THE ORIGINS OF SURFACE-TO-AIR GUIDED MISSILE TECHNOLOGY

German flak rockets and the onset of the Cold War

By
JAMES MILLS

CASEMATE

Philadelphia & Oxford

Published in the United States of America and Great Britain in 2022 by
CASEMATE PUBLISHERS
1950 Lawrence Road, Havertown, PA 19083, USA
and
The Old Music Hall, 106–108 Cowley Road, Oxford OX4 1JE, UK

Hardcover Edition: 978-1-63624-277-4
Digital Edition: 978-1-63624-278-1

A CIP record for this book is available from the British Library

Printed and bound in the United Kingdom by TJ Books

Typeset by DiTech Publishing Services

For a complete list of Casemate titles, please contact:

CASEMATE PUBLISHERS (US)
Telephone (610) 853-9131
Fax (610) 853-9146
Email: casemate@casematepublishers.com
www.casematepublishers.com

CASEMATE PUBLISHERS (UK)
Telephone (01865) 241249
Email: casemate-uk@casematepublishers.co.uk
www.casematepublishers.co.uk

Cover image
An engineering drawing of a C-2/W-3 prototype flak rocket, 6 September 1943, captured by the US Army in May 1945. Source: *Projekt Wasserfall*," Deutsches Museum, https://www.digipeer.de/index.php?id=385942567

Contents

List of Abbreviations

American

AAF	United States Army Air Forces
AMC	Air Materiel Command
APL/JHU	Applied Physics Laboratory, Johns Hopkins University
ATSC	Air Technical Service Command
BOMARC	Boeing Michigan Aeronautical Research Centre
BTL	Bell Telephone Laboratories
BRL	Ballistic Research Laboratory, Aberdeen Proving Ground
BuAer	US Navy Bureau of Aeronautics
BuOrd	US Navy Bureau of Ordnance
CIT	California Institute of Technology
CTV	Control Test Vehicle
DAC	Douglas Aircraft Corporation
DNI	Director of Naval Intelligence
GAL	Guggenheim Aeronautical Laboratory
GAPA	Ground to Air Pilotless Aircraft
GE	General Electric
HAWK	Homing All-the-Way Killer
JATO	Jet-Assisted Take-Off
JIOA	Joint Intelligence Objectives Agency
JPL	Jet Propulsion Laboratory
NACA	National Advisory Committee for Aeronautics
NASA	National Aeronautics and Space Agency
NASM	National Air and Space Museum
NavTechMisEu	United States Naval Technical Mission in Europe
NOL	Naval Ordnance Laboratory
OCO	Office, Chief of Ordnance (US Army)
OSRD	Office of Scientific Research and Development
RV	Research and Test Vehicle
USAF	United States Air Force
USFET	United States Forces European Theatre
USSTAF	United States Strategic and Tactical Air Forces

WSPG	White Sands Proving Ground
XSAM	Experimental Surface-to-Air Guided Missile
XSSM	Experimental Surface-to-Surface Guided Missile

British

ADD	Armaments Design Department
ADE	Armaments Design Establishment
ADI (K)	Assistant Directorate of Intelligence (Prisoner Interrogation), Air Ministry
ADI (Ph)	Assistant Directorate of Intelligence (Photographic Reconnaissance), Air Ministry
ADI (Sc)	Assistant Directorate of Intelligence (Science), Air Ministry
AGE	Admiralty Gunnery Establishment
AI2 (g)	Air Intelligence, Technical, Air Ministry
AI3 (e)	Air Intelligence, Enemy Order of Battle, Air Ministry
AP/WIU	Air Prisoner-of-War Interrogation Unit
ARD	Armaments Research Department
ARE	Armaments Research Establishment
ASE	Admiralty Signals Establishment
BAFO	British Air Forces of Occupation
BIOS	British Intelligence Objectives Subcommittee
CCG (BE)	Control Commission for Germany (British Element)
CEAD	Chief Engineer, Armaments Design
CRDD	Chemical Research and Development Department
DCOS	Deputy Chiefs of Staff
DGP	Director of Guided Projectiles
DRPC	Defence Research Policy Committee
DTPA	Directorate of Technical and Personnel Administration, Ministry of Supply
ERDE	Explosives Research and Development Establishment
GAP	Guided Anti-aircraft Projectile
GC&CS	Government Code and Cypher School
GPE	Guided Projectiles Establishment
JIC	Joint Intelligence Committee/Subcommittee
LOP/GAP	Liquid Oxygen-Petrol/Guided Anti-aircraft Projectile
MAP	Ministry of Aircraft Production
MI5	Military Intelligence Section 5, War Office
MI6	Military Intelligence Section 6, War Office
MI15	Military Intelligence Section 15, War Office
MoD	Ministry of Defence

MoS	Ministry of Supply
MOSEC	Ministry of Supply Establishment Cuxhaven
NPL	National Physical Laboratory
PDE	Projectile Development Establishment
RAE	Royal Aircraft Establishment
RAF	Royal Air Force
RPD	Rocket Propulsion Department
RRDE	Radar Research and Development Establishment
RTV	Rocket Test Vehicle
SAGW	Surface-to-Air Guided Weapon
SRDE	Signals Research and Development Establishment
STIB	Scientific and Technical Intelligence Branch
TNA	The National Archives
TRE	Telecommunications Research Establishment

British and American

AAM	air-to-air missile
ADRC	Air Documents Research Centre
CAFT	Consolidated Advance Field Team
CIC	Combined Intelligence Committee
CIOS	Combined Intelligence Objectives Subcommittee
CIPC	Combined Intelligence Priorities Committee
DGDN	Diethylene glycol dinitrate
FIAT	Field Information Agency, Technical
LOX	Liquid Oxygen
LPRE	Liquid Propellant Rocket Engine
NATO	North Atlantic Treaty Organisation
SAM	surface-to-air missile
SCAEF	Supreme Commander Allied Expeditionary Force
SHAEF	Supreme Headquarters Allied Expeditionary Force
SPRE	Solid Propellant Rocket Engine
SSM	surface-to-surface missile

French

BEE	*Bureau d'Études d'Emmendingen*
CEPA	*Centre d'Études des Projectiles Autopropulsés*
CIEES	*Centre Interarmées d'Essais des Engins Spéciaux*
DCA	*Défense Contre Avions*
DCCAN	*Direction Centrale des Constructions et Armes Navales*
DEFA	*Direction des Études et Fabrications d'Armement*

DGER	*Direction Générale des Études et Recherches*
DTCAN	*Direction Technique des Constructions et Armes Navales*
DTIA	*Direction Technique et Industrielle de l'Aéronautique*
EA/G	*Groupe Engins-autopropulsés-guidage*
EA/P	*Groupe Engins-autopropulsés-propulsion*
GANES	*Groupe d'Aéronautique Navale d'Engins Spéciaux*
GOPA	*Groupe Operationnel des Projectiles Autopropulsés*
LRBA	*Laboratoire de Recherches Balistiques et Aérodynamiques*
MARUCA	*Marine Ruelle Contre Avions*
MASURCA	*Marine Supersonique Ruelle Contre Avions*
MATRA	*Société Générale de Mécanique, Aviation, Traction*
PARCA	*Projectile Autopropulsé Radioguidé Contre-avions*
SCIT	*Service Central d'Information Technique*
SDECE	*Service de Documentation Extérieur et de Contre-espionnage*
SEPR	*Société d'Études de la Propulsion par Réaction*
SFENA	*Société Française d'Équipements pour la Navigation Aérienne*
SNCAC	*Société Nationale de Constructions Aéronautiques du Centre*
SNCASE	*Société Nationale de Constructions Aéronautiques du Sud-Est*
SSM	*Service de la Securité Militaire*
STAé	*Service Technique Aéronautique*
STCAN	*Service Technique des Constructions et Armes Navales*
STRIM	*Société Technique de Recherches Industrielles et Mécanique*

German

A-4/V-2	*Aggregat-4/Vergeltungswaffen Zwei*
AG	*Aktiengesellschaft* (open stock company)
AIA	*Aerodynamische Institut der Technische Hochschule Aachen*
AVA	*Aerodynamische Versuchsanstalt*
BMW	*Bayerische Motoren Werke Flugmotorenbau GmbH*
DFS	*Deutsche Forschungsanstalt für Segelflug Ernst Udet*
Dipl.-Ing.	*Diplom-Ingenieur*
Dr.-Ing.	*Doktor-Ingenieur*
Dr. Phil.	*Doktor Phil. Experimentelle Physik* (Doctor of Philosophy, Experimental Physics)
Dr. rer. nat.	*Dr. rerum naturalium* (*Doktor der Naturwissenschaften*, Doctor of Natural Sciences)
DVL	*Deutsche Versuchsanstalt für Luftfahrt*
EMW	*Elektromechanische Werke GmbH*
FuG	*Funkgerät*
GBN	*Generalbevollmächtigte für technischen Nachrichtenmittel*

GEMA	*Gesellschaft für Elektroakustische und Mechanische Apparate GmbH*
GmbH	*Gesellschaft mit beschränkter Haftung* (limited liability company)
HFW	*Henschel Flugzeug-Werke AG*
HWA	*Heereswaffenamt*
HWK	*H. Walter KG*
KG	*Kommanditgesellschaft* (limited stock company)
LFA	*Luftfahrtforschungsanstalt Hermann Göring*
LFM	*Luftfahrtforschungsanstalt München*
OKL	*Oberkommando der Luftwaffe*
OKM	*Oberkommando der Marine*
NSDAP	*Nationalsozialistische Deutsche Arbeiterpartei* (National Socialist German Workers' Party; Nazi Party)
RLM	*Reichsluftfahrtministerium*
SS	*Schutzstaffel*
TH	*Technische Hochschule*
TLR	*Technische Luftrüstung*
WaPrüf	*Waffenamt-Prüfwesen*
Wasag	*Westfälisch-Anhaltische Sprengstoff AG*
WVA	*Wasserbau-Versuchsanstalt Kochelsee GmbH*
ZWB	*Zentralstelle für Wissenschaftliche Berichterstattung*

Russian

KB	*Konstrukturskoye byuro* (Design Bureau)
NII	*Nauchno-issledovatelskiy institut* (Scientific-Research Institute)
NKGB	*Norodny komitet gosudarstvennoy bezopasnosti* (People's Commissariat for State Security)
OKB	*Opytno-konstrukturskoye byuro* (Experimental Design Bureau)
SB	*Spetsialnoye byuro* (Special Bureau)
SKB	*Spetsialnoye konstrukturskoye byuro* (Special Design Bureau)

'...a well planned development of the art of rockets will have revolutionary consequences in the scientific and military spheres ... much in the same way as the development of aviation has brought revolutionary changes in the last 50 years.'

Dr. Wernher von Braun
Former technical director, Peenemünde-East, 1945[1]

'For the future defense against hostile aircraft, it seems clear that supersonic guided missiles will be used, propelled either by rockets or more probably by a ramjet. The fully automatic radar beam-guiding methods of control of the type suggested, but not experimentally tried, by the Germans will probably be used for guiding, supplemented by simplified heat-homing devices and proximity fuses.'

Dr. Theodore von Kármán
Director, Army Air Forces Scientific Advisory Group, August 1945[2]

Introduction

All surface-to-air guided missile systems that have entered service since the early 1950s can at least in part be traced back to pioneering research and development that was carried out in Germany, and to a lesser extent in the United Kingdom and the United States, in the years immediately leading up to and during World War II. Germany indeed holds many firsts. The German Air Ministry was the first government department to conduct an extensive surface-to-air missile (SAM) development programme; German scientists and engineers were the first to integrate the various sub-systems and assemblies into functional guided missile designs; and the *Luftwaffe* had the distinction of being the first armed service in the world to test-launch an experimental surface-to-air guided missile. Unbeknownst to the Germans, their contemporaries in the UK and the US were also interested in the weapons, mainly prompted by the German attainments with guided weapons and the use of suicide aircraft by the Japanese in the Pacific theatre. In the Soviet Union and France, there was little, if any, progress with the technology during World War II, and the defence industries in both countries essentially had to start from scratch in 1945, which necessitated a heavy reliance on the wartime German research and development.

At the end of war, amidst the ruins of Hitler's once-mighty Third Reich, in the bomb-damaged and dispersed industrial infrastructure and arguably the world's leading aeronautical research complex, Allied troops found the shattered remains of the world's first comprehensive SAM development programme. It very quickly became apparent to technical and counter-intelligence officers from the Allied countries who found evidence of the German programme that as with the V-1 flying bomb and the V-2 ballistic missile, Germany had made incredible advances with SAM technology, and clearly had a significant technological lead over the Allies. Scientific and technical experts from the four Allied countries then set about documenting the German progress, interrogating the German scientists, engineers, and technicians (collectively referred to as specialists) who were involved in the programme, removing hardware for examination and analysis, and recruiting selected German specialists to exploit their scientific and engineering knowledge.

This book is an account of the Allies' transfer and exploitation of German SAM technology after World War II, with a particular emphasis on surface-to-air guided missiles. In exploring the subject, several questions had to be posed. What was the comparative progress in research and development of SAM technology by Germany and the Allies during World War II? What was the quantity and composition of the German knowledge about SAMs, from both human and material sources, that was captured by the Allies shortly before and after the final German surrender in May 1945? What were the processes through which the German SAM technology was transferred to the Allies? Which German specialists with expertise from the German SAM development programme were recruited by the Allies after the war? And how did German scientific and engineering knowledge and expertise that appertained to SAMs generally contribute to the postwar SAM programmes of the Allies up until about 1960?

The Allies' seizure of documentary material and hardware from the German aeronautical research and experimental establishments, as well as industrial contractors, which were involved in the development programme proved to be very successful, particularly for the Anglo-Americans. It can be reasonably concluded that almost the entire archives on all the major SAM projects were captured by British and American investigators. Certainly, the quantity and quality of the documents were very high, and collectively constituted an extraordinary intelligence windfall for the Allies. Take for instance the seizure of the archives of Peenemünde-East, which contained the development histories of the *Wasserfall* and *Taifun* missiles among over 500,000 documents, all of which went to the US and the UK; or the archives of arguably the world's best aeronautical research establishment in 1945, the *Luftfahrtforschungsanstalt Hermann Göring* near Völkenrode, which contained documents on all aspects of German SAM research and development among its 4,900 volumes and reports, also acquired by the US and the UK. The scientific and technical knowledge from these two establishments would have been considered a successful haul for the Anglo-Americans, yet much more was recovered at other locations throughout Germany up until 1947. Hundreds, and probably thousands, of intelligence reports and monographs about the German guided missile programmes, and about the careers and work of the German specialists involved with the programmes, were written between 1945 and 1948. These documents not only summarised the progress that was made with SAM technology in Germany, but also provided the investigators with views on future developments and information about the scientific and engineering credentials of German specialists who were potentially considered for employment in the Allied countries after the war. Broadly speaking, the German specialists who did go to work in the Allied nations specialised in fields such as rocket engines, electronic and mechanical equipment in control and guidance systems, aerodynamics, and ancillary technologies such as computer simulators.

How the defence industries in each country benefited from the advances made in Germany with SAM technology differed. There were several factors – the amount of scientific and technical intelligence that was brought back from Germany; the calibre of the German specialists who were recruited; where the specialists were employed; what projects the specialists worked on; and the scope of the postwar guided missile programmes in each country. Furthermore, American and British departments shared all of the intelligence from Germany and exchanged the results of its evaluation and exploitation after the war. Of the Western Allies, the US gained the most out of the technology transfer. The Americans' vast economic and industrial resources enabled the world's newest superpower to be in the best position to exploit the German knowledge. The most visible legacy of the German SAM development programme in the US was in the design concepts of several missiles that were developed by the US Army Ordnance Department. The UK also significantly benefited from the technology transfer, with support from the Australian government in the provision of personnel and development and testing infrastructure in South Australia as part of the Anglo-Australian Joint Project. But Britain was not in an economically or technologically strong position to fully exploit the knowledge compared to the US, a lack of supersonic wind tunnels particularly handicapping their effort. In France and the Soviet Union, the industrial contractors often resumed from where the Germans left off in 1945, and as a result, German ideas and technology were highly influential in the first experimental SAMs that were built and tested in both countries during the late 1940s and early 1950s.

In the second half of the 1940s, while the threat of armed conflict between the Western powers and the communist bloc was relatively low, there was no great urgency to bring into service first-generation surface-to-air guided missile systems that would exceed the limit of German advances in World War II. In the meantime, anti-aircraft artillery and small, unguided solid propellant rockets, using radar fire control, remained in service, although it was realised that the rapid developments in jet aircraft would soon render those weapons obsolete, at least at high altitudes. The outbreak of the Korean War in 1950 and increased military spending hastened the development of surface-to-air guided missile systems, which led to the first American system, the Nike (later Nike-Ajax), becoming operational in December 1953. This was soon followed by the first Soviet system, the S-25, in 1955, while the first British system, the ramjet-powered Bloodhound, did not enter service with the Royal Air Force until 1958. France, meanwhile, could not successfully develop its own first-generation system in time, and thus had to purchase or manufacture under licence American SAM systems before the realisation of an indigenously developed system.

The terminology used to describe what are usually referred to in the English language today as SAMs, and less commonly as surface-to-air guided weapons (SAGWs), has altered since World War II. Guided missiles are any self-propelled projectiles with a guidance system, whereas guided weapons are a much broader field, including armaments with a guidance system but not necessarily self-propelled, for instance precision-guided munitions. Until 1945, there was no standard term in use to describe the weapons. In Germany, SAMs were usually described as *Flak-Raketen* (flak rockets), an abbreviation for *Flugabwehrkanone Raketen*, which translates as air defence cannon rockets. All of the ground-launched anti-aircraft missiles in development in Germany during the war, from the smallest solid propellant rockets right up to the most advanced guided missiles, were described as such. The description is, however, something of a misnomer, because technically the abbreviation *Flak* was used to describe conventional anti-aircraft artillery. SAMs were less commonly described with the more appropriate terms *Flugabwehrraketen* (air defence rockets), abbreviated as *FLA Raketen*, or *Flug-Raketen* (flying rockets). In the UK and the US, the nomenclature varied, from anti-aircraft torpedoes, anti-aircraft rockets and anti-aircraft guided missiles to guided anti-aircraft projectiles. Since 1945, SAM has become the standard term in NATO and the US, whereas in the UK SAGW is also used. In France, the weapons were referred to as *fusées de DCA* or *projectiles de DCA* (*Défense Contre Avions*, anti-aircraft defence rockets or projectiles), or with the general term to describe all guided weapons, *engins spéciaux* (special machines). In the French language, *engin* is a general engineering term used to describe any machine that has a particular function, such as a crane for example. *Fusées de DCA* remains in use, while *engins spéciaux* has gradually been replaced with *missiles tactiques* to describe all ground-, ship- and aircraft-launched tactical guided missiles in the categories of *sol-air* and *surface-air* (SA, ground-to-air, in the case of ship-launched missiles surface-to-air), *sol-sol* (SS, ground-to-ground), *air-sol* (AS, air-to-ground) and *air-air* (AA, air-to-air). When referring to the development of the weapons in Germany, both the acronym SAM and the term flak rocket are used, while everywhere else SAM is used.[1]

Chapters 1 and 2 constitute a comparative history of the development of SAM technology in Germany, the UK and the US between 1939 and 1945. France and the Soviet Union are omitted due to the relative lack of development in those two countries during the war. German specialists undertook a vast amount of research and development at research and experimental establishments, armaments firms, universities and institutes of technology with the intention of producing an operational surface-to-air guided missile system. The aim was to design a weapon system that could provide more effective ground-based air defence than anti-aircraft artillery against the British and American strategic bomber aircraft attacking Germany, an objective the Germans were unable to achieve during the war. The many technical modifications that were made to each missile system and the various vicissitudes in the research and development are, for the most part, not dealt with, as these details are so numerous and specific in nature that they would detract from the overall account. Of the Allies, the UK held the lead in SAM development in 1945 and was the first to test-launch a SAM. The US was, however, in front of the UK in the field of liquid propellant rocket engine (LPRE) development. By the end of hostilities, the US Navy and US Army had initiated large-scale, long-range research and development programmes with funding and resources that outweighed what the British government could commit to its own programmes on a relative scale.

Chapter 3, 'Anglo-American investigations of intelligence targets linked to the German SAM development programme, 1944–48', is focused on the capture and postwar investigation and documentation by British and American investigators of intelligence targets that were directly involved or associated with the German SAM development programme. The targets included individual German military and civilian specialist personnel. In most situations, the German specialists assisted the British and American investigators in the technology transfer process by writing reports and monographs about their careers and wartime work. These documents aided the Anglo-Americans in the future recruitment and employment of their erstwhile enemies. The Germans also assisted in the search for hidden documents. They built samples of technology to be dispatched to the UK and the US for evaluation, as what took place at the facilities of BMW in Munich, for example.[2]

Chapters 4, 5 and 6 account for the transfer and exploitation of German SAM technology by the US, the UK and France after 1945. The reader will see that while there was an interruption between the wartime and postwar research and development, German work on SAM systems resumed in the three countries by the late 1940s. These chapters address the next stages of the technology transfer process, which were the physical removal of captured German documents, hardware and specialist personnel, and how the Western Allies derived benefits from the German knowledge. The processes through which the transfers were executed were not the same in each case, due to the differences in the organisational structures of the responsible government and military entities.

On the American side, German specialists were recruited by the technical branches of the armed services through government-run programmes, Project Overcast and afterwards Project Paperclip. Concerning the specialists with expertise in SAM technology, the rocket and guided missile specialists formerly employed in the Peenemünde-East organisation provides a useful indicator as to their capabilities. Between 140 and 150 scientists and engineers from that one organisation were recruited, with the largest group (between 125 and 135) recruited by the US Army Ordnance Department. The US Army Air Forces (US Air Force from 1947) and the US Navy each recruited between six and 12 from the organisation. While most of these specialists brought experience from the A-4/V-2 programme, a certain number had or also worked on flak rocket projects. I estimate that around 80 per cent of the German specialists who brought expertise with SAM technology, or who were subsequently connected with SAM projects in the US after the war, came from the Peenemünde-East organisation. Also notable was the recruitment of three out of the four technical directors of the major SAM projects at the prime contractors. They were Dr. Wernher von Braun, the technical director of the *Wasserfall* project at Peenemünde-East, by the US Army Ordnance Department; *Dr.-Ing.* Werner Fricke, the head of the *Rheintochter* flak rocket development group at *Rheinmetall-Borsig AG*, by the US Air Force; and *Dr.-Ing.* Herbert Wagner, the chief designer of guided missiles at *Henschel Flugzeug-Werke AG*, the firm that developed the Hs 117 (also known as the *Schmetterling*), by the US Navy. *Dipl.-Ing.* Klaus Scheufelen, the designer of a small, unguided flak rocket developed at Peenemünde-East called the *Taifun,* was also recruited by the US Army Ordnance Department. The Germans were initially employed as consultants to military agencies and defence contractors, and afterwards almost all of them went on to work for the US government (such as with NASA), for universities, or in private industry on grander projects, such as ballistic missiles, space rockets and advanced aircraft.

On the British side, a total of 162 German specialists were recruited through a government employment programme for the British defence sector, the Deputy Chiefs of Staff (DCOS) scheme. Their employment in the UK was sponsored by research and experimental establishments under the administrative control of two government departments, the Ministry of Supply and the Admiralty, and in a few cases by private firms. As early British postwar rocket and guided weapons programmes were primarily focused on the development of test vehicles and tactical missiles (for close attack and defence in land, sea and airborne warfare scenarios) rather than strategic missiles (for area defence and long-range cruise and ballistic missiles), the recruitment of German guided missile and rocketry specialists was specifically directed towards those objectives. Generally speaking, the German specialists who brought knowledge of SAM technology to the UK, or who were involved in postwar British SAM research and development, can be placed into five categories: 1) the design and development of rocket engines, including turbopumps, valves, combustion chambers, propellant feed systems, propellant injection systems and technology suitable for the two preferred oxidizing agents that were mixed with rocket fuels, hydrogen peroxide and nitric acid; 2) rocket propellant, combustion and heat transfer research; 3) technology in missile control and guidance systems; 4) electronic analogue and electromechanical computers/guided weapon simulators; and 5) miscellaneous fields, such as specialist welding and field instrumentation.

The French government recruited hundreds more German specialists for the country's defence sector than the British government did through the DCOS scheme, perhaps over 1,000. After the crippling effects of the Occupation, the French armed forces needed to rebuild their capabilities with the most modern weaponry, and for this purpose selected groups of German specialists from fields where Germany had made great advances up until 1945 – for example in jet engines, guided missiles, and wind tunnels – were sought after. German specialists were recruited by the technical services of the government directorates that were responsible for the research, development and procurement of armaments and equipment for the French armed forces. They were employed by a mixture of organisations – the government directorates themselves, research and experimental establishments, and state- and privately-owned companies. Commensurately, a greater number of guided missile and rocketry specialists were recruited. From the Peenemünde-East organisation came between 70 and 80 specialists – including a number with research and/or development experience from the SAM development programme – divided between wind tunnel, propulsion and guidance specialists.

The transfer and exploitation of German SAM technology by the Soviet Union – which significantly contributed to the first Soviet surface-to-air guided missile system to enter service, the S-25, in 1955 – is treated separately in an appendix. This has been done because the historical events have, for the most part, been documented in a number of English-language works on early Soviet rocket and guided missile development.[3] My research has added some previously unpublished (as far as the author is aware) details of a technical nature that have been sourced from British and American intelligence reports, in particular concerning the Soviet modifications to the control and propulsion systems in the *Wasserfall* SAM system, from 1946 to 1948.

This book goes some way to try to solve the enduring mystery and address the speculation concerning whether the Germans ever tested one of the experimental SAMs against an airborne target. There is circumstantial evidence to suggest that the Germans planned such an experiment, most likely using a *Rheintochter* and the Fi-103/V-1 as a target drone. The infrastructure was in place on the Baltic coast in Pomerania by late 1943, and a year later both missiles could be steered with enough accuracy to permit such a test, although probably not at a very high altitude. However, there is not a single shred of evidence in any of the American, British and French intelligence reports, or in the many reports and monographs that were written by German specialists after the war, to indicate that an attempt was ever made. Then there is the question of whether any of the German missiles were tested against American or British bomber aircraft. I am certain that none of the experimental guided flak rockets were ever fired against Allied aircraft, but there is certainly enough circumstantial evidence to suggest that the small, unguided *Taifun* was probably fired against Allied aircraft in the last weeks of the war.

Lastly, the intention of this book is not to weaponise or politicise the history of the technology transfer as part of an anti-Marxist or anti-communist agenda, nor is its intention to surreptitiously promote and perpetuate Nazi ideological mysticism. Rather, this account is a sober reckoning of the consequences for the German SAM programme, the personnel involved and the German attainments in the field – which overall were more advanced than the Allies – as a result of the outcome of World War II, the greatest conflagration in history.

CHAPTER ONE

German *Flak-Rakete* Research and Development During World War II

During World War II, the *Reichsluftfahrtministerium* (RLM, German Air Ministry) commissioned six major flak rocket projects as part of an advanced flak armament development programme. The histories of these projects have been accounted for to a certain extent since 1947. Therefore, although some of the content of this chapter is already historical knowledge, the *raison d'être* is original; its purpose is twofold. The chapter introduces the reader to World War II German flak rocket technology, the places where research and development were carried out, and the leading German scientists and engineers who were involved in research and development. It also establishes the general progress of SAM research and development in Germany during the war and forms the basis upon which the concurrent developments in the US and the UK are compared. The reader can therefore trace the objects and people more easily throughout the account and during the transnational technology transfer processes by the Allies.

The origins of the German SAM development programme

Guided and non-guided flak rockets were developed by the RLM to eventually replace anti-aircraft artillery (*Flugabwehrkanonen*, abbreviated as flak) as ground-based anti-aircraft armaments in the German air defence network. By the end of 1941, this network – which was considerably expanded following the defeats of Denmark, Norway, the Netherlands, Belgium, Luxembourg and France by Germany during the summer of 1940 – stretched from Denmark to France. The network comprised surveillance and tracking radars, day- and night-fighter aircraft, searchlights, and heavy and light flak artillery. The heavy guns were designed and developed by two of Germany's great armaments manufacturers: *Krupp AG* supplied three guns, the 88mm *Flak 18, 36* and *37*, which were designed before the war; and *Rheinmetall-Borsig AG* supplied the 88mm *Flak 41*, the 105mm *Flak 38* and *39*, and the largest-calibre gun in the *Luftwaffe* anti-aircraft arsenal, the 128mm *Flak 40*, which entered service in 1942. The *Flak 41* was the standard heavy flak gun in service with the *Luftwaffe*, a multi-purpose weapon that could also be deployed as an anti-tank gun and field artillery piece.

The *Luftwaffe* anti-aircraft artillery organisation was divided into formations that were analogous to an artillery branch in an army. There were corps *(Flakkorps)*, of which there were seven, each sub-divided into divisions, brigades, regiments, battalions *(Abteilungen)* and batteries *(Batterien)*. A heavy flak regiment normally comprised three or four battalions of heavy guns and several battalions of light flak guns. A heavy flak battalion comprised four batteries of heavy guns, while a heavy flak battery featured combinations of heavy and light guns, manned by a flak troop *(Zug)*. The composition of the personnel and the number of guns in a battery varied; each battery was usually armed with four or six heavy guns. Batteries could be mobile, transported by road or rail, semi-mobile or static. A battery with more than six guns was referred to as a *Großbatterie*.

As the *Luftwaffe* air defence network reached the peak of its strength in 1942, proponents of flak rocket development in the RLM began to push for the development of the weapons, including *Generalleutnant* Kurt Steudemann, the *Inspekteur der Flakartillerie*, who in February 1941 had called for anti-aircraft missile development because of the failure of anti-aircraft artillery against British bombers.[1] By this time, Germany had become a world leader in the embryonic field of guided missile technology. The lead that Germany possessed with the technology can be traced back to efforts by the German Army to circumvent the conditions of the Versailles Treaty of 1919, which restricted Germany's military strength by prohibiting certain categories of armaments, such as aircraft. The numerous constituent technologies in guided missiles, such as rocket propulsion, electronics for guidance and construction techniques, as well as the aerodynamics, were still largely experimental. Research and development with liquid and solid propellant rocket engines (LPREs and SPREs) prior to mid-1942 laid the groundwork for the

Table 1. The specifications of the *Rheinmetall-Borsig* 128mm *Flak 40* anti-aircraft gun

Muzzle velocity:	2,890ft per second
Maximum horizontal range:	68,650ft
Ceiling:	48,500ft
Estimated effective ceiling:	35,000ft
Weight of projectile:	26kg (high explosive) or 27kg
Estimated lethal radius of burst:	65ft
Rate of fire:	8–12 rounds per minute
Weight in action:	16.75 tons
Weight in draught (pulling):	26.5 tons
Weight of static equipment:	12.75 tons
Elevation:	-3° to +88°

Source: *TNA, AIR 40/1151, MI 15 Periodical Intelligence Summary No. 18, 12.3.1945*

propulsion systems that were designed for each of the prototype flak rockets in the programme. At Peenemünde-East (*Heeresversuchsanstalt Peenemünde*), a LPRE was being developed for the A-4/V-2 ballistic missile; *Rheinmetall-Borsig* designed powder and solid propellant artillery rockets and assisted take-off (ATO) units for aircraft; *Wilhelm Schmidding AG* designed both SPREs and LPREs; *H. Walter KG* (HWK) developed hydrogen peroxide propulsion systems for torpedoes and submarines, a turbopump for the A-4/V-2, ATO units for aircraft, a LPRE for the *Messerschmitt Me 163 Komet* interceptor aircraft and a LPRE for the Hs 293 air-to-surface guided missile that was being developed by the aircraft and guided missile manufacturer *Henschel Flugzeug-Werke AG* (HFW), one of a number of subsidiaries in the large, family-owned *Henschel* industrial complex.

The Hs 293 was one of two tactical anti-shipping weapons that were in development for the *Luftwaffe* by 1941. *Dr.-Ing.* Max Kramer of the *Deutsche Versuchsanstalt für Luftfahrt* (DVL, German Experimental Establishment for Aviation) at Berlin-Adlershof (later of *Ruhrstahl AG* from 1943–45, and who emigrated to the US in 1946 to work for the US Navy Bureau of Aeronautics) designed a precision-guided bomb designated the X-1 (otherwise referred to as 'Fritz X'). Advances that were being made in Germany with radio remote control technology were of fundamental importance to the design of each weapon. Development of the new electronic technology commenced in the central laboratories of *Siemens und Halske AG* in 1936. In 1939, the RLM issued an order to another electronics firm, *Stassfurter Rundfunk GmbH* of Stassfurt, south of Magdeburg in Anhalt, to design and construct a radio-control system for guided bombs. In early 1940, the firm designed a suitable radio receiver, and in combination with a radio transmitter that was designed by *Telefunken GmbH* – a subsidiary of the giant electrical engineering firm *Allgemeine Elektricitäts Gessellschaft* (AEG) and one of Germany's largest electronics companies – the radio remote control system was code named *Kehl-Strassburg*. By the end of 1940, the feasibility of the system was proved in tests that were run with both the X-1 and the Hs 293. Basically, the system worked by the transmission of radio signals from the parent aircraft to the receiver in the weapon, which moved control surfaces to steer it towards the target. The *Kehl-Strassburg* system formed the basis of all subsequent radio remote control systems that were developed for German tactical guided missiles during World War II.[2]

The aerial war and current technological developments in Germany continued to stimulate interest in anti-aircraft rocket weapons. On 13 May 1941, a research programme was initiated by the Department for Aeronautical Research in the RLM to gather supersonic aerodynamic, stability and steering data for flak rocket development. The programme was co-ordinated from *Abteilung* LC 1 (*Forschung*, research) in the *Technisches Amt* (technical office) of the *Generalluftzeugmeister* (GL, Aircraft Master General) department in the RLM. The GL department was established in Berlin in 1938 to manage all research and development, supply, production and procurement for the expanding *Luftwaffe*. Until 1941, the GL was World War I fighter ace *Generaloberst* Ernst Udet, who was replaced in that year by *Generaloberst* (later *Generalfeldmarschall*) Erhard Milch following the former's suicide. According to Walter Wernitz, a former employee in the Department for Aeronautical Research, initially there was a problem acquiring a suitable projectile. The intention was to utilise missiles that were designed by the aircraft industry, but none could be supplied. The aerodynamic shapes of others that were in development were considered unsuitable for the velocities that were sought.[3]

Subsequently, the programme was contracted to the *Luftfahrtforschungsanstalt Hermann Göring* (LFA, Hermann Göring Aeronautical Research Establishment), a new facility south of the village of Völkenrode near Braunschweig in central Germany. The specialists at the LFA already had some expertise with missiles – an experimental, rocket-propelled air-launched test vehicle had recently been designed there, code named *Hecht* (Pike). The new research programme was code named *Feuerlilie* (Fire lily) and was led by *Dr.-Ing.* Gerhard Braun of the LFA Institute for Aerodynamic Research. The *Feuerlilie* programme was undertaken in association with *Rheinmetall-Borsig*, which supplied the propulsion system, its standard RI-502 ATO unit that was designed for heavily laden aircraft. After 1942, *Rheinmetall-Borsig* designed more powerful rockets that built upon the firm's previous work with ATO designs and research (and later development) that was undertaken during the *Feuerlilie* programme. Meanwhile, in 1941, *Rheinmetall-Borsig* proposed a flak rocket to the RLM, but was unsuccessful. Elsewhere, the *Heereswaffenamt* (HWA, Army Ordnance Office) began a study of a flak rocket in May 1941, and the following month, *Dr.-Ing.* Herbert Wagner, the chief designer of the Hs 293 at HFW, and Dr. Theodor Sturm of *Stassfurter Rundfunk* also proposed a flak rocket, designated the Hs 297. However, the HFW missile was also rejected, on account of the preference for offensive rather than defensive weapons by the *Oberkommando der Luftwaffe* (OKL, *Luftwaffe* High Command).[4]

By the first half of 1942, the calls for anti-aircraft missile development had grown louder in the RLM. German flak guns were reaching the limits of their capability – the weapons were becoming increasingly uneconomical because the guns expended ammunition at an unsustainable rate, and future aircraft would soon be too fast and would operate at altitudes beyond the ranges of the guns then in service. Faced with this situation, on 22 June 1942, the *General der Flakwaffe, Generalmajor* Walter von Axthelm, who had recently succeeded Steudemann as the *Inspekteur der Flakartillerie*, produced a report for the RLM entitled *Übersicht über den Entwicklungsstand und die Entwicklungsabsichten der Flakartillerie* (Review of the state of development and development designs for flak artillery). In this report, Axthelm proposed a new flak armament development programme that called for more powerful anti-aircraft guns and guided anti-aircraft missiles, the latter of which were the natural technological evolution in ground-based anti-aircraft armaments.[5]

Up until the first half of 1942, the RAF had only inflicted negligible damage on the German war economy and had not yet had an appreciable effect on armaments production. However, the British strategic bombing capability had recently been enhanced with the entry into service of three new makes of four-engine heavy bombers with RAF Bomber Command. In the second half of 1941, the Short Brothers Stirling and the Handley Page Halifax entered service, followed in March 1942 by the Avro Lancaster, the foremost British heavy bomber of the war. The British were also in the process of developing effective radio navigational and target identification aids (the 'Gee' and H2S systems), as well as blind bombing radar devices to assist bomber pilots to reach and identify their targets (the 'Oboe' system). When used successfully by mass concentrations of hundreds of bomber aircraft against area targets, these technologies enabled the RAF to cause massive destruction, as was demonstrated in the first of the three 'thousand bomber raids' against Cologne on 30 May 1942, and by the 630 aircraft that attacked Dusseldorf on the night of 31 July 1942. Furthermore, the aerial war had escalated in 1942 with the entry of the US Army Air Forces (AAF) into the European theatre. The US Eighth Air Force began bombing operations from British bases against German targets outside of Germany proper on 17 August 1942, when 12 Boeing B-17 Flying Fortresses attacked the marshalling yards at Rouen in France. Subsequently, on 1 September 1942, *Reichsmarschall* Hermann Göring, the commander in chief of the *Luftwaffe*, was forced to authorise the development of SAMs as part of a programme to improve Germany's ground-based air defences in accordance with von Axthelm's recommendations. The RLM produced theoretical specifications for a guided missile with a ceiling of 8,000 metres (around 26,500ft).[6]

The German SAM development programme started with two missile design concepts, one fuelled by liquid propellants and the other by solid propellants, to be radio-controlled and to travel at supersonic velocities. Initially, two collaborative projects between the RLM and HWA were based at Peenemünde-East. The first missile, designated C-1, was planned to have a SPRE. However, this project was cancelled in the spring of 1943, possibly out of preference for a flak rocket with a LPRE being designed at the same establishment, the C-2, later code named *Wasserfall* (Waterfall), a scaled-down version of the A-4/V-2. Another possible reason for the cancellation of the C-1 was preference by the RLM for a solid propellant flak rocket under development by *Rheinmetall-Borsig*. This project, which commenced towards the end of 1942, was code named *Rheintochter* (Rhine Maiden) and was based at the firm's Berlin-Marienfelde facilities. The *Rheintochter* was named after the three Rhine maidens in Richard Wagner's opera *Das Rheingold*, the first in the tetralogy *Der Ring des Nibelungen* (The Ring of the Nibelung). The cancellation of the C-1 left the *Rheintochter* as the sole solid propellant missile under development. A third joint project at Peenemünde-East was a thin, unguided rocket designed to be fired in barrages, code named *Taifun* (Typhoon), which was initiated later, in 1944.

These projects at Peenemünde-East were under the direction of the civilian technical director of development, the young, charismatic and talented Dr. Wernher von Braun.[7]

The continued inability of the *Luftwaffe* air defence network to defend against the Anglo-American air raids on industrial and civilian targets during the Combined Bomber Offensive – in particular the devastating bombing of Hamburg from 25 July to 2 August 1943 – stimulated the RLM to urgently expand the SAM development programme. By the end of August 1943, the HFW Hs 297 that Wagner and Sturm proposed to the RLM in 1941 was reactivated, re-designated the Hs 117 and code named *Schmetterling* (Butterfly). The Hs 117 was developed by a team under Wagner in the Development Department for Guided Missiles at the firm's Berlin-Schönefeld facilities, who drew on experience from the development of the Hs 293. In the autumn of 1943, *Messerschmitt* initiated a project code named *Enzian* (Gentian, an alpine plant), headed by Dr. Hermann Wurster, a former test pilot at the firm and holder of the 1936 world speed record in a *Messerschmitt* Bf 109 (Me 109) fighter. The *Enzian* project was initially based at the Augsburg plant in Bavaria, but after the workshops there were destroyed in a double air raid by RAF Bomber Command and the US Eighth Air Force on the night of 25 February 1944, the project was moved to a client firm, *Holzbau-Kissing KG* at Sonthofen, 100km to the south in Upper Bavaria.[8]

The German SAM development programme did not develop any weapons for the *Kriegsmarine* because the programme was directed towards the realisation of a missile system only for the *Luftwaffe*. However, from as early as May 1944, the *Kriegsmarine* sought from *Rheinmetall-Borsig* the requirements of a *Rheintochter*-type guided missile, most likely for use as a surface-to-surface anti-ship weapon. In the domain of small, ship-launched anti-aircraft rockets, there was at least one in development for the *Kriegsmarine* from as early as 1943, designated the R 42, which had a calibre of 214mm. To place the diameter into perspective, the planned production version of the *Taifun* had a calibre of 100mm.[9]

Research and development structure

Once the SAM development programme was established, the RLM nominated prime contractors to meet the specifications. These organisations were responsible for the technical direction of each project. They formed development groups to design and plan for the manufacture of prototype missiles in concert with German industrial firms. The military direction of the development programme was, by 1943, the responsibility of *Abteilung 5* in the *Flak Entwicklung* division of the GL department (GL/Flak-E 5) under *Major* Dr. Friedrich Halder, who was a champion of SAM development during the war.[10] Simultaneously, the aeronautical research establishments and industrial firms carried out research and development work with various aspects of the technology to assist the tasks of the prime contractors. Many aspects of the technology were still theoretical in 1942, and unsolved questions required continuous and time-consuming applied research.

Aeronautical research in Germany was, up until May 1942, under the control of the already-mentioned Department for Aeronautical Research that was attached to the *Technisches Amt* in the RLM. The department was headed by Dr. Adolf Bäumker, who was one of Germany's most prominent scientists in the field of aeronautical research. Due to the war situation, applied research was prioritised over basic or fundamental research at the aeronautical research establishments. Structural and organisational problems in the German aeronautical research complex, and the need for aeronautical research to have a free hand, led in May 1942 to the separation of the Department for Aeronautical Research from the RLM. The department became an independent government organisation (although the RLM continued to fund

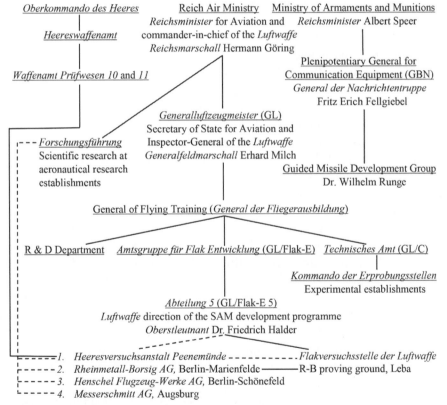

Research and development structure of the German SAM development programme, 1943. A dotted line indicates association. (Sources: 'General Report on Guided Missiles', Intelligence Report GDM-2, Signal Corps, USFET, 11.7.1945, RG 38, National Archives College Park; TNA, AIR 40/1310, 'Section IV – Part 4 – Controlled Missiles', MAP, 7 August 1945; Boog et al., *Germany and the Second World War, Volume VII: The Strategic Air War in Europe and the War in West and East Asia 1943–1944/5*, p266)

its activities and the research establishments) under the authority of a four-man directorate, the Aviation Research Command of the State Minister of Aviation and Supreme Commander of the *Luftwaffe* (*Forschungsführung*, or *FoFü* for short).[11]

The chairman of the *FoFü* was Professor Dr. Ludwig Prandtl, 'the father of aerodynamics', who was the former director and then-current chairman of the *Aerodynamische Versuchsanstalt* (AVA, Aerodynamic Test Establishment) at Göttingen. Bäumker became one of the directors, while also holding the directorships of the *Akademie der Luftfahrtforschung* (a society that promoted aeronautical research) and the new *Luftfahrtforschungsanstalt München* (LFM) in Bavaria. The third director was Professor Dr. Walter Georgii, who was the director of the *Deutsche Forschungsanstalt für Segelflug Ernst Udet* (DFS, Ernst Udet German Research Institute for Gliding) at Ainring. He was appointed deputy chairman of the *FoFü* in November 1943. The fourth director was Professor Dr. Friedrich Seewald, the recently appointed head of the *Aerodynamische Institut der Technische Hochschule Aachen* (AIA). Seewald later became chairman of the DVL, which was the oldest aeronautical research establishment in Germany, founded in 1912. The main role of the *FoFü* was to decide what research should be done at each establishment. The fields of research that were applicable to flak rockets included flight, engines, electro-physics (radio), armaments, flak, materials and chemistry. The *FoFü* also oversaw the *Feuerlilie* research programme at the LFA. The establishments could enlist universities or *Technische Hochschulen* (TH, institutes of technology) to work on particular research problems, although these institutions were under the control of the German education ministry. Hirschel has observed that the increased autonomy of aeronautical research freed it from adverse political and financial constraints, and subsequently led to positive outcomes, which he attributed to Bäumker, who successfully exploited his relations with Hermann Göring.[12]

Most of the theoretical and practical research undertaken for the SAM development programme was done at three locations. Two were civilian aeronautical research establishments under the control of the *FoFü*, the LFA and the DFS; the other was Peenemünde-East. The construction of the LFA began in 1935 as part of the National Socialist regime's programme to reverse the defeat of 1918 and the humiliation of Versailles by rebuilding the *Luftwaffe*. By 1937, the establishment was operational. The LFA employed about 1,500 people and had the most modern facilities in the world. There were four institutes at the establishment, of which two – the Aerodynamics Institute, which had several sub- and supersonic wind tunnels, and the *Institut für Motorenforschung* (Institute for Engine Research), where rocket propellant research was carried out – were concerned with guided missile research and some aspects of development.[13]

The DFS at Ainring emerged from an organisation called the *Forschungsinstitut der Rhön-Rossitten-Gesellschaft* (Research Institute of the Rhoen-Rossitten Society).

Prime contractors/prototype assembly
 Heeresversuchsanstalt Peenemünde
Henschel Flugzeug-Werke AG, Berlin
 Rheinmetall-Borsig AG, Berlin
 Messerschmitt AG, Augsburg

Experimental and testing installations
Flakversuchsstelle der Luftwaffe, Karlshagen
Erprobungsstelle der Luftwaffe, Karlshagen
Rheinmetall-Borsig proving ground, Leba

Aeronautical research establishments
 Deutsche Versuchsanstalt für Luftfahrt, Berlin-Adlershof
 Luftfahrtforschungsanstalt Hermann Göring, Völkenrode
 Deutsche Forschungsanstalt für Segelflug Ernst Udet,
 Ainring
 Aerodynamische Versuchsanstalt, Göttingen
 Wasserbau-Versuchsanstalt Kochelsee GmbH
 Technische Hochschule Aachen

*Assembly of prototype missiles was moved to *Holzbau-Kissing KG* in Sonthofen in 1944.

The Greater German Reich in 1943 with the locations of the key sites in the SAM development programme. (Source: Edmaps, https://www.edmaps.com/html/germany_3.html)

Its original main purpose was to study gliding. Early on, the institute was located at Darmstadt in Hessen, then moved to Braunschweig in 1939, where it remained only until the summer of 1940, when it was transferred to Ainring, an alpine location better suited to high-altitude gliding research. There were five institutes in the establishment, plus two departments and one laboratory. Broadly speaking, the fields of inquiry included infra-red research, research on compasses in steel aircraft, model studies for solving remote control procedures, the development of an apparatus for disturbance-free control signal transmission, the development of gyroscopic control for missiles and aircraft, theoretical research of stabilisation of pilotless flying bodies, control procedures and the connection of seekers to the automatic control of missiles. The DFS was an important centre of theoretical research for the SAM development programme. A great deal of work was carried out in the *Institut für Flugausrüstung* (Institute for Flight Equipment) under Prof. Dr. Eduard Fischel. Fischel specialised in three areas: model techniques for studying trajectories, the determination of the magnitudes of required acceleration and the optimum relative speeds between missiles and targets. In *Abteilung G 5* (theoretical investigations section) under *Dipl.-Ing.* E. Stinshoff, the research problems related to SAMs included mathematical calculations of pursuit curves and hit probabilities based on the aerodynamic shape and speed of a projectile; the gearing of an infra-red seeker based on the flight mechanics of rockets; the design and construction of simulators to train ground crews in the control of flak rockets (primarily the C-2); beam-rider guidance; and, during 1944 and January 1945, the development of the stabilisation and control of the *Enzian*, which had been contracted out to the DFS by *Holzbau-Kissing*.[14]

Peenemünde-East was an experimental establishment founded in 1937 to develop liquid propellant rockets for the German Army. The installation was under the administrative control of the *Entwicklungs- und Prüfwesen* (development and testing division) in the HWA, which was responsible for carrying out research, development and testing of new ordnance. The work at Peenemünde-East was overseen by the *Heereswaffenamt Prüfwesen 11* (Testing Division 11, abbreviated as *Wa Prüf 11*), which was tasked with liquid and solid propellant rocket development. *Wa Prüf 11* was headed by *Oberst* (later *Generalmajor*) *Dr.-Ing.* Walter Dornberger until September 1943 (from 15 December 1943, liquid propellant rocket development was overseen by a new section, *Wa Prüf 10*). The establishment, renamed *Heimat-Artillerie Park 11* (HAP 11) on 17 May 1943, was divided into two sectors. The *Entwicklungswerk* (EW, developments works, also known as Peenemünde-East) under Wernher von Braun comprised departments concerned with design, guidance and control, propulsion systems and testing. The other sector was the *Versuchsserienwerk* (VW, pilot production plant, otherwise known as Peenemünde-South), where test missiles were manufactured and assembled. The VW was under the management of von Braun's deputy since 1940, *Dipl.-Ing.* Eberhard Rees.[15]

In late 1942, the RLM set up a special experimental section at the HAP to develop and test-launch flak rockets, called the *Flakversuchsstelle der Luftwaffe*. It was placed under the command of a *Luftwaffe* officer, *Hauptmann* (later *Major*) Dr. Hermann König from GL/Flak-E 5 and made use of the existing facilities at the establishment. Men with engineering and scientific degrees who were serving in the *Wehrmacht* were withdrawn from front-line service and sent to Peenemünde-East to work in the *Flakversuchsstelle* as civilians, or as was the case with *Luftwaffe* personnel, in a military capacity. A number of the personnel who were working on the A-4/V-2 project were also transferred to the *Flakversuchsstelle*. Most of the design and development of the C-2 and *Taifun* projects was done at Peenemünde-East and the *Flakversuchsstelle*, with some research and design work contracted out to other establishments, including the DFS, the LFA and the TH Darmstadt. An *Abteilung* (No. 224) with a staff of around 120 was created within the guidance and controls department at Peenemünde-East to carry out research and development work in connection with the steering and remote control system in the C-2. Until April 1944, the section was headed by Dr. Oswald Lange, who was replaced as *Abteilungsleiter* that month by Dr. Theodor Netzer. *Abteilung* 224 was divided into three sub-sections, each headed by a deputy section leader. One sub-section was concerned with steering (such as servomotors and servomechanisms); the second with remote control (ground and on board electronic guidance equipment, proximity fuses and seekers); and the third with proving and test-launches, and also comprised workshops.[16]

The components of the LPRE in the C-2 were built and tested in the propulsion department. Manufacturing of parts of test missiles was contracted out to private industry except for the tail section of the fuselage, which was fabricated in the VW. The contractors sent the parts to the HAP for final assembly in the VW under the direction of personnel from a special army unit, the *Versuchskommando Nord* (VKN, Northern Experimental Command), which was set up in late 1941 by the commander in chief of the German Army, *Generalfeldmarschall* Walther von Brauchitsch, to provide serving engineering and technical personnel for the HAP. The other flak rocket that was developed at the HAP, the *Taifun*, was designed by *Luftwaffe Leutnant Dipl.-Ing.* Klaus Scheufelen, who was attached to the *Flakversuchsstelle* as a range officer.[17]

On the subject of wind tunnel testing of models of the various prototypes, Bernd Krag produced an analysis of the aerodynamic development of each missile (in English) in 2010, so the technical details shall not be repeated here.[18] Aerodynamic and thermodynamic research for the A-4/V-2, C-2 and *Taifun* projects, and some minor work for the *Rheintochter* project, was undertaken in two 40cm x 40cm (working area) supersonic Mach 4.4 wind tunnels in the Aerodynamics Institute at Peenemünde-East under Dr. Rudolf Hermann. From 1939, the institute carried out aerodynamic and exterior ballistics research for the *Wehrmacht* and the armaments industry. The Aerodynamics Institute had customers including *Rheinmetall-Borsig*, *Krupp*, *Westfälisch-Anhaltische Sprengstoff AG* (*Wasag*, where Hermann Oberth, a pioneer of rocket technology in the 1920s, was an employee), *Skoda* and *Röchling*. Design work was done for numerous HWA projectile development programmes, which included field and long-range artillery, anti-tank weapons and multiple rocket launchers (*Nebelwerfer*). The institute also helped to design projectiles for the *Luftwaffe*, including flak artillery and rocket armaments for aircraft. Following an RAF raid against Peenemünde in August 1943, wind tunnel experiments for the C-2 project were disrupted for 10 months, from January to October 1944, while the two supersonic wind tunnels were transferred to a safer location at Kochel, 60km south of Munich in Upper Bavaria, where they operated under the cover of an innocuous front company called *Wasserbau-Versuchsanstalt Kochelsee GmbH* (WVA, Hydraulic Engineering Research Establishment)

As already mentioned, Peenemünde-East contracted out research and development work for the C-2 project to several other concerns. Engineers and physicists at the TH Darmstadt worked in close association with specialists at Peenemünde-East on projects related to the guidance and control system. In the Mathematics Institute at the TH Darmstadt, which was a computing institute for the *Wehrmacht*, a Dr. Czerniakorski and a junior staff of about 20 people carried out trajectory computa-tions.[19] Aerodynamic experiments in the subsonic region were carried out in a 10ft diameter wind tunnel at the firm *Luftschiffbau Zeppelin GmbH*[20] at Friedrichshafen, near Lake Constance in Baden, southern Germany, under *Dr.-Ing.* Max Schirmer. Aerodynamic research was also done in the low-speed wind tunnel at the LFM at Otterbrunn near Munich.[21]

Wind tunnel experiments for the *Rheintochter* project were mostly carried out in the Mach 3.7 wind tunnel at the AVA in Göttingen under Professor *Dipl.-Ing.* Otto Walchner, a swept-wing and high velocity ballistics expert, and at the AIA in Aachen. Experiments for the Hs 117 project were primarily carried out in the wind tunnels of the DVL in Berlin under the supervision of Dr. Bernhard Gothert, with some research also done at the AVA. Experiments for the *Feuerlilie* and *Enzian* projects were done in the subsonic wind tunnels at the LFA (one tunnel, the A-3, was 8 metres in diameter and could accommodate full-scale models of the *Enzian*). The scientific knowledge about the sub-, trans- and supersonic aerodynamics of guided missiles that was accumulated at these establishments was one outcome of the large sums of money that the Nazi regime spent on wind tunnels for aviation research and development.[22]

The research establishments and educational institutes could choose to publish the results of their research for distribution to, and exchange with, other establishments. Soon after the Nazi seizure of power in January 1933, an agency was created for this purpose, the *Zentrale für wissenschaftliches Berichtswesen über Luftfahrtforschung* (ZWB, Centre for Scientific Publications on Aeronautical Research). The ZWB was originally attached to the DVL but was detached in 1940. The president of the ZWB was Hermann Göring and its director was Adolf Bäumker. The agency played a very important role in effecting the dissemination of knowledge, stimulating research and reducing duplication in Germany. Once published, the reports were distributed to an approved list. The reports had a standard cover which consisted of a dark blue central part with light blue at the top and bottom. The ZWB maintained a card index that contained the title and an abstract of each report. Peenemünde-East was one organisation that had its reports regularly published and distributed by the ZWB. For example, a report produced in the Aerodynamics Institute entitled 'The aerodynamic development

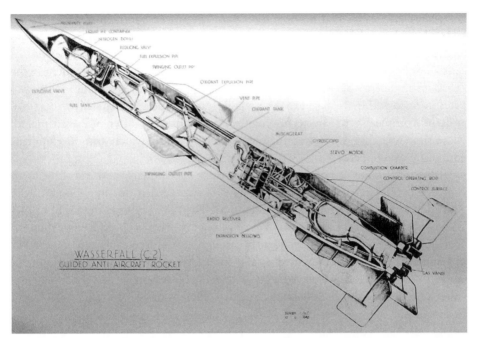

Internal diagram of a C-2 *Wasserfall* flak rocket. The first C-2 prototype, the W-1, had four trapezoidal shaped fins at the mid-section with a leading edge of just over 90° to the fuselage. The fins at the mid-section of the second, third and fourth prototypes (W-2, W-3 and W-4) were of a parallelogram shape with a low aspect ratio. Attached to the trailing edge of each of the four rear fins was a control surface. Gas vanes provided control immediately after launch. (Source: TNA, DEFE 15/216)

of the flak rocket *Wasserfall* by Dr. Hermann Kurzweg, dated March 1945, was indexed in HAP *Archiv* 66 as report number 171 and published by the ZWB with the reference ZWB/Pa/Re/66/171. In another example, a 1943 report entitled 'Paper on propellant questions for A-4 and *Wasserfall*' by Dr. Adolf Thiel and *Dipl.-Ing.* Gerhard Heller was indexed in HAP *Archiv* 110 as report number six, then published by the ZWB with the reference ZWB/Pa/Re/110/6. Access to the disseminated results of the scientific and engineering research no doubt assisted the development work by the prime contractors, who were allowed latitude to innovate and invent within the boundaries of the required specifications.[23]

Weapons and technologies intended for service use with the *Luftwaffe* usually underwent testing or evaluation at experimental establishments called *Erprobungstellen* (*E-Stellen*) prior to series production or delivery to operational *Luftwaffe* formations. While some research and testing related to flak rockets was carried out at the *E-Stelle Karlshagen*, otherwise known as Peenemünde-West because it was adjacent to Peenemünde-East, test-launching of complete prototypes was done at purpose-built missile-firing ranges. The solid propellant *Rheintochter* was test-launched at Leba in eastern Pomerania, and all the liquid propellant missiles were test-launched at the *Flakversuchsstelle*. Engineering staff from the prime contractors were present at test-launchings in order to evaluate the performance of the technology. It was envisaged that a test-launching programme would, if a missile system performed satisfactorily, eventually lead to series production and operational deployment in the *Luftwaffe* air defence network. Due to the increasingly urgent war situation, there was pressure from within the RLM to get a missile into production as quickly as possible, and the complexity of the technology was consequently often underestimated. More time was always required to solve problems, and research and development constantly overlapped.

The general structure of aeronautical research in Germany and the three main centres where research and development work was carried out in support of the SAM development programme – Peenemünde-East, the LFA and the DFS – along with a number of other key research establishments, have been dealt with thus far. Peenemünde-East will appear frequently throughout the account because a number of German scientists and engineers who undertook research and/or development work for the C-2 and *Taifun* projects at that establishment went to work for the Allies after the war. In the following two parts, there is more information about the

An *Enzian* flak rocket being hoisted onto its launcher on the Greifswalder Oie. The *Enzian*, which resembled the Me 163, had two swept wings with ailerons for control surfaces, and spoilers at the rear. (Source: TNA, AVIA 40/4812)

A Hs 117 *Schmetterling* flak rocket on its launcher at Peenemünde. (Source: TNA, AIR 40/2532)

test-launching programmes and the work of the scientists and engineers at the German industrial firms engaged in the development of missile subsystem technologies for the prime contractors.

Propulsion systems and test-launching

At least seven flak rocket propulsion system design concepts were conceived during the SAM development programme. Among the guided missiles, there was a single-stage missile with a LRPE (C-2); a two-stage missile with a solid propellant booster rocket and a solid propellant sustainer rocket in a tandem configuration (*Rheintochter* R-I); three two-stage missiles with solid propellant booster rockets and a liquid propellant sustainer rocket in a parallel configuration (Hs 117, *Enzian* and *Rheintochter* R-IIIf [*flüssig*, liquid]); and a two-stage missile with solid propellant booster rockets and a solid propellant sustainer rocket in a parallel configuration (*Rheintochter* R-IIIp [*Pulver*, powder]). The *Taifun* was a single-stage rocket in liquid and solid propellant versions. The general layout of the LPREs in the guided missiles was fairly standard. At the forward end of the missile behind the warhead and proximity fuse (and in future a seeker), there was a compressed gas bottle to force the propellants into the combustion chamber. In the centre of the missiles were forward and rear propellant tanks which contained the fuel and the oxidizer, with room for control system equipment either forward of the propellant tanks or between the rear propellant tank and the combustion chamber. The general layout of the solid propellant *Rheintochter* R-I was different – in the sustainer rocket, the proximity fuse and the control and guidance equipment were situated at the forward end of the missile, but the warhead was situated behind the SPRE, which had splayed exhaust nozzles to force the exhaust gases out of the sides of the missile.[24]

SPREs are technologically simpler propulsion systems than LPREs because they have no moving parts, no propellant tanks, and no injection system; they require no cooling; and they are not fuelled prior to launch. Thrust is produced by ignition of the propellants in tubes, with the exhaust expelled through a nozzle or nozzles. *Rheinmetall-Borsig* supplied its own rockets for the booster and sustainer rockets in the R-I, and also the booster rockets for the R-III which comprised two RI-503 ATO units. *Wilhelm Schmidding* at Bodenbach on the Elbe River in the Sudetenland (present-day Podmokly, part of the city of Děčin in the Czech Republic) manufactured the booster rockets (RLM designation 109-553) for the Hs 117 (two) and *Enzian* (four). Three German firms supplied the solid propellants for these rockets: *Wasag* at Rheinsdorf near Zwickau in Saxony; *Wolff AG* at Walsrode, south-east of Bremen; and *Dynamit AG*, formerly *A. Nobel* (named after the founder of the Nobel Prizes), which was 51 per cent owned by the giant conglomerate *IG Farben*.[25]

By comparison, the LPRE technology had four main components: a propellant feed system, propellant tanks, a thrust cylinder consisting of a combustion chamber and a nozzle, and in some cases a control mechanism to regulate propellant flow. German chemists and rocket engineers considered liquid propellants as more suitable

A *Rheintochter* R-III flak rocket with booster rockets attached, presumably being hoisted onto its launcher at the Leba proving ground. (Source: TNA, AIR 40/2532)

for guided missiles than solid propellants because of a better specific consumption (the impulse, or thrust, that is created per unit of volume of propellant). By 1945, thousands of liquid propellants had been researched in Germany, but only one oxidiser was considered suitable for LPREs in flak rockets, and that was nitric acid. Nitric acid is an inorganic, highly corrosive substance that was first experimented with as an oxidiser in Germany in 1942. All of the LPREs in the flak rockets used nitric acid in various concentrations, with one exception, the HWK RI-209 ATO unit (109-502) that was used in the early prototypes of the *Enzian* (it was fitted to the *Junkers* Ju 88 and *Heinkel* He 111 medium bombers), which used hydrogen peroxide. Evidently the RI-209 was unsatisfactory, as it was replaced with the RI-210b ATO unit from the same company, which used a mixed acid of about 92 per cent nitric acid with 8 percent sulphuric acid as the oxidiser – the latter chemical was added to protect against corrosion in the tanks – and petrol as the fuel, which was cheap and easily available. Further planned prototypes of the *Enzian*, a subsonic E-4 and a supersonic E-5, were to be fitted with a nitric acid/petrol LPRE that was being developed at the *Vierjahresplan-Institut für Kraftfahrzeug* (VfK, Four-Year-Plan Institute for Motor Vehicles) in Berlin under the direction of *Dr.-Ing.* Hans-Joachim Conrad from the TH of Berlin.[26]

The LPREs that were designed for the C-2, *Taifun* and *Rheintochter* R-IIIf (the latter also developed at the VfK under the direction of Conrad) used fuels known as *Visols*. *Visols* are a group of vinyl-ether compounds that are hypergolic, in that they ignite spontaneously upon contact with an oxidiser in the combustion chamber. The research with *Visols* – which were found to be particularly suitable for flak rockets – was carried out in the Institute for Engine Research at the LFA under Dr. Wolfgang Noeggerath. A *Visol* named vinyl-isobutyl-ether was selected for use in the C-2 engine, but after the supply of propellants became disrupted by bombing raids *Visols* were blended with other propellants. Research with hypergolic propellants was also done at *BMW Flugmotorenbau GmbH* (BMW Aircraft Engine Construction) under chief chemist Dr. Hermann Hemesath, who found that a group of aromatic amino compounds, code named *Tonkas*, were suitable as fuels with concentrated nitric acid. BMW entered the rocket engine field in the autumn of 1939 under *Dipl.-Ing.* Helmut von Zborowski. The firm had three locations where rocket engine development was carried out. Projects and basic design calculations, some design work, experimental development, and chemistry were done at Allach near Munich; development was also carried out at Zühlsdorf in Berlin; and some design work and manufacturing were done at Bruckmühl, south of Munich. Series production was reportedly at the *Gerätewerk* at Stargard in Pomerania (present-day Stargard Szczeciński in Poland). BMW was contracted to supply a LPRE for the Hs 117, the 109-558, a nitric acid/*Tonka* engine that was designed by Hans Ziegler. HWK was contracted to supply an alternative LPRE for the Hs 117, the 109-729, a nitric acid/petrol engine that was designed by *Dr.-Ing.* Johannes Schmidt, the director of the rocket engine development department at the firm. The 109-729 was almost identical to the 109-558 in layout, weight and performance. The differences were a pyrotechnic igniter in the combustion chamber, a propellant regulation device and, unlike most German combustion chamber designs, it was not regeneratively cooled. Almost all the liquid propellants for these engines were supplied by *IG Farben*.[27]

Moving on to the subject of test-launching, Germany has the distinction of being the first country to test-launch a prototype guided SAM, when a *Rheintochter* R-I was fired on 12 October 1943 from a *Rheinmetall-Borsig* proving ground near Leba in eastern Pomerania on the Baltic Sea coast, about 200km east of Peenemünde. The SPRE in the test missiles was static-tested nearby. During the *Rheintochter* test-launching programme, the missile was launched from an inclined ramp mounted on an 88mm gun platform supplied by *Krupp*, a technique that was also adopted by the British and the Americans.[28]

It is interesting to observe that the Leba proving ground was located adjacent to the target area for tests of the *Fieseler* Fi-103/V-1 flying bomb, which was fired from Zempin, to the south of Peenemünde. The utilisation of the proving ground at Leba suggests that the RLM may have been planning to attempt an interception of a Fi-103/V-1 using radars and a *Rheintochter* R-I. Some authors (such as Pocock in 1967) have mentioned that the *Luftwaffe* did designate the Fi-103/V-1 as a *Flakzielgerät* (anti-aircraft target apparatus), but only in order to deliberately mislead Allied intelligence organisations. However, there is evidence that the *Blohm und Voss* Bv 246, an experimental, long-range, air-launched glider bomb, was considered for use as a *Flakzielgerät*, which suggests the designation was not misleading. Further evidence pointing to the use of the Fi-103/V-1 as a target drone was the creation, in February 1945, of a department within the headquarters of the *General der Flakwaffe* that dealt with plans, tactics and weapons for defence against flying bombs and rockets. Also, after World War II, the French government research establishment *Arsenal de l'Aéronautique* actually designed two target drones based on the design of the Fi-103/V-1, the Ars 5.501 and the CT-10.[29]

The development of the Fi-103/V-1 began in the spring of 1942, about six months before the commencement of the *Rheintochter* project. The missile was launched from the ground for the first time on 24 December 1942, and flew at altitudes between 1,000ft and 4,000ft at a maximum speed of around 720 km/h. To track and plot the trajectories of test missiles, during the autumn of 1943 the 14th company of the *Luft Nachrichtung Versuchs Regiment* (Air Signals Experimental Regiment) installed tracking stations equipped with *Würzburg* radars, which could determine altitude, distance and bearing, along the Baltic coast at Peenemünde, Zempin, Greifswalder Oie (a small island about 10km off the coast of Peenemünde), Horst, Jershöft, Stolpmünde and Leba. When the Fi-103/V-1 moved out of the range of one radar station, the missile would be picked up by the next station. R. V. Jones revealed in *Most Secret War* in 1976 that the Government Code and Cypher School at Bletchley

Park in Buckinghamshire intercepted the plots of Fi-103/V-1 trajectories from the radar stations. A sample taken from December 1943 shows the projectiles falling across a broad area about 100km wide to the west, south-west, north-west and north of the Leba station. Another sample taken from 5–10 May 1944 shows the projectiles falling within an area approximately 30km across to the west and north-west of the station. The most distant impact was about 50km away, the nearest about 15km. Despite the increased accuracy of the Fi-103/V-1, if an interception was planned, it could only have taken place after July 1944. A summary of launches 16 to 29 in the third series of the *Rheintochter* test-launching programme, which ran until late July 1944, describes several of the missiles exhibiting a number of major problems, such as erratic steering and control, and sometimes the wooden rudders and wings broke off during flight.[30]

All of the liquid propellant flak rockets were test-launched on the Greifswalder Oie. The island was ideally suited for the purpose – the necessary infrastructure was already in place and the location was highly secure. Prior to being utilised for the SAM development programme, the island was used from early December 1937 to test-launch the precursors of the A-4/V-2, the A-3 and A-5 test vehicles. The main drawbacks of the location were that the island was accessible only by ferry or barge, and the water between the mainland and the island was vulnerable to mine warfare. Germany also has the distinction of being the first country to test-launch a liquid propellant SAM – a C-2 on 29 February 1944. The C-2 was launched vertically from a stationary platform of tubular construction with wheels for easy transportation to and from the launching site. The commencement of the Hs 117 and *Enzian* test-launching programmes soon followed in May 1944, but tests of the former missile were initially carried out at the *E-Stelle Karlshagen* with dummy rounds of an air-launched variant (Hs 117H). Subsequently, due to delays in the delivery of the LPRE, the experimental BMW 109-558, only booster-powered rounds could be tested. The Hs 117 was launched from a zero-length launcher, while the *Enzian* and a version of the *Feuerlilie*, the F-55, were like the *Rheintochter* launched from an inclined ramp mounted on a converted 88mm gun platform that was supplied by *Krupp*, although the launcher for the *Enzian* and the *Feuerlilie* comprised two rails which were 6.8 metres in length. The *Taifun* was fired in barrages from a multi-barrelled projector.[31]

To recapitulate, German engineers and technicians innovated to apply the scientific and technical knowledge of solid and liquid propellants and rocket propulsion, already partly embodied in the various applications of rocket engine technology in experimental and operational armaments, to each of the flak rocket projects. Closely associated with the design of the propulsion systems were the launching methods, which were also a combination of existing technology and new ideas. A corollary of these developments was the creation of infrastructure to possibly attempt an interception of an airborne target. Such a test almost certainly did not take place, but there is enough circumstantial evidence to suggest that the objective was at the very least planned for. Specialists from several of the abovementioned German industrial firms would become of interest to the Allies because of their scientific and engineering knowledge in the field of rocket propulsion.

Guidance and control systems, proximity fuses and seekers

The co-ordination of radio, electrical and radar developments for German guided missile projects was the responsibility of a development group composed of representatives from the industrial and commercial sectors, the *Luftwaffe* and the German Army. At the end of 1942, a series of development groups were formed under the control of a *Generalbevollmächtigte für technischen Nachrichtenmittel* (GBN, Plenipotentiary General for Communication Equipment) in the Reich Ministry of Armaments and Munitions under *Reichsminister* Albert Speer. The Plenipotentiary was *General der Nachrichtentruppe* Fritz Erich Fellgiebel, who would later be involved in the 20 July plot to assassinate Hitler. The development group concerned with guided missiles was led by Dr. Wilhelm Runge of *Telefunken* (from 1944 of the DVL). This group was divided into several committees – *Hauptkommission Elektrotechnik* (electrical technique), headed by Dr. Lüschen of *Siemens*; *Sonderkommission Funkmeßtechnik* (radar technique), led by Dr. Rottgardt of *Telefunken*; *Arbeitskommission Fernlenktechnik* (remote control technique), headed by Theodor Sturm of *Stassfurter Rundfunk*, who was also the *Luftwaffe* representative on the committee (Hans Schuchmann from *Siemens und Halske* was the Army's representative); and the *Sonderkommission für Elektrische Munitionausrüstung* (proximity fuses and seekers), which was jointly headed by Runge and Professor Dr. Friedrich Gladenbeck of AEG. These committees had the power to decide which firms should be contracted to undertake research and development, to monitor the progress of developments and to make judgements on the performance of the work done at each firm.[32]

During the war, German scientists and engineers developed two electronic methods to control and guide SAMs, namely the command-link (also referred to as command-control or command-guidance) and beam-rider techniques. The problems of guiding the flak rockets then in development were much different and more complicated than those encountered with the Hs 293 and A-4/V-2. Since a SAM has to manoeuvre quickly towards a rapidly moving airborne target, the technology had to be custom designed for flak rockets. The very complex nature of the problems was made even more challenging because Germany was badly retarded in the field of centimetre-wave (microwave) radio research and development, a domain in which the Anglo-Americans reigned supreme during the war. There were a number of reasons why Germany was so far behind, which have been set out by Boog *et al.* in 2001.[33] To summarise, a lack of scientific knowledge of microwave frequencies from 3–30 GHz led to the conviction by German radar scientists that

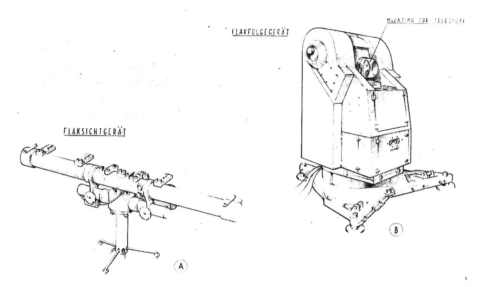

Equipment in the *Burgund* command-link guidance system for the Hs 117. (A) *Flak Sicht Gerät* and (B) *Flak Folge Gerät*. (Source: NAA: D879, DB16)

research in that area would be prohibitively expensive and therefore unprofitable. Consequently, *Telefunken* closed down its microwave laboratory in November 1942 after a suggestion to do so was made by Runge to Fellgiebel, despite disagreements by certain scientists and engineers. The decision did not last very long; after a British H2S centimetric airborne radar was recovered from a downed Stirling bomber in February 1943, microwave research recommenced in mid-1943. Another possible reason why there was a lack of scientific and engineering progress in the field was the forced closure of amateur radio clubs by the National Socialist regime, which consequently suppressed their innovative and inventive potential.[34]

The contractors at first adapted the *Kehl-Strassburg* command-link guidance system that was originally designed for the X-1 and the Hs 293. The system for the Hs 117 which utilised this technology was code named *Burgund*. This system represented the earliest technique to track an airborne target and the missile during its trajectory, which was by optical means without the use of radar. Basically, there were six main parts of the *Burgund* system: an optical target sighting and tracking apparatus (*Flak Sicht Gerät*) manned by one operator; a parallax calculator unit that determined the direction of the missile launcher, developed by *Kreiselgeräte GmbH* of Berlin-Zehlendorf (the same firm that was contracted by the HWA to design the guidance systems in the A-3 and A-5 test vehicles and the A-4/V-2); the launcher itself, supplied by *Krupp*; an optical missile tracking and steering apparatus (*Flak Folge Gerät*) that was developed by *Askania-Werke AG* of Berlin-Fridenau (also manned by one operator, who fired the missile after receiving an order from the gunnery officer by phone); the *Bodo* unit, which generated signals to direct the missile towards the target, developed by *Telefunken*; and to transmit the signals, the *Kehlheim* unit that housed the radio transmitter, antenna and ancillary equipment. Testing of the radio technology in the Hs 117 system was done at the *E-Stelle Karlshagen*, in *Abteilung E 4* under Dr. Josef Dantscher. The main drawback of the optical tracking technique was that it could only be used during the day and in clear weather. There was also the likelihood that the radio frequencies would be susceptible to jamming.[35]

The RLM intended to eventually replace the optical tracking and sighting technique with two radars, which had the advantage of being able to operate independently of meteorological conditions. The first command-link guidance system to be designed for this purpose was being developed by *Telefunken* for the supersonic *Rheintochter* and C-2, and was code named *Elsass*. Because the C-2 was launched vertically, a computer had to be designed to steer the missile onto its correct trajectory during the first 200 metres of flight before a radar could take over and steer the missile toward the target aircraft. This apparatus was called an *Einlenkrechner* (also referred to as the *Einlenk-Gerät*), a computer for guidance onto a line-of-sight trajectory. The design and development of the *Einlenkrechner* was carried out in Section 224 at Peenemünde-East and by German industrial firms. An experimental C-2 ground

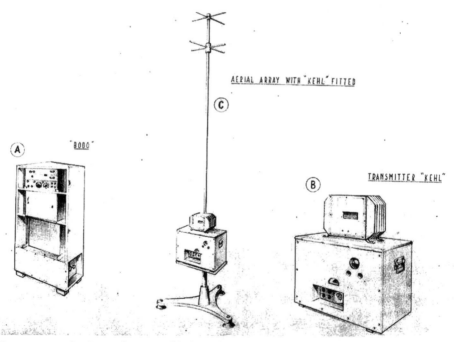

Equipment in the *Burgund* command-link guidance system for the Hs 117. (A) *Bodo* apparatus, developed by *Telefunken*, which generated signals to direct the missile towards the target, transmitted to the receiver in the missile by the *Kehlheim* radio equipment (B) and (C). (Source: NAA: D879, DB17)

installation on the Greifswalder Oie consisted of a firing platform, a *Flak Sicht Gerät*, *Flak Folge Gerät*, *Einlenkrechner* and the *Kehlheim* unit, in various positions grouped together within a few metres of one another. The radar that was initially intended for experiments was the *Telefunken Würzburg Riese* (Giant *Würzburg*). Later in the war, the 3-metre diameter *Telefunken Mannheim* gun-laying radar was also used, which entered service with flak units during 1944. The *Mannheim* had a range of around 40km (approximately 24.9 miles), which was less than the *Würzburg Riese*, but nevertheless it was more accurate.[36]

Having two supersonic missiles in development did not necessarily equate to a waste or duplication of guidance system technology, although trial and error did result in constant modifications. For example, servomotor technology designed for the A-4/V-2 and by the aircraft industry was adapted to the C-2. In the design of the first C-2 prototype, the W-1, engineers at Peenemünde-East utilised *Askania-Werke* servomotors that were originally designed to move the control surfaces of the A-4/V-2 which, like the C-2, were four rudders in the tail. However, these proved unsatisfactory because they could not move the rudders in the C-2 design – which were larger than those in the A-4/V-2 – quickly enough to ensure stability. Also adapted from the A-4/V-2 was a *Mischgerät* (mixing device), designed to modify the radio transmissions from the ground and the gyroscope signals in the missile. This device was also modified for use in the *Enzian*. In the C-2/W-2 prototype, hydraulic servomotors manufactured by *Luftfahrtgerätewerk Hakenfelde GmbH* (LGW) of Berlin-Spandau, a subsidiary of *Siemens*, were used instead. LGW also supplied the C-2 project with three standard aircraft autopilot gyroscopes for stabilisation, which were mounted in the tail section. The C-2/W-3 prototype, which was test-launched about 35 times, was equipped with the LGW K12 electro-hydraulic autopilot servomotor, standard equipment for *Luftwaffe* aircraft including the *Junkers* Ju 88 medium bomber. The K12 servomotor was also replaced, this time with an electric servomotor designed in Section 224, which was used in about five test-launches. A simplified hydraulic servomotor being built by LGW for a C-2/W-4 prototype was never tested in flight.[37]

The development of the *Burgund* and *Elsass* command-link guidance systems accompanied the early stages of the development of a fully automatic guidance system code named *Rheinland*, which was being designed to operate independently of weather conditions. For the *Rheinland* system to be effective, each missile required at least a proximity fuse so the warhead could detonate within range of a target aircraft (if a direct hit could not be achieved), but ideally be equipped with a seeker. From 1943 onwards, there was an intensive effort across German aeronautical research establishments and industry to design and develop both of these technologies. There were between 30 and 50 proximity fuse projects in development in Germany during the war. Proximity fuse technology worked on the principles of electronics, acoustic signals and infra-red radiation. The RLM made at least seven contracts with German

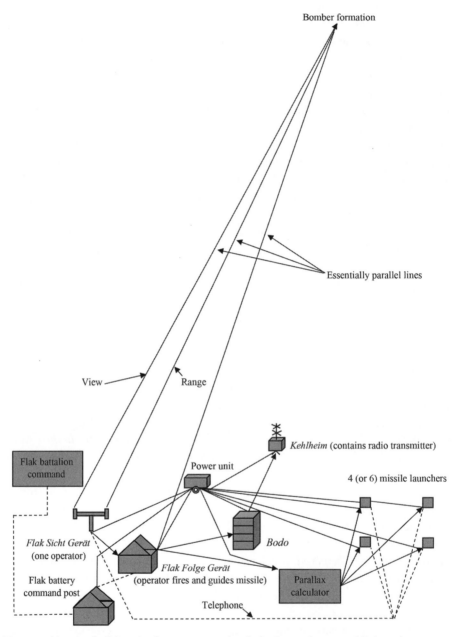

The general layout of a flak rocket battery equipped with the *Burgund* command-link guidance system. Not to scale. (Source: '*Burgund* Control Equipment for the Rocket *Schmetterling*', Intelligence Report GDM-1, Intelligence Branch, Technical Liaison Division, US Army, 28.6.1945, RG 38, National Archives College Park)

industry to develop a proximity fuse for the C-2 and the Hs 117, and at least one for the *Taifun*.[38]

The realisation of a seeker to automatically guide a missile in the terminal phase of the trajectory towards a target aircraft was arguably one of the most complicated tasks for the Germans from both a scientific and engineering perspective. Science had already proven that a seeker could detect light, acoustic signals, high-frequency radio waves or infra-red radiation. The detection techniques to choose from were: 1) passive; 2) semi-active, where radio signals transmitted from the ground towards the target are reflected back to the missile; 3) active, where the missile transmits a radio signal towards the target and detects the deflection; or 4) the use of television. The technology did not progress far beyond the laboratory, however. The possibility of developing a seeker for the subsonic Hs 117, designed to intercept bomber aircraft flying at speeds of up to 500 km/h, was impossible in 1944 and 1945. A seeker for the C-2, which was designed to travel at supersonic velocities of up to Mach 2.3, was an even more complicated undertaking. The only realistic solution to the problem was the use of a radio-operated fuse which would be fired by command from the ground on the basis of optical observation, or a proximity fuse controlled by radar coincidence.[39]

Taking a look at some of the seeker projects, in the high-frequency category there was a passive short-wave device in the laboratories of *Blaupunkt GmbH* in Berlin under Dr. Güllner; an active device code named *Dackel* (*dachshund*, sausage dog) in the laboratories of the *Forschungsanstalt der Deutschen Reichspost*; a passive

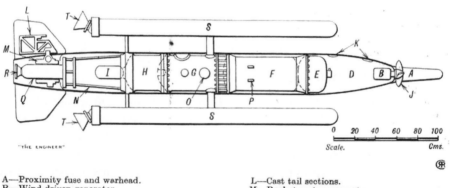

A—Proximity fuse and warhead.
B—Wind-driven generator.
D—Control compartments, radio, &c.
E—Pressure air bottle.
F—Fuel tank.
G—Centre section.
H—Rear fuel tank.
I—Speed control valve.
J—Windmill for generator.
K—Inspection ports.

L—Cast tail sections.
M—Rocket motor support.
N—Thrust rods.
O—Wing supports (main spar).
P—Forward wing supports.
Q—Reaction chamber.
R—Expansion nozzle.
S—Booster rockets.
T—Expansion nozzles of booster rockets.

Internal diagram of the Hs 117 *Schmetterling*. (Source: Eric Burgess, 'German Guided and Rocket Missiles', *The Engineer*, no. 184 (10 October 1947): 332–333)

device code named *Windhund* (greyhound); and *Moritz*, a semi-active device. In the acoustical category, there was a device being developed by *Telefunken* code named *Dogge* (mastiff). Regarding infra-red seekers, the technique had the advantage of not being detectable like the transmissions from radar devices, but was susceptible to some weather conditions. An experimental infra-red seeker design concept which utilised lead sulphide cells as the receiver element perhaps showed the most promise. One of the leading German firms that developed this technology was *Electroacoustic AG* at Kiel, work that was led by the firm's director of infra-red research and development, Dr. Edgar Kutzscher. Among Kutzscher's projects were a device for the C-2 that was designed in conjunction with Dr. Helmut Weiss at Peenemünde-East, and one for the Hs 117. Dr. Richard Orthuber at AEG was also developing an infra-red device for the C-2, code named *Netzhaut*. In the *Institut für Flugausrüstung* at the DFS under Prof. Fischel, the staff were developing a device for the *Enzian* in conjunction with its inventors, Baron von Pfieffer and the Dr. Alfred Kepka firm of Vienna, code named *Madrid*. *Rheinmetall-Borsig* was developing a device for the *Rheintochter*. In the television category, Dr. Werner Rambauske, a physicist, had a television seeker in development for the *Enzian*. Rambauske's project began in 1939 with *Gallnow und Sohn* at Stettin in Pomerania, then from 1941–43 with *Askania* in Berlin, and thereafter at the *Institut für Physikalische Forschung* at Stargard in Pomerania after the *Askania* plant was bombed.[40]

Postwar, like their counterparts in the aerodynamics and rocket propulsion fields, the leading German scientists and engineers who undertook research and/or development work with remote control and guidance technologies would draw the attention of Allied scientists and engineers who sought to exploit their knowledge.

Reorganisation of the SAM development programme, August 1944–January 1945

During 1944, the destructive effects of the Combined Bomber Offensive on German cities and industry, with the accompanying demise of Hermann Göring's prestige within the leadership of the Nazi regime and the *Luftwaffe*, in addition to the success of a combined Armaments Ministry/RLM fighter aircraft production programme in the first half of 1944, led to major changes in the air armaments industry. In June 1944, Albert Speer, the *Reichsminister* for Armaments and War Production – who had succeeded in increasing the output of new armaments in his domain despite the gradual destruction of German industry – convinced Hitler to transfer air armaments production from the RLM to the Armaments Ministry. Hitler's decision resulted in the resignation of *Generalfeldmarschall* Erhard Milch, the Secretary of State for Air in the RLM and *Generalluftzeugmeister* (GL), because he lost his posts as both the chief of supply and procurement. These two positions were eliminated on 29 July, and the transfer was officially complete on 1 August. The ramifications for the SAM development programme and other *Luftwaffe* experimental weapons programmes were that the RLM lost control over series production prior to deployment. The OKL reconstituted the reduced functions of the GL department as the *Technische Luftrüstung* (TLR, technical air armament). The TLR was established on 1 August 1944 and was based in Berlin. It was headed by the *Chef der Technischen Luftrüstung*, *Oberst* (later *Generalmajor*) Ulrich Diesing, until he was killed in a car accident on 17 April 1945. The TLR was divided into several directorates (*Amtsgruppen*) and had a staff of about 13,700 when it was created. The acute manpower shortage in Germany resulted in successive staff cuts, so by the time Germany surrendered nine months later, the number of personnel had been reduced to 4,500.[41]

The creation of the TLR did not result in any significant changes to the co-ordination of the SAM development programme by the RLM. Flak armament development continued in the TLR (TLR/Flak-E). Dr. Friedrich Halder, now an *Oberstleutnant* and *Abteilungschef* of TLR/Flak-E 5, retained overall military responsibility for the programme in collaboration with other sections within the TLR, including TLR/Fl.-E (aircraft development, for items such as electronic components) and TLR/Fl.-B (supply). The technical direction of each flak rocket project remained the responsibility of the respective prime contractors. The aeronautical research establishments remained under the control of the *Forschungsführung*, and the *Luftwaffe* experimental establishments continued to be controlled by the *Kommando der Erprobungsstellen* (KdE) in the RLM.

The summer of 1944 also saw changes at Peenemünde-East as the SS sought to gain control over the HWA rocket programmes. *SS-Reichsführer* Heinrich Himmler coveted Peenemünde-East and the A-4/V-2 programme, and since the previous year had gradually acquired a greater stake in the programme through the provision of slave labourers and the construction of the *Mittelwerk* factory. After the attempted *putsch* on 20 July 1944 by a group of German Army officers led by *Oberst* Claus von Stauffenberg, the chief of staff to *Generaloberst* Friedrich Fromm, the commander of the Replacement Army and overall head of armaments supply for the Army, Himmler took over Fromm's position. In order to prevent the SS from also gaining control of Peenemünde-East, on 1 August 1944 the Armaments Ministry converted the establishment into a state-owned company, *Elektromechanische Werke GmbH* (EMW).'

The damage that the Allies inflicted on German industry had a more significant effect on the conduct of the programme than these reorganisations. Factories and establishments were often partially or completely destroyed or dispersed to other parts of Germany. To take one example, bombing affected the supply of chemicals from the *IG Farben* plant at Leuna from January 1944, although the conglomerate's plant at Ludwigshafen-Oppau continued to operate under normal conditions until it too was bombed during September 1944. By 15 December 1944, the production capacity at the latter plant had slowed to 70 per cent, then an air raid on the night of 15/16 December 1944 made the plant inoperable. In other examples, *Siemens* was late on the delivery of all its guidance technology contracts in the second half of 1944, partly due to air raids, and three daylight AAF air raids on Peenemünde on 18 July, 2 August and 25 August 1944 caused such damage to the manufacturing facilities there that C-2 assembly operations were forced to relocate to Zwickau. These attacks did not, however, affect the test-launching operations on the Greifswalder Oie – by September 1944, the *Flakversuchsstelle* was launching six C-2 missiles on the island per month. Anglo-American bombing raids also hindered the development of the Hs 117. Attacks on BMW facilities caused delays in the delivery of the experimental 109-558 LPRE for the missile, the first of which was reportedly not delivered to HFW until August 1944.[42]

By October 1944, the RLM had selected the Hs 117 and C-2 for series production, and each missile was assigned a RLM designation, 8/117 and 8/45 respectively. The RLM decided to rush the Hs 117 into production commencing in March 1945 even though the missile had yet to be test-launched with the sustainer rocket – according to records, the first launch with the BMW LPRE did not take place until November 1944. The planned supply and series production processes for the Hs 117 were as follows.

Industrial firms contracted to manufacture components of the missile would send the parts to three factories, under the control of the Armaments Ministry, for final assembly. At full capacity, 3,000 missiles would be assembled monthly, 1,000 in each factory. From these factories, the assembled missiles would be sent to nine supply points manned by *Luftwaffe* personnel, where the fuelling of the missiles would take place with propellants supplied by *IG Farben*. Each supply point would handle a maximum of 300 missiles per month. From there, the missiles would be forwarded to the missile batteries, around 10 of which would be serviced by every supply point. TLR/Flak-E decided to split a normal flak battery into two half-batteries, with each half-battery armed with between four and six missile launchers. Ninety batteries armed with four missiles apiece would total 360 missiles ready for firing at any one time (or 540 missiles with six missiles per battery). On 22 October 1944, TLR/Flak-E 5/II B produced a manufacturing timetable that scheduled an output of 150 missiles per month beginning in March 1945, rising to the maximum output of 3,000 missiles per month by November 1945 through to March 1946. Series production of the C-2 was not planned to start until October 1945, at an output of 50 missiles per month, which would rise to 900 missiles per month by March 1946. These numbers correlate with 70 Hs 117 batteries and three C-2 batteries that the RLM estimated would be in service by the end of 1945, although well short of the 2,170 batteries that the *Luftwaffe* believed were needed to defend Reich airspace, which would have taken many more months to realise. The radar and guidance system equipment in the ground installations of the Hs 117 and C-2 were a combination of technology in service and experimental development and were planned to be standardised as much as possible. Considering the war situation, the state of German industry and other armaments priorities, the production targets in relation to the Hs 117 were very unlikely to be achieved; those in relation to the C-2 were completely unrealistic.[43]

Meanwhile, the development priority of the *Taifun* was actually raised above that of the C-2 as the RLM considered the barrage rocket to be a more promising weapon in the immediate term. In September 1944, the design of the *Taifun* was submitted to the RLM and was frozen the following month. The first batch of prototypes were test fired by November 1944, with another 10,000 ordered for further ballistics trials. The underground *Mittelwerk* factory near Nordhausen would end up manufacturing these rockets, and the projector was manufactured by *Skoda*. Although the rocket's performance was still unsatisfactory, the tests indicated the weapon to be potentially more effective than flak artillery. Compared to the standard 88mm flak shell, the *Taifun* took 14 seconds to reach an altitude of 32,800ft compared to the 28 seconds the 88mm shell took to reach 29,500ft.[44]

As for the *Enzian* and the *Rheintochter*, both missiles failed to attain the required ceiling during their respective test-launching programmes, and guidance remained a problem. In the case of the former, there were problems with the HWK LPRE. Although *Rheinmetall-Borsig* made more test-launches of the R-I than the other guided SAMs, the performance of the SPRE remained unsatisfactory. The RLM continued to finance solid propellant research – the *Versuchsanstalt Grossendorf* firm, situated about 80km east of the *Rheinmetall-Borsig* proving ground at Leba on the Hela Peninsula near Danzig, was contracted to develop a booster rocket for the *Rheintochter* called the Firework Z. Amazingly, in tests the Firework Z successfully burned 250kg of propellant in less than one second before the sustainer rocket would take over. In late 1944, the approaching Red Army forced the evacuation of the project personnel and operations to the state-owned company *Waffen-Union Škoda-Brünn* at Příbram in the Reich Protectorate of Bohemia-Moravia, where the director, SS officer Rolf Engel, had taken up a position in the test division of the company.[45]

In November 1944, the production of anti-aircraft artillery was awarded top priority by Hitler, despite the desperately needed resources for the SAM development programme. The situation was summarised in a speech on 1 December 1944 by Albert Speer at the *E-Stelle Rechlin* at Rechlin on Lake Müritz in Mecklenburg. While discussing the so-called *Wunderwaffen* (wonder weapons) – highly advanced and revolutionary armaments, including flak rockets, that the Nazi regime hoped would turn the tide of defeat and used propaganda to promote – Speer referred to the series production of the Hs 117 planned for the following spring. He appealed for additional assistance for the missile's development, since it was his view that:

> It is most important to obtain a weapon that can be used against the enemy through cloud cover in bad weather because the enemy is enabled by his radio technical equipment to fly even in bad weather. The development therefore of Flugraketen [flying rockets] reaching a height of 11 to 12,000 metres [approximately 36,000–39,350ft] is a most important problem which must have special priority.

Clearly, even with the planned introduction into service of the Hs 117, there were almost insurmountable challenges still facing the development of the experimental *Rheinland* guidance system.[46]

Through all these efforts, Germany was not only the first country to pioneer guided SAM technology, but the *Luftwaffe* was the first air force in the world to plan an air defence network armed with guided SAMs. The accelerated and desperate development efforts during the second half of 1944 pushed forward the technological evolution of the experimental weapons systems to the point where, like the fate of the Third Reich, the outcome for Germany would be all or nothing. The German scientists, engineers and technicians who were involved in the SAM development programme were either going to produce a weapons system that would help save Germany – practically impossible by the end of 1944 – or risk the products of their years of research and development falling into the hands of the enemy.

Evacuation of operations to central Germany and the end of World War II

As 1945 arrived, World War II in the European theatre entered its final stages, with Winston Churchill's strategic and ideological vision of defeating Hitler and Nazi Germany by sleeping with the devil – the alliance with the Soviet Union – about to be realised. On 12 January 1945, the Red Army began a winter offensive in Poland. Meanwhile, on the same day, Albert Speer authorised the creation of a

Table 2. The number of test-launches of each guided flak rocket that had taken place by early 1945

	Total launches	Without control	Pre-set control	High freq. control
Rheintochter R-I	88	42 (*16)	21 (8)	25 (5)
Rheintochter R-IIIf and p	7		(no data available)	
Wasserfall	28	1	–	27 (10)
Enzian	24	24 (15)	–	–
Schmetterling	80			
(a) launched from aircraft	21	–	1 (1)	20 (6)
(b) launched from ground	59	2 (1)	–	57 (31)

* The numbers in parentheses indicate tests at which stability, control or the propulsion system failed.
Source: TLR/Flak-E. 5, Az. 140 Nr. 194/45, 1945, in The Story of Peenemünde, 524

joint Armaments Ministry/RLM working group to consolidate all guided missile development. The following day Speer approved its formation, and it was called the *Arbeitsstab Dornberger*, after *Generalmajor* Walter Dornberger, who initially had the role of technical director, while von Braun was the managing director, chairman, advisor to Dornberger and head of A-4/V-2 and C-2 development. As Dornberger was overburdened with other tasks, he delegated technical management to von Braun. Von Braun's deputy was Albin Sawatzki, who had distinguished himself at *Henschel und Sohn* for his management of Tiger tank production and had been appointed the chief production planner of the A-4/V-2 in 1943. But he was also known to brutalise the slave labourers in the underground *Mittelwerk* factory. The secretary of the working group was Johannes Thiry from LGW, with 10 ordinary members who were prominent early figures in the field of guided missile development in Germany. Meanwhile, selected staffs from the German Army, the *Luftwaffe*, Peenemünde-East, the *Flakversuchsstelle* and the operations of several firms including HFW, *Kreiselgeräte*, *Ruhrstahl* and *Telefunken* were in the process of evacuating to Thuringia in a grouping of concerns known collectively as the *Entwicklungsgemeinschaft Mittelbau* (Central Construction Development Co-operative).[47]

On 27 January 1945, the *Arbeitsstab* met for the first time in Berlin. The evacuation of companies and research establishments was discussed and an assessment of the missile projects determined which remained viable for further development and production with the limited facilities in central Germany. It was decided at this meeting to reduce the tactical guided missile developments down to the three most promising projects: the C-2, Hs 117 and the *Ruhrstahl* X-4 AAM, while the *Rheintochter* and *Enzian* were cancelled. The C-2 remained the most complex design, but nevertheless still showed considerable development potential, while the Hs 117 was selected due to its simple design and ease of manufacture. The *Taifun* (which now included a solid propellant version, the *Taifun P*) was also retained, along with a ground-launched version of the 55mm solid propellant air-to-air R4M aircraft rocket, code named *Orkan* (Hurricane), which was also under development at the *Waffen-Union Škoda-Brünn* at Příbram in Bohemia-Moravia. The *Orkan* was designed for use against low-flying aircraft, and in an equally remarkable performance as the Firework Z, could apparently attain an altitude of 5km in three seconds.[48]

As already indicated, the control of German guided missile development was the object of rivalry between the SS, the German Army and the *Luftwaffe*. Himmler prevailed when the *Arbeitsstab* was subordinated under the authority of SS *Obergruppenführer Dr.-Ing.* Hans Kammler, who oversaw the construction of the *Mittelwerk* complex using slave labour and built the gas chambers at a number of death camps, including Auschwitz-Birkenau. Kammler had Göring, the now ineffective chief of the *Luftwaffe*, appoint him to lead a *Programm Brechung des Lufterrors* (Programme for Breaking the Air Terror) on 26 January 1945. All German guided missile development was now effectively under the control of the SS. On 6

February 1945, Kammler officially ordered the *Arbeitsstab* to immediately terminate the *Rheintochter* and *Enzian* projects. By this time, the total number of test-launches of the *Enzian* was reported to be 38. Kammler also ordered the termination of the HFW Hs 298 AAM project. The development of the C-2 and the Hs 117 was shut down at a later date. HFW had looked beyond the Hs 117 and planned at least two flak rockets to supersede the missile, *Projekt S2* and the *Zitterrochen* (Electric ray). By the end of the war, only aerodynamics experiments with models of the latter were done in the subsonic and supersonic regions in the wind tunnels of the AVA.[49] Also cancelled in February 1945 was the *Feuerlilie* research programme by the *Forschungsführung*. According to an Anglo-American intelligence report, one reason for the cancellation was that the factory of the firm which produced the fuselages, *Ardelt-Werke GmbH* at Breslau in Lower Silesia (present-day Wrocław in Poland), which also manufactured the fuselages of the *Rheintochter* R-I, was about to fall into the hands of the Red Army.[50]

On 31 January 1945, von Braun received an order from Kammler to commence the evacuation of the facilities and personnel (including families) of EMW to central Germany. The evacuation began on 17 February with the transfer of the personnel on the A-4/V-2 and *Taifun* projects, followed by those on the C-2 and A-4b (a winged derivative of the A-4) projects. The new arrangements in the Harz Mountains were only of a makeshift nature. The main office of EMW was set up in a former agricultural school at Bleicherode, a small cotton milling town in Halle-Merseburg. The *Flakversuchsstelle* was transferred to the village of Neubleicherode, 13km east of Bleicherode; the C-2 development group was installed in a carriage

Table 3. The organisation of the *Arbeitsstab Dornberger*, February–April 1945

SS Obergruppenführer Hans Kammler, *Sonderbeauftragter II* (Special Emissary II)
Generalmajor Walter Dornberger, *Inspekteur V-2*
Arbeitsgruppe

Chairman:	Prof. Dr. Wernher von Braun, EMW (A-4/V-2 and C-2)
Deputy:	*Dipl.-Ing.* Albin Sawatzki, *Mittelwerk* (A-4/V-2 production)
Secretary:	Johannes Thiry (LGW)
Members:	

- *Oberstleutnant* Dr. Herbert Axster, HWA (Dornberger's chief of staff)
- *Flieger-Oberstabsingenieur* Rudolf Brée, *Abteilungschef* TLR/Fl. E 9 (Hs 117, Hs 298 and *Ruhrstahl* X-4 air-to-air missiles, and the X-7 anti-tank missile)
- *Oberstleutnant* Dr. Friedrich Halder, *Abteilungschef* TLR/Flak-E 5 (*Rheintochter* and *Enzian*)
- Dr. Georg Weiss, director of the *Forschungsanstalt der Deutschen Reichspost* (television research)
- Prof. *Dr.-Ing.* Herbert Wagner, HFW (Hs 117 and Hs 298)
- *Dr.-Ing.* Ernst Steinhoff, EMW (guidance and controls)
- Dr. Wilhelm Runge, DVL (proximity fuses and seekers)
- *Dipl.-Ing.* E. Stinshoff, DFS (infra-red seekers)
- Dr. Theodor Sturm, *Stassfurter Rundfunk* (radio controls)
- Heinz Kunze, *Reichsministerium für Rüstung und Kriegsproduktion* (A-4/V-2 production).

Source: TNA, FO 1031/12, 'Brief Interrogation Report on Prof. Dr. Wernher von Braun', 8.3.1947

shop in the village; and Section 224 was housed in surface workings of a nearby mine. The headquarters of the *Arbeitsstab Dornberger* was established at Bad Sachsa, a hot springs resort about 20km north of Bleicherode. Several large test areas were set up for the A-4/V-2 and C-2, and smaller ones for the Hs 117, *Taifun* and X-4. Elsewhere, the *WVA Kochelsee* in southern Bavaria was not in imminent danger and remained in place, and aerodynamic and thermodynamic research continued. The *E-Stelle Karlshagen* was evacuated westwards to the *Seefliegerhorst Wesermünde*, a seaplane base about 4.5km north by north-west of Bremerhaven. The evacuation of the inter-service installation at Peenemünde was completed by the third week of March.[51]

By February 1945, the overall coordination of the SAM development programme was in a state of confusion and showing signs of complete breakdown. *Rheinmetall-Borsig* was unaware of the cancellation of the *Rheintochter* project by the *Arbeitsstab Dornberger* and the evacuation of Peenemünde-East with the associated loss of the test-launching facilities on the Greifswalder Oie. In around mid-February, the firm was still making arrangements for further test-launches of the *Rheintochter* at Peenemünde-East using the *Kogge* radio transmitter with the *Rheinland* guidance system; and on 20 February the company released a memorandum announcing that a further 20 *Rheintochter* R-I and 15 R-III test-launches were planned to take place.[52]

The status of the Hs 117 project was as follows. The design of the sixth prototype version, the A2 type VI, was completed by January 1945, although aerodynamic research at the DVL in Berlin continued until as late as March 1945. Series production was scheduled to commence that month in an underground factory called *Hydra*, at Himmelberg near the village of Woffleben, 3km north by north-west of the *Mittelwerk* factory at Niedersachswerfen. The original intended purpose of the *Hydra* factory was the production of *Junkers* aircraft, and a floorspace of 123,000 square metres was planned. Tunnelling work at the site, referred to by the code name *Anhydrit* (after the anhydrite gypsum in the hills), commenced in mid-April 1944 using slave labourers that were supplied by the SS from concentration camps. When HFW moved its guided missile production operations there in early 1945, the tunnelling at the site was incomplete. The estimated time to assemble one complete Hs 117 in the *Hydra* factory was about 780 hours – the body, wings and tail about 80 hours; the booster rockets about 50 hours; the LPRE about 250 hours; and the electrical apparatus about 400 hours. The planned production of the missile was also affected by the impact of Allied bombings raids and technical issues. Two serious setbacks were further delays in the delivery of the BMW 109-558 LPRE and the delivery of the booster rockets from *Schmidding*. BMW had difficulties in manufacturing the oxidiser tank, which was considered too heavy for the missile. The problem led the designers of the Hs 117 to consider building a whole new missile with a lighter tank. A warhead had been designed and appeared to be relatively successful. Experiments had shown that a device which contained about

40kg of powder explosive detonated at a distance of 7 metres could take down a Boeing B-17 Flying Fortress. In addition to all these technical problems, a suitable proximity fuse and seeker were not realised.[53]

By February 1945, the command-link guidance system for the Hs 117 remained the *Burgund* using the optical tracking technique, while the development of the *Elsass* system for the C-2 proceeded more slowly and was still unfinished. Although experimental work on the *Rheinland* system had clearly made some initial progress, there was a considerable amount of further research and development that needed to be accomplished before the entire system could be built and tested. The evacuation of EMW and *Telefunken* to the Harz Mountains, and the *E-Stelle Karlshagen*, effectively ended any further development of the system.[54] Anglo-American electronic counter-measures would have been a major problem for the Hs 117 missile batteries. For example, there was the use of radio jamming equipment or the deployment of metallised strips of paper, code named 'window', against the German radar network that had so successfully rendered the *Luftwaffe* anti-aircraft artillery defences ineffective. These tactics would have presented significant challenges for the operation of the *Burgund* command-link system, as Herbert Wagner, the chief designer of the Hs 117, remarked in 1956:

> Based on the test launchings with the Hs 117, and on experience with the Hs 293, all concerned felt certain that this weapon would be successful. Large scale production was initiated and in progress, in spite of the fact that the sustainer rocket was not yet operating at the desired degree of efficiency. However, in spite of the use of the 40-centimetre wavelength for the radio link, there is scarcely any doubt that enemy jamming would soon have made the weapon unsuccessful.[55]

Amid the chaos of the rapidly collapsing fronts and the shattered industrial sector, along with the general difficulties that were being experienced with transportation, development and production, the complete termination of the Hs 117, C-2 and X-4 projects took place by the end of March 1945. The development of small rockets such as the *Taifun* and *Orkan* continued for another week or two beyond that date. The manufacturing of the additional 10,000 *Taifun* rounds and the projectors were undertaken in the *Mittelwerk* factory, and further test-launches were reportedly carried out at a site near Woffleben. The *Taifun* had been scheduled for introduction into service by 1 September 1945; 400 batteries of 12 projectors per battery were planned, with each projector mounted on a converted 88mm gun carriage that could hold 30 rockets.[56]

Faced with inevitable defeat, the staff of the research establishments, experimental and testing establishments, armaments manufacturers and government departments made difficult but necessary decisions about how to preserve the archives in their possession. The outcomes of these decisions – after the concerns that were connected to the SAM development programme in western, northern, central and southern Germany surrendered to the Allies – are discussed in Chapter 3. The staff at some places simply retained the archives in the original location(s), which occurred at the LFA near Völkenrode; in other instances, the staff chose to microfilm the archives before possibly burning the originals, as happened at HWK at Kiel; the staff at a number of establishments buried the archives in sealed containers, which was done by the staff at the *Messerschmitt AG Oberbayerische Forschungsanstalt* at Oberammergau in Upper Bavaria, the DFS at Ainring and the *E-Stelle Karlshagen*; while some establishments deposited their archives in caves or mines, as was done by the personnel at Peenemünde-East with the approximately 14 tons of documents that contained all the scientific and technical data on the A-4/V-2, A-4b, C-2 and *Taifun* projects.

The approach of American troops towards central Germany forced the evacuation of around 450 guided missile specialists from their ephemeral enterprise in the Harz Mountains south to Upper Bavaria on 6 April. The *Taifun* project was evacuated to the town of Gmunden, about 15km south-west of Linz in Austria. Gmunden was close to the *Zement* underground factory, which had been planned as a possible site for the production of the A-4/V-2. The evacuation to Upper Bavaria meant that the German SAM development programme was essentially at an end. There were no longer any central authorities or organisations to co-ordinate missile and rocket development, and no manufacturing and testing infrastructure. In the Harz Mountains, undamaged infrastructure remained in place, while around Bleicherode, mechanical components and equipment for the Hs 117 and C-2 were left behind, burned, damaged or hidden in the surrounding area.[57]

Conclusion

The German SAM programme thus did not complete the development of any missile systems. Overall, the late start of the programme, the experimental technology, the vicissitudes of compressed research and development phases, and the exigencies of the supply and production situations were all factors that made it impossible for the *Luftwaffe* to introduce an effective first-generation missile system within a timeframe that was constantly being reduced as the war gradually turned against Germany. When the programme commenced on 1 September 1942, the Third Reich was at the peak of its power. As the conflict progressed and Germany continued to lose territory, the many complex problems that required solving, the delays caused by bombing raids, lack of priority and the increasing perilousness of the strategic situation forced the RLM to seek simpler and more immediate solutions. The military and civilian specialists who were involved in the programme were well aware that if an operational weapon did enter service, it was likely to have many technical shortcomings in the near term. A compromise had to be made between a missile system that was suitable for the air defence network and the knowledge that constantly updated scientific developments and enemy counter-measures may have soon rendered the technology obsolete. If the RLM authorised the development of the HFW Hs 297 in 1941, the missile system would almost certainly have become operational by 1945.

It can definitely be argued that the decision by the RLM to have four major flak rocket projects in development by the end of 1943 led to an unnecessary waste of material and human resources and a duplication of effort. However, the personnel in GL and TLR/Flak-E clearly chose to standardise the technology in order to consolidate and economise the resources across the programme. The radio remote control technology supplied by *Telefunken* and *Stassfurter Rundfunk* were two examples, as were the equipment in the ground apparatus of the *Burgund* and *Elsass* guidance systems for the Hs 117 and C-2 respectively.

The original scientific and engineering advances that were made in Germany during the war were considerable and very significant. The accomplishments by an industry that was under constant aerial bombardment, with the many complicated problems – both theoretical and practical – that required solving in compressed research and development phases, were remarkable. The RLM and the *Luftwaffe* sought to counter the numerical superiority of the Allies' aircraft production with revolutionary technologies, and the conduct of the aerial war forced the RLM to rapidly develop more advanced anti-aircraft weapons. Due to the successful work by German scientists, engineers, and technicians in areas such as missile design and construction, supersonic aerodynamics, telemetry, flight behaviour, rocket engine design and liquid propellant research, by May 1945 Germany was, overall, several years ahead of the Allies in the development of the technology, despite Germany's relative lack of progress in microwave research and development. The eventual transfer and exploitation of the German scientific and technical knowledge by the Allies were direct consequences of the outcome of the war and the ground-breaking advances that were made in Germany between 1941 and 1945. In the next chapter, the comparative British and American progress with SAM technology during the war is discussed and analysed and elucidates for the reader the significant gap in SAM research and development between the belligerents by the end of the war.

CHAPTER TWO

Comparative British and American SAM Research and Development up to 1945

The development of what were referred to in the UK as 'guided anti-aircraft projectiles' and in the US as 'anti-aircraft guided missiles' was not seriously considered until 1944. Intelligence on German rocket development at Peenemünde, in addition to the operational use by the *Luftwaffe* of the HFW Hs 293 and *Ruhrstahl* X-1 in the Mediterranean theatre, stimulated research and development of surface-to-air guided missile systems in both countries. Another factor that stimulated research and development was the introduction into combat in 1944 of the British Gloster Meteor and German *Messerschmitt* Me 262 first-generation jet fighter aircraft. It was realised that the increased speed and manoeuvrability of these aircraft would soon render anti-aircraft artillery obsolete. Military planners in both countries were also aware that the first generation of jet-powered bomber aircraft being designed (the English Electric Canberra light bomber and the Boeing B-47 heavy bomber) would also necessarily require more advanced anti-aircraft defences. A third factor that stimulated research and development – albeit in the short term – was the need to develop effective counter-measures against the Japanese use of suicide aircraft to attack US Navy vessels in the Pacific theatre. This chapter discusses the background to the decisions made in the UK and the US to initiate SAM research and development programmes, focuses on the exchanges of scientific and technical intelligence between the two allies and compares the progress made with SAM research and development in both countries and with Germany.

British 3-inch rockets, the CP 3 and the creation of the GAP organisation

The origins of British SAM technology can be found in the mid-1930s, when the Ordnance Research Department (ORD) initiated a 3-inch diameter (76mm) solid propellant anti-aircraft rocket project. To undertake this task, a special unit was created in July 1936 within the ORD under the Director of Ballistics Research, Dr. Alwyn Crow. Around the same time, there was also a 2-inch anti-aircraft rocket in development. These rockets were designated with the security name Unrotated Projectile (UP). The first test firings of the 3-inch rocket took place during the spring of 1939 in the British colony of Jamaica. To further develop and test-fire the new rocket, in the summer of that year the newly created Ministry of Supply (MoS) founded the Projectile Development Establishment (PDE) at Aberporth in Wales. The operational ceiling of the 3-inch rocket was 18,000ft, 6,000ft higher than shells fired by the standard Swedish Bofors 40mm light anti-aircraft gun. It was constructed of lightweight materials with stabilising fins and a screwed ring to receive the warhead at the forward section, and was fuelled with cordite, a double-base propellant invented by Sir Frederick Abel and Sir James Dewar in 1889, consisting of 58 per cent nitroglycerin (by weight), 37 per cent guncotton and 5 per cent vaseline. The rocket was a multiple-purpose and versatile weapon which could be fired from Royal Navy warships as either an anti-aircraft or shore bombardment weapon, from RAF fighters against air or ground targets, and used as an air-defence weapon by the British Army's Anti-Aircraft (AA) Command. The 3-inch rocket, also referred to as the 'Z' projectile, entered service with the British armed forces in 1940, with the Admiralty given preference because the weapon was badly needed for the protection of merchant shipping. In November 1940, the first Z battery in AA Command became operational, under the command of Major Duncan Sandys (pronounced 'sands') of the Royal Regiment of Artillery, who was the son-in-law of the British Prime Minister Winston Churchill. Each Z battery was armed with 10 rounds per projector. Multiple projectors were designed, with twin, nine and 20 barrels. A battery contained 128 barrels and was manned by 274 personnel, which from 1941 also included members of the Home Guard. By the end of March 1941, 18,600 rockets had been delivered, of which 8,400 were allocated to AA Command. The security name UP was dropped after 1942.[1]

Following the armistice between Germany and France on 22 June 1940 – which left the British Commonwealth to fight Nazi Germany – the UK sought to make a new alliance with the US. Technical information about the 3-inch rocket was passed on to the Americans after Churchill approved the exchange of classified scientific and technical information with the US through the Tizard Mission in August 1940. In 1941, the exchanges with the Americans extended to other technologies that later in the war would be important to American SAM development, including industrial techniques to manufacture dry-extruded Ballistite propellants (Ballistite was a smokeless propellant invented by Alfred Nobel in 1887), microwave radio research and radar technology. As the war continued, the technical exchanges extended into the domain of LPRE research and development that was being undertaken in both countries.[2]

Also in 1941, the MoS awarded a £10,000 research contract to the Shell International Petroleum Company to design a rocket assisted take-off (ATO) unit using any fuel except cordite, which was then in short supply. The man who designed the rocket under the contract was Isaac Lubbock, an engineer with the Asiatic Petroleum Company (a subsidiary of Royal Dutch/Shell). Lubbock designed an ATO unit that used a combination of aviation fuel, water and liquid oxygen (LOX) as the oxidising agent, which he dubbed 'Lizzie'. The idea of using LOX in an ATO unit was also being explored in Germany. In January 1939, the RLM and the HWA commenced an inter-service project to develop a 1,000kg-thrust LOX/petrol ATO unit for heavily laden bombers; however, the development ran into problems in late 1941 and the technology never came to fruition. With Lubbock's engine, a five-second run was achieved on 15 August 1942, and by the end of September 1942 the engine could run for 30 seconds. On 7 May 1943, a visiting delegation of scientists observed a demonstration run of the engine using petrol instead of aviation fuel, and on this occasion the duration of the run was 23 seconds. The Lizzie was apparently performing better than nitric acid and aniline ATO units that were in development in the US in the Jet Propulsion Laboratory (JPL) of the Guggenheim Aeronautical Laboratory at the California Institute of Technology (GALCIT), under contract to the US Army Air Forces. In the summer of 1943, Lubbock visited the US and observed the American progress with the technology. He was informed about the American experiments with nitric acid and aniline, and the development of a pump as an alternative to the use of heavy compressed air bottles to force the propellants into the combustion chamber, two technologies that were also currently being developed in Germany. The information that Lubbock brought back to the UK considerably influenced the investigations that were being conducted by British experts into guided projectiles on the basis of intelligence on German progress at Peenemünde.[3]

While in Germany it was the Anglo-American bombing campaign that stimulated research and development of guided SAMs for use against high-altitude targets, in the UK the introduction into combat of the HFW Hs 293 and *Ruhrstahl* X-1 by the *Luftwaffe* in the Mediterranean theatre during the second half of 1943 prompted the first serious investigation into the development of surface-to-air guided missiles for the British armed forces. These German weapons were deployed from the parent aircraft at distances, altitudes and positions from the target vessel that made naval anti-aircraft guns ineffective as counter-measures against the parent aircraft, although both weapons could be shot down by naval guns while in flight. The Hs 293 could be launched at altitudes from as low as 1,000 metres up to 7,000 metres (3,280–22,960ft), and at a distance from the target vessel of between 12km and 16km. The X-1 could be dropped from an altitude of between 4,000 and 7,000 metres (13,120–22,960ft). The Hs 293 was introduced into service on 25 August 1943, and claimed its first casualty two days later, when one of the missiles sunk the Royal Navy corvette HMS *Egret* in the Bay of Biscay. The X-1 went into service two days later, and claimed its first casualty on 9 September 1943 with the sinking of the Italian battleship *Roma*, which was on its way to the British naval base on Malta after Italy surrendered to the Allies. The *Luftwaffe* continued to use both weapons in the Mediterranean for the next six months. Although the use of a number of effective counter-measures by the Royal Navy and the RAF against the Hs 293 and X-1 – including fighter aircraft, radio jamming and smokescreens – forced the *Luftwaffe* to quit the anti-shipping campaign by 4 March 1944, the deployment of the *Fieseler* Fi-103/V-1 against Britain beginning in June 1944 necessitated continued research into the fledgling technology.

The Admiralty commenced an investigation into the various technological problems of shipborne anti-aircraft missiles during the autumn of 1943. As the *Luftwaffe* air offensive against Britain had ceased the previous year, improvements to anti-aircraft defences by this stage of the war were not a very high priority for British military planners; nevertheless, the preliminary studies by the Admiralty led to the creation of an inter-service Guided Anti-aircraft Projectiles (GAP) Committee in early 1944. The first meeting took place on 16 March 1944 and was chaired by Canadian chemist Sir Charles Goodeve, the Assistant Controller for Research and Development at the Admiralty, with Dr. Herbert Gough, the Director-General of Scientific Research and Development at the MoS, as vice-chairman. The other attendees were senior representatives from the Admiralty, the War Office, the MoS and the Ministry of Aircraft Production (MAP). The requirements of the Royal Navy and the British Army were indicated at the meeting, and it was agreed that these were very similar and a joint investigation was desirable. The terms of reference of the committee therefore were:

> To be responsible for the initiation and direction of all research required to assess the prospects of meeting the Admiralty and War Office requirements for guided anti-aircraft projectiles.

To study the technical aspects in more detail, four subcommittees were created. The subjects of the subcommittees and their chairmen were as follows: Propulsion, under Dr. (later Sir) Alwyn Crow, Controller of Projectile Development, MoS; Radio and Radar, Dr. Paris, Controller of Physical Research and Signals Development, MoS; Aerodynamics, (later Sir) Ben Lockspeiser, Director of Scientific Research, MAP; and Stabilisation and Servos, Colonel A. V. Kerrison, of the Admiralty Gunnery Establishment. In 1944 and 1945, a number of other committees, subcommittees and panels were created by the GAP organisation to study the various subsystems in SAMs.[4]

At the second meeting of the GAP Committee on 27 April 1944, the subcommittees presented their reports and the Naval and General Staffs tabled more specific requirements. The Admiralty requirements, similar to those of the Army except with a restriction on the weight of the missile, were for a:

> GAP whose maximum performance is to be capable of engaging directly approaching aircraft flying at heights of up to 40,000 feet and speeds of 500 miles per hour [800 km/h], so that there is a good prospect of destruction before the aircraft reaches the point of release of free-falling guided bombs.

The size and weight of the projectile and associated projection and control apparatus should be such that they can be employed in any warship of the size of a destroyer or larger. This limits the weight of the projectile to a maximum of 500 lbs [227 kilograms]. If the weight is much in excess of this, the weapon will be limited to cruisers and above.

The projectile must be such that it can be guided to the target within functioning distance of its proximity fuse and carry such a charge as to be certainly lethal at the maximum functioning distance of the fuse.

The British had clearly identified the general principles and technical specifications of a surface-to-air guided missile system – albeit largely theoretically – around three years after the Germans. As a result of the deliberations at this meeting, the GAP Committee made its first report to the respective ministers. In this report, the committee concluded that while there was a reasonable prospect of developing a GAP to meet Admiralty and War Office requirements, the required research and development effort would be very great, and in their opinion the minimum period of development under the most favourable circumstances would be two years.[5]

In 1944, the GAP organisation considered turbojet, ramjet, SPREs and mono- and bi-propellant LPREs as the methods of propulsion, with the ramjet considered as outstandingly the most promising. Ramjets are air-fed jet engines which rely on the forward motion of the missile for a supply of oxygen as the oxidising agent to mix with the fuel. The air intake collects and diffuses the air to a low velocity before it enters the combustion chamber. Once the air has travelled through the intake ducts, it is mixed with solid or liquid fuel in the combustion chamber which then ignites, propelling the missile forward. A ramjet therefore requires the missile to be propelled to a sufficient velocity for the ramjet to function, provided there is sufficient ram pressure (air pressure obtained through forward motion) and a sufficiently high temperature in the combustion chamber for the ramjet to produce a net forward force or thrust, which ejects a high-velocity jet in the process. Ramjets are most effective at speeds between Mach 2 and 4. Ramjet propulsion could not be selected for the GAP due to the lack of experimental development, but studies of the technology continued.[6]

A LPRE was determined to be the most feasible propulsion system in the meantime. During March and April 1944, one of four subcommittees of the GAP Main Committee, the GAP Propulsion Subcommittee – of which Lubbock was a member – assessed the possible liquid propellants for the propulsion system, including nitric acid and aniline, and hydrogen peroxide. In the subcommittee's first report to the GAP Main Committee on 26 April 1944, nitric acid and aniline were considered as the most likely propellants, the supply of which was satisfactory but posed potential problems on board a ship. A two-stage missile was considered essential. The booster stage (solid propellant), estimated to weigh about 500lbs (227kg), would bring the missile to a supersonic velocity of over Mach 1.3 in five seconds, then the sustainer, powered with a nitric acid/aniline LPRE and estimated to weigh about 600 lbs (272kg), would attain a velocity of around Mach 2 and bring the missile to an altitude of 40,000ft. The idea for a nitric acid/aniline LPRE was undoubtedly influenced by the information Lubbock had brought back from the US, as at this stage the subcommittee did not recommend a LOX engine, despite the technical knowledge acquired from Lubbock's LOX/petrol ATO unit. It appears the success of the A-4/V-2, first fired against the UK on 8 September 1944, strongly influenced the decision on the choice of propellants. By 26 September 1944, the date of the first meeting of the GAP Working Committee (set up to proceed with research into a GAP on a scale fixed by the Main Committee) Technical Panel C (Liquid Fuels) – set up on 18 September 1944, composed of representatives from the MoS Armaments Research Department (ARD) and Armaments Design Department (ADD) and the Asiatic Petroleum Company – a nitric acid and aniline LPRE had been rejected in favour of LOX/petrol, hydrogen peroxide/alcohol or nitromethane. On 1 November 1944, Technical Panel C decided that the LPRE should use LOX and alcohol, which was changed to LOX/petrol about two weeks later. As a result, the GAP was dubbed the Liquid Oxygen-Petrol/Guided Anti-aircraft Projectile (LOP/GAP). Experience would later show – and validate the German and American preference for nitric acid as an oxidiser in LPREs for SAMs – that LOX was totally unsuitable as an oxidiser in LPREs for any tactical guided missiles.[7]

Meanwhile, the British Army AA Command – responsible for the air defence of the UK – had a rudimentary solid-propellant SAM in development (compared to the *Rheintochter* prototypes), the CP 3. The provenance of the project was a proposition in a paper written in 1942 by a British Army officer, Captain H. B. Sedgfield of the Royal Electrical and Mechanical Engineers (REME). Some initial research was done that year by a small team at the electronics firm A. C. Cossor Ltd. under the director of research, L. H. Bedford, on a projectile to be guided by a radio beam. A. C. Cossor manufactured the radio receiver equipment in the Chain Home radar stations that were constructed in 1938 and helped the RAF win the Battle of Britain in 1940. At some point during 1943 or early 1944, the CP 3 project was expanded after the commanding officer of AA Command, General Frederick Pile, authorised AA Command to work in conjunction with A. C. Cossor to undertake further development. A. C. Cossor undertook all the theoretical analysis, design, development and construction of the guidance system, while AA Command provided the propulsion, servo system, a servo simulator, assembly facilities and a firing range at the experimental establishment at Walton-on-the-Naze on the coast of Essex. The method of propulsion was rather simplistic – six 3-inch rockets were wrapped around the fuselage. The contemplated launching method resembled the technique used by the Germans to launch the *Rheintochter* and the *Enzian* – the platform of a standard British 3.7-inch (94mm) heavy anti-aircraft gun. The dimensions of the CP 3 were 2 metres in length and 267mm in width, with a single pair of aerodynamic surfaces with a wingspan of 838mm and a launch weight of 145kg. The purpose of the CP 3 project was not to be a long-term research and development programme, but to build a suitable prototype missile as soon as possible and to demonstrate the practicability of the confinement of a self-guided projectile within a conical scan radar beam.[8]

The British LOP/GAP programme

On 12 December 1944, the ADD, with assistance from the National Physical Laboratory (NPL) at Teddington in Middlesex, completed a paper on a proposed design and specifications of the LOP/GAP.[9] The contrast between the scientific and engineering knowledge that was embodied in the German missiles then being developed and the British combination of theoretical and practical knowledge that would be in the LOP/GAP was very great. For instance, in the UK there was little experience with LPRE technology besides Lubbock's engine and the scientific and technical intelligence on the A-4/V-2, while the breadth and depth of rocket propellant research in Germany was far more extensive.[10] Perhaps the most significant handicap was in the domain of supersonic aerodynamics, despite the British historical technological attainments in wind tunnel technology. The world's first supersonic wind tunnel was built at the NPL by Sir Thomas E. Stanton in 1921, and by 1945 the High-Speed Laboratory in the Aerodynamics Division of the NPL had two supersonic wind tunnels, one with a circular working section 1ft in diameter, the other 5-inches x 2-inches.[11] However, the working sections of these tunnels were too small to accommodate scale models for testing, which then had to be followed by test firings to verify the results. In contrast, the supersonic wind tunnels in Germany could accommodate scale models. The problem had been summed up in the first report of the GAP Aerodynamics Subcommittee on 24 April 1944:

> Supersonic wind tunnels are required to get the necessary control data. If work is to proceed now on a GAP it must be of the nature of an inspired guess.[12]

At that time, the most suitable tunnel was the subsonic High-Speed Wind Tunnel at the Royal Aircraft Establishment (RAE), Farnborough. Built between 1939 and 1942, it had a 10ft x 7ft working section which could accommodate complete models of fighter aircraft and could generate a top speed slightly higher than Mach 0.8.[13] Plans were being made to construct a larger supersonic wind tunnel in the UK, but it would not be operational for several years. In the meantime, the best way to acquire data at supersonic velocities was by launching prototype missiles, and then only from several seconds of flight, which was an inefficient, time-consuming and expensive process.

From late 1944 to mid-1945, British engineers designed a missile that was mostly based on British ideas, innovations, and technology – there was no intelligence on German SAMs – that did not rely very much on the known German techniques. The LOP/GAP was designed as a supersonic, two-stage SAM with a solid propellant

Table 4. Tentative specifications of the LOP/GAP sustainer rocket, 12 December 1944

Ratio of LOX to petrol by weight:	1.39: 1
Specific Impulse:	215 seconds
Reaction temperature:	2,400 degrees Celsius
Required thrust:	1,075lbs (488kg)
Warhead:	125lbs (56.75kg)
Propellants:	125lbs
Body and propellant tanks:	50lbs (22.7kg)
Radio and batteries:	50lbs
Servomotors and power sources:	25lbs (11.35kg)
Wings and control surfaces:	25lbs
Combustion chamber and igniter:	13lbs (5.9kg)
Cordite fuel expulsion gear:	6lbs (2.72kg)
Pipelines and fittings:	6lbs
Total mass excluding solid propellant booster:	425lbs (193kg)

Source: TNA, AVIA 6/15499, 11-12

booster and liquid propellant sustainer rocket in a tandem configuration. According to one set of figures, the weight of the booster stage was 609lbs, and the sustainer, fully fuelled, weighed 469lbs, which brought the total weight to 1,075lbs, or 488kg, at launching. Another set of figures puts the weight of the sustainer at 425lbs. The aerodynamic sustainer had four wings of a rhomboidal shape at the mid-section of the fuselage in a cruciform configuration, with a similar arrangement of the control surfaces at the rear, near the centre of gravity. The radio-control equipment and servo gear were to be positioned between the petrol tank and the combustion chamber near the control surfaces, and the warhead was to be situated at the front of projectile.[14]

Like the Germans, British (and American) engineers found that ATO units could be modified as propulsion systems for the experimental weapons. The companies that developed and produced these rockets found a new purpose for their designs. The LOP/GAP booster was powered with seven British 5-inch (12.7cm) solid propellant ATO units. The rockets were arranged in a cylinder 15 inches (38.1cm) in diameter, and 4ft 6 inches (137.2cm) in length and could produce 11 'g' (gravitational force) of thrust for four seconds. The most obvious German influence in the LOP/GAP design – from technical intelligence on the A-4/V-2 at the RAE, where recovered parts of the German missile were sent for analysis and reconstruction – was in the LPRE. The Asiatic Petroleum Company, in collaboration with the ADD, designed a combustion chamber that was a scaled-down version of the regeneratively cooled design in the A-4/V-2. However, the engineers who designed the LPRE for the LOP/GAP chose to stick with the propellant combination of LOX and petrol that had been successfully tested in Lubbock's ATO unit, rather than LOX with ethanol or methanol that were the preferred fuels at Peenemünde-East. Another departure from the German design was the positioning of the fuel and oxidiser tanks, which were placed the opposite way around as in the A-4/V-2. Also, the propellant feed system used cordite to force the propellants into the combustion chamber instead of pressurised gas or steam systems that were favoured by German designers. Later on, the launcher was based on the standard 3.7-inch heavy anti-aircraft gun platform, the same mounting that was used to launch the CP 3. For guidance, the GAP organisation had firmly decided by 1945 to use the beam-rider technique with an American Bell Telephone Laboratories (BTL) SCR-584 microwave gun-laying radar and predictor on S-band (Super High Frequency, between 2 and 4 GHz). Experiments had already been done in the UK using the SCR-584 to track 3-inch rockets fired from the ground; by October 1944, the radar had successfully tracked one of the projectiles to an altitude of almost 30,000ft.[15]

By early February 1945 – over two years since German engineers at *Rheinmetall-Borsig* and Peenemünde-East began drawing up plans of the *Rheintochter* and the C-2 respectively – the design of the first prototype missile, minus guidance system equipment, was complete. The materials to be used in the LOP/GAP were fixed at a meeting held on 31 January 1945. Production orders for a first series of parts for 12 LOP/GAP prototypes were placed with British industrial firms. The LOP/GAP was originally planned to be fabricated as far as possible out of a light alloy metal – except for the combustion chamber, which was to be made of steel – but a decision was later made to use mild steel instead of the light alloy. The contractors included the Aluminium Plant and Vessel Company Ltd., which manufactured the propellant tanks; Rubery-Owen, which manufactured the combustion chamber and burners; and the radio-control system was designed by A. C. Cossor in conjunction with the Radar Research and Development Establishment at Great Malvern in Worcestershire. On 6 June 1945, construction of the prototypes commenced, which were dummy rounds to begin with. After the final design stage had been completed, the first two test-launches of boosted dummy rounds of the LOP/GAP, both uncontrolled, took place on 1 August 1945 – almost 22 months since the first *Rheintochter* R-I was test-fired at Leba in Pomerania on 12 October 1943 – at the Ynyslas Experimental Gunnery Establishment range in Cardiganshire (now Ceredigion), Wales. During the

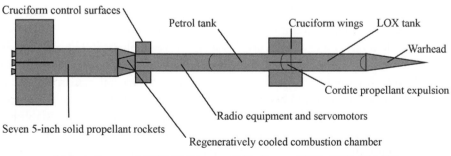

9.5-inch diameter LOP/GAP, February 1945. (Source: TNA, AVIA 6/15499)

Table 5. The British government departments and research and experimental establishments that were involved in the LOP/GAP project during World War II

• Admiralty Gunnery Establishment (AGE), Portland, Dorset:	servomechanisms
• Admiralty Signals Establishment (ASE), Haslemere, Surrey:	radio control
• Armaments Design Department (ADD), Fort Halstead, Kent:	overall missile design
• Armaments Research Department (ARD), Fort Halstead, Kent:	propulsion
• National Physical Laboratory (NPL), Teddington, Middlesex:	aerodynamics
• PDE, Aberporth, Pembrokeshire:	propulsion and test firings
• Radar Research and Development Establishment (RRDE), Great Malvern, Worcestershire:	radio control and radar
• RAE, Farnborough, Hampshire:	aerodynamics and propulsion
• Signals Research and Development Establishment (SRDE), Christchurch, Dorset:	telemetry equipment

Source: TNA, AVIA 6/15499

test firings, radar was used to track the missiles, but in each instance there was bad 'clutter' (attenuation of the radio signal) that resulted in the failure of the radar to track both rounds. A positive result was the satisfactory separation of the sustainer from the booster at an altitude of about 900 metres (2,950ft).[16]

Meanwhile, the development of the CP 3 was progressing. By June 1945, between 12 and 20 men at AA Command and a greater number at A. C. Cossor were engaged on the project. The first test-launch probably took place at the Walton-on-the-Naze experimental establishment during either June or July 1945. By the end of the war against Japan, two prototypes had been test-fired. The data obtained from these launches provided some basic information on aerodynamics, missile behaviour and the performance of the solid propellant rockets. At some point in 1945, the project was assigned the code name 'Brakemine'. When AA Command was dissolved in late 1945, the GAP organisation took over the project.[17]

As these two guided missile projects slowly progressed in the UK, and scientific and technical exchanges with the US deepened, British government officials were sent to the US to collect information on the latest American research and development. The MAP dispatched a Guided Missiles Mission to the US, which was subsequently joined by an attached representative from the GAP organisation. In December 1944, the GAP representative forwarded information on American developments in fields including control systems, telemetry, aerodynamics, propulsive ducts (work done at the California Institute of Technology [CIT] and by the National Advisory Committee for Aeronautics [NACA]), liquid propellant rockets, brief details on current American projects and a draft General Staff specification for a GAP. Formal arrangements were subsequently made between both countries to exchange information about guided missiles, and a committee was established during February and March 1945 for this purpose, the British-Washington Guided Missiles Committee. Aspects of the LOP/GAP design subsequently influenced SAM development for the US Army.[18]

American departments likewise sent representatives to the UK. They included the Office of Scientific Research and Development (OSRD), a civilian government department that co-ordinated rocket and guided weapon research and development for the US Army and US Navy; the US Army Ordnance Department; and the US Navy Bureau of Ordnance (BuOrd). The American personnel liaised with the GAP organisation in order to exchange technical information and monitor the current progress of projects and facilitated the exchange of liaisons and visitors. American liaison officers also brought back knowledge on the A-4/V-2 from the RAE. The man whom Vannevar Bush, the director of the OSRD, nominated to liaise with the GAP organisation was Dr. Guy Stever, a physicist and scientific liaison officer attached to Section T in the OSRD. Section T had successfully overseen the development of the first radio-activated proximity fuse for the US Navy in collaboration with the Carnegie Institute in Washington DC and later with the Applied Physics Laboratory of the Johns Hopkins University (APL/JHU) at Silver Spring in Maryland. In the spring of 1944, Bush sent Stever to London to obtain information on the GAP programme and other British and German scientific developments. After D-Day on 6 June 1944, Stever went to France on an intelligence-gathering mission as a representative of the OSRD to examine German radar and guided missile technology, including as a member of a joint British-American Combined Intelligence Objectives Subcommittee (CIOS) team which investigated targets in the Paris area soon after the city was liberated in August 1944. In October 1944, he returned to the US to report to Bush on the GAP programme and other guided missile developments, and subsequently returned to Europe to participate in a number of other investigations.[19]

US Army projects: ORDCIT, Project Nike and GAPA

Following the entry of the US into World War II on 7 December 1941, the first major rocket project for the American armed forces did not commence until mid-1943, with the development of an aircraft-launched rocket for the BuOrd. The BuOrd contracted the research and development of a potential weapon to the CIT, which was headed by Dr. Charles C. Lauritsen, who had campaigned in 1941 for the armed forces to accelerate rocket development using the British example of dry-extruded Ballistite propellant. The outcome was the 3.5-inch aircraft rocket. In the US Army, rocket weapons had become recognised as a distinct category of armaments like artillery and tanks. The Technical Division of the Office, Chief of Ordnance (OCO) therefore set up a Rocket Branch in September 1943 to undertake research and development with the technology. The first chief of the new branch was Colonel Dr. Gervais Trichel, an electrical engineer by training. The rocket programme for the US Army was originally located at the Naval Powder Factory at Indian Head in Maryland, just south of Washington DC, and later transferred to the new Allegany Ballistics Laboratory in West Virginia with contract support from George Washington University. The idea of developing a SAM system for the US Army can be traced back to as early as February 1944, when the Army Ground Forces sent the Army Service Forces (with the AAF the three main organisational elements of the US Army) an inquiry about the development of a direction-controlled, anti-aircraft rocket torpedo. Twelve months passed before serious consideration was given to exploring the problem.[20]

Meanwhile, in early 1943, British intelligence reports concerning the development of guided missiles at Peenemünde prompted the US Army Ordnance Department to commence research into long-range, rocket-propelled surface-to-surface missiles (SSMs). Subsequently, two missile projects, ORDCIT and Project Hermes, would help to lay the foundation for future SAM development for the US Army. The work commenced on 1 June 1944, when the Rocket Branch contracted the Guggenheim Aeronautical Laboratory (GAL) at the CIT to begin a study for that purpose, an undertaking that was dubbed the ORDCIT project. The first phase of the ORDCIT project was the design and firing of a series of solid propellant test vehicles. The first of these, the XF10S1000 – better known as the 'Private A' – was a two-stage projectile in a tandem arrangement. It was boosted by four 4.5-inch artillery rockets filled with a double-base propellant that produced 10 tons of thrust for 0.18 seconds. The sustainer rocket was powered by an Aerojet 30AS1000 solid propellant engine of 30 seconds duration that generated 1,000 lbs (about 450kg) of thrust (the early Aerojet rockets were designated in this fashion based on the propellant used and the thrust produced). The dimensions of the Private A were 2.34 metres in length, 24.4cm in diameter and a span of 85cm, with a weight of 240kg at launch. Four symmetrical fins in a cruciform shape were fitted to the tail for stabilisation. Aerodynamics experiments with full-scale models of the sustainer rocket commenced in the 10ft-diameter subsonic wind tunnel at GALCIT in October 1944. During the first test-launching programme at the Leach-Spring/Leach Lake area of Camp Irwin Reservation near Bartow, California, from 1–16 December 1944, 24 rounds were fired from a 36ft ramp at inclined angles between 50 and 80.5 degrees. The average range attained was 16.5km, with the maximum range about 18.3km. Based on the results, the tests were considered successful.[21]

By early 1945, work on the ORDCIT project was focused on aerodynamics research with a reduced scale model of the Private A and a subsequent version, the 'Private F'. From 9 February to 2 March 1945, experiments were conducted in the new Mach 1.72 wind tunnel at the Supersonic Wind Tunnel Test Laboratory at the Ballistic Research Laboratory, Aberdeen Proving Ground (BRL/APG), near Baltimore in Maryland. The fuselage of the Private F was identical to the Private A except that the four symmetrical fins were replaced with a single fin and two horizontal lifting surfaces fitted towards the forward end. Another alteration was the installation of a 9kg charge at the front end of the combustion chamber to provide a continuous smoke trail to enable better observation. A 5.5kg black powder explosive charge was also fitted to detonate on impact. Seventeen rounds were fired during the next test-launching programme at the Hueco Firing Range at Fort Bliss, Texas, from 1–13 April 1945. The purpose of these test-launchings was to acquire data on the behaviour of an uncontrolled projectile with wings; there was no guidance, control, radar tracking or telemetry instrumentation used during these tests. As with the Private A, the Private F lacked aerodynamic stability during flight and had wayward trajectories. It did not perform as well as the Private A, although a maximum range of 4,500 metres was attained. The experience with the Private rockets enabled the personnel in the Rocket Branch to obtain basic experimental data to build upon on subsonic and supersonic aerodynamics, flight behaviour, launching techniques and technical knowledge. The first phase of the ORDCIT project concluded in April 1945. It is not necessary to compare the modest attainments of the Private rockets with the Fi-103/V-1 and the A-4/V-2; the differences are self-evident.[22]

Phase two of the ORDCIT project called for the development of an experimental missile with a LPRE that could produce a thrust of 2,000lbs (approximately 900kg). This missile, which was designated 'Corporal', would eventually become the US Army's first short-range tactical SSM. In December 1944, Trichel authorised a feasibility study of a liquid propellant rocket for use by the Army Signal Corps as a high-altitude sounding rocket to carry instruments for meteorological research purposes. The altitude requirement from the Signal Corps was 100,000ft or higher. The feasibility study theorised that a two-stage missile powered with a solid propellant booster and LPRE sustainer could reach the desired altitude. The booster stage was planned to consist of a cluster of solid propellant rockets to impart a velocity of 400ft/second. The sustainer rocket was powered by a LPRE that used a propellant combination of 80 per cent aniline and 20 per cent furfuryl alcohol mixture as the fuel, and red fuming nitric acid (RFNA) as the oxidiser, which produced 1,500lbs (680kg) of thrust for between 40 and 50 seconds. The rocket would be fired from a launching apparatus 100ft high. For this project, the original design of the Corporal was scaled down and re-designated the 'WAC Corporal' (without altitude control). The WAC Corporal was also used to study a liquid propellant missile generally and as a first step in the development of a guided SAM system.[23]

Meanwhile, the Rocket Branch entered into a contract with the General Electric (GE) company for a separate guided missile research and development programme, partially in response to the operational use of the Fi-103/V-1 and A-4/V-2 by the Germans, but primarily to design and build these new weapons of war.[24] In addition to long-range SSMs, the US Army Ordnance Department sought guided SAMs for use against high-altitude aircraft. The contract commenced on 20 November 1944, and the programme was subsequently code named Project Hermes after the Greek mythological god.[25] GE contemplated three phases to the project:

1. the collection and evaluation of information on guided missiles and associated apparatus;
2. a scientific mission to Germany to survey German developments and obtain samples of weapons and components; and
3. the submission of designs for ground and anti-aircraft missiles desired by the army and when approved, the construction of prototypes for test firing and possible preparation of manufacturing plans.[26]

In early 1945, the US Army Ordnance Department requirement for an operational SAM system as part of Project Hermes was cancelled when a separate contract for that purpose was negotiated with the Western Electric Company. An amendment to the prime contract with GE reoriented Project Hermes towards research and development only, leading towards a family of SSMs for the US Army. The origin of what would eventually become the US Army's first guided SAM system has been unofficially credited to a young, unnamed engineer from Bell Telephone Laboratories (BTL), the research and development arm of the Western Electric Company, who was on active service with the Ordnance Department in the autumn of 1944. BTL was an industry leader in communications systems and electronics, particularly with telephone systems, radar, and computers, with facilities in New York and New Jersey. During the war, Western Electric and BTL produced about half of the radar equipment for the American armed forces. While stationed at Frankford Arsenal in Philadelphia, Pennsylvania, the unnamed engineer made a proposal to the Ordnance Department on how to improve the existing fire-control radar technology used with anti-aircraft artillery, which was soon to become obsolete with the advent of high-altitude supersonic jet aircraft. Officially, by January 1945 the Ordnance Department was in discussions with the Army Service Forces (ASF) concerning SAMs. On 26 January, the ASF sent a letter to the Office, Chief of Ordnance (OCO) that gave approval to the Ordnance Department to develop a guided SAM system. On 31 January, the OCO authorised contract negotiations with BTL for a feasibility study of the technical characteristics of a system, which led to Western Electric signing a contract with the New York Ordnance District on 8 February for BTL to undertake the study over the next three months. The Ordnance Department requirements were for a missile to be capable of engaging manoeuvrable jet aircraft travelling at 600mph (966 km/h) at an altitude of 60,000ft, well beyond the range of anti-aircraft artillery. The contract was jointly sponsored by the ASF, the Ordnance Department and the AAF Air Technical Service Command (ATSC) at Wright Field in Dayton, Ohio. As a result of this contract, the planned Project Hermes SAM was relegated to a test vehicle.[27]

Six days after the final German surrender, on 14 May 1945, staff from BTL made a preliminary verbal presentation of the feasibility study to about 70 civilians and officers of the US Army in Washington DC. The Ordnance Department approved the study, and on 30 May, Western Electric signed a research and development contract with the Ordnance Department for BTL to proceed with the design and development of the missile system. Just prior to the signing of the contract, it was estimated that the cost of the project would be approximately $1.5 million per year. The comparative annual costs of the German SAM projects in *Reichsmarks* would be difficult to determine. It was also estimated that approximately 60 BTL scientists and 60 miscellaneous BTL personnel would be directly involved with the project, with another 60 BTL staff indirectly involved. These figures do not represent the total number of personnel on the project. Later, there were the Ordnance Department personnel attached to the project and staff from the other major contractors. To place the number of people involved into perspective, in early 1945 over 1,200 personnel at Peenemünde-East were working on the C-2 project, in addition to the numerous staff of the industrial firms who fabricated parts of the test missiles.[28]

The AAF, meanwhile, sought to develop its own guided SAM system. In the field of guided weapons, the AAF had quickly acquired experience and expertise. During the war, the AAF embarked on numerous missile projects as part of its missile development programme. The use by the *Luftwaffe* of the X-1 and the Fi-103/V-1 in particular prompted intensive efforts by the AAF to reproduce these weapons. The results were the two best-known projects – the VB-1 (or Azon, azimuth-only; azimuth is an angle measured clockwise from the south or north), a controlled vertical bomb which was similar to the X-1 and was used in combat in France, Italy and the Burma–India theatre with some success; and the JB-2 flying bomb, in essence a reverse-engineered Fi-103/V-1, which reached the experimental production stage by the end of the war. On 16 May 1945, Boeing Aircraft Corporation, under contract to the ATSC, commenced a three-month study to design a ground-to-air pilotless aircraft (GAPA). The distinction between the SAM system that the US Army sought and the GAPA – which were essentially for the same purpose – was a matter of semantics. The development of ground-launched guided missiles was the prerogative of the Ordnance Department, whereas aircraft development was the exclusive domain of the AAF. The AAF regarded any unmanned vehicle with wings or other aerodynamic surfaces as pilotless aircraft, not guided missiles, so although technically the GAPA was planned as an aircraft, in reality it was a SAM. As a result of the AAF contract, the Ordnance Department, in agreement with the AAF, assumed full sponsorship of the BTL contract on 16 June 1945. The information contained in the BTL document helped to form the foundation for the Boeing study.[29]

On 15 July 1945, BTL released the formal version of the feasibility study, entitled 'AAGM Report'. The report formed the basis for examination and experimental verification of the many problems which faced the designers but was produced before all of the scientific and technical data from the German SAM development programme was fully evaluated. Notable were the similarities with the British 9.5-inch LOP/GAP design of February 1945 that fulfilled the Admiralty requirement. Like the LOP/GAP, the tentative American design concept was also a supersonic two-stage missile with a solid propellant booster and liquid propellant sustainer in a tandem configuration. The overall weight of the American missile was slightly less, and the length of both missiles was around 19ft. Another similarity with the British design concept was the tentative manoeuvrability requirement – both missiles were to be capable of making 5 'g' turns up to an altitude of 40,000ft. There were, however, key differences in the designs that were the product of research and technical developments in each country. For instance, the American booster comprised eight 5-inch solid propellant rockets instead of the seven 5-inch ATO units in the LOP/GAP booster. Although both sustainer rocket designs were powered by LPREs with a regeneratively cooled combustion chamber (not necessarily of British influence, as the design was prevalent in German LPREs and had been investigated in the US), American engineers chose not to use LOX in the engine but a propellant combination of aniline and RFNA, building on the American wartime research with the propellant combination in the JPL at GALCIT.[30]

Other differences in the designs took into account the intended applications. The guidance system technology was one example – the British missile system was to use the beam-rider technique, which was more suitable at sea, whereas the Americans chose command-link, more suitable for land-based missile systems. One radar would track the missile and the other the target, with the flight and trajectory data entered into a computer for conversion into steering and tracking information on the ground. The two radars were to be combined into a single apparatus with two separate lens antennae mounted on a rotatable beam structure on top of a van designed for radar equipment. The azimuth of the radar beam that tracked the target would be adjustable by moving the beam structure. The differences between the target and missile azimuths were planned to be adjustable by moving the missile radar antenna.[31]

Following the publication of the AAGM Report, BTL produced a four-year development plan on 27 July 1945 for submission to the Ordnance Technical Committee in accordance with the recommendations in the report. In the event the plan was approved by the committee, the overall authority to direct, co-ordinate and supervise the development was the Ordnance Rocket Branch. BTL would subcontract the Douglas Aircraft Corporation (DAC) of Santa Monica, California (famous for the DC series of transport aircraft), to conduct the aerodynamic studies, to engineer and manufacture the missile structure and the launcher devices, and to conduct test firings at the new Ordnance Department firing range at White Sands Proving Ground (WSPG) at Las Cruces in New Mexico. The Aerojet Engineering Corporation of Pasadena and Azusa near Los Angeles would be subcontracted to supply the solid propellant booster and the LPRE for the sustainer rocket. The JPL/CIT agreed to act as a consultant to DAC and Aerojet regarding the propulsion system. The BRL at the Aberdeen Proving Ground in Maryland – where there was a suitable supersonic wind tunnel – would undertake wind tunnel experiments, in addition to the development of a fragmentation warhead in association with a number of other US Army establishments. BTL would retain the overall technical leadership of the project and would be in charge of the design and manufacture of the command-link technology in the guidance system. In many aspects of the technology, such as aerodynamics, thermodynamics, rocket propulsion, guidance, warheads, construction, handling, launching procedures, flight testing, telemetry and trajectory calculations, fundamental and applied research to solve these problems had already commenced in Germany. However, the results of the German experiments had to be validated after the war, a task which necessitated a duplication of effort.[32]

Nevertheless, the Americans had a considerable lead over Germany in radar and microwave techniques, as well as proximity fuse technology. From 1941–43, the APL/JHU, with British assistance, successfully designed a radio-activated device for use in anti-aircraft shells. The idea of developing a proximity fuse for anti-aircraft projectiles was originally British, after the introduction into service of the 3-inch rocket in 1940. The first British proposal, which was presented in April 1940 during discussions between the Air Defence Research and Development Establishment, the PDE and the War Office, was for a proximity fuse that worked on the Doppler reflection principle from an aircraft. The following year, a radio-triggered fuse was proposed for the 3-inch rocket. However, the Americans successfully completed the development of the technology. At the APL/JHU, a small radar was designed to fit into an anti-aircraft shell that could withstand the concussion from being fired from an artillery piece or naval gun. The first test firing, by the US Navy, took place in January 1942, and the technology was introduced into service by the end of the year. The first casualty of the device was in the Pacific theatre on 5 January 1943. The US (and the British) led Germany in this technological field despite the vast amount of German research and development of far more sophisticated anti-aircraft weapons.[33]

One month after the final Japanese surrender and nearly two weeks after VJ Day, the BTL development plan was recommended to the Ordnance Department by the Ordnance Technical Committee on 13 September 1945 and was subsequently approved on 4 October 1945. The AAF, meanwhile, was not prepared to enter into any contract negotiations with industry concerning a surface-to-air guided missile system until Boeing completed the GAPA study. Boeing was scheduled to complete the study on 27 August 1945, but the final report was not presented to the ATSC until 20 September. The need to evaluate the German scientific and technical intelligence may have been a factor in the delay. With regard to the specifications of the envisaged missile system, the particulars were still vague at this point:

> This pilotless aircraft will probably be designed to operate at supersonic speeds. It is expected to employ a jet-assisted take-off unit of some type and will undoubtedly be rocket or jet propelled after launching. Various types of control systems are under consideration; however, some form of radar control looks to be the most promising … The GAPA will be suitable for use against high altitude aircraft [40,000–60,000ft].

Six days later, the report was presented to the headquarters of the AAF. Following the completion of the study, the AAF contracted Boeing to commence the first phase of the development of a GAPA.[34]

The US Navy Bumblebee programme and anti-kamikaze weapons

The US Navy was actually the more active of the armed services in the pursuit of a surface-to-air guided missile system. Initially, the Bureaus of Aeronautics (BuAer) and Ordnance (BuOrd) clashed over which technical service had the prerogative to develop and employ the weapons. The bureaus were two of the three that acquired the US Navy's major weapons systems, along with the Bureau of Ships. The BuAer was responsible for aircraft and aviation equipment; the BuOrd, generally speaking, was responsible for all offensive and defensive arms and armaments, such as guns, torpedoes, bombs and rockets. Since guided missiles could potentially be launched from ships as well as aircraft, there was no clear delineation concerning the prerogatives. Each bureau embarked on its own projects or research and development, so by the end of the war, three surface-to-air guided missile projects had been initiated by the US Navy. The first was the 'Lark' under development by the BuAer. The Lark was conceptualised in October 1944 as a two-stage subsonic missile powered by a solid propellant booster and a liquid propellant sustainer. The BuAer approved the development on 6 February 1945 as a potential counter-measure against Japanese suicide aircraft in the Pacific theatre. Two versions of the missile were planned – a SAM was contracted out to the Fairchild Engine and Airplane Corporation, which was designated KAQ-1, and an AAM version was under development by the Consolidated Vultee Aircraft Corporation (Convair), designated KAY-1. The programme was assigned a high priority in 1945, but development proceeded slowly. No prototype missiles had been test-launched by the end of the war.[35]

The BuAer had a second project, a missile proposed by its Air Material Unit on 10 May 1945, as a consequence of the slow progress with the Lark and the continued kamikaze attacks in the Pacific theatre. Originally called 'Baby Lark', it was subsequently renamed 'Little Joe'. The Little Joe was designed as a short-range, two-stage radio-controlled missile. The intended range was about 4km, with a ceiling of 2.4km. The booster stage was powered by four 8.2cm calibre AN Mk 7 solid propellant rockets mounted around the fuselage, in a configuration which appears to have resembled the British 'Brakemine'. These rockets provided a thrust of around 900kg for one second, less than other American and British rockets. The sustainer rocket was an Aerojet 8AS-1000E ATO, the same rocket that was fitted in the McDonnell 'Gargoyle' air-to-surface missile that was also being developed by the BuAer. The first prototypes of the Little Joe were constructed by the end of June 1945 and the first powered test-launch was made on 9 July. In these firings, nine second 'smoky' tests (to denote the visible exhaust trail) were assessed as successful 'beyond expectation'. By VJ Day, 15 missiles were built for testing, but the project was terminated in December 1946.[36]

The third enterprise was the 'Bumblebee' programme by the BuOrd. The origins of the project can be traced back to as early as July 1944, when for a similar purpose as the British Admiralty 10 months earlier, the Chief of the BuOrd, Rear Admiral George F. Hussey, requested from the OSRD an analysis and evaluation of how to protect a naval task force against guided missiles launched from enemy planes beyond the range of conventional guns. During the second half of 1944, the director of the APL/JHU and leader of Section T in the OSRD, renowned physicist Dr. Merle Tuve, was tasked with producing the proposal for the BuOrd. The proposal was subsequently approved and the APL was awarded a development contract in January 1945. The importance of the programme to the BuOrd and future naval operations is contained in guidelines that were issued that month to the APL by Rear Admiral Hussey, which are summarised below:

> A comprehensive research and development programme shall be undertaken, embracing all technical activities necessary to the development of one or more types of rocket-launched, jet-propelled, guided anti-aircraft missiles ... This programme shall include pertinent basic research, investigations, and experiments, and the design, fabrication and testing of such missiles, their component parts, and supplementary equipment ... The Bureau of Ordnance does not expect by this request and in the immediate future to obtain an ideal or ultimate aircraft weapon and is aware that the actual results of these efforts cannot be guaranteed or accurately predicted. The Bureau believes, nonetheless, that an immediate attack must be made on this problem and expects that this will result at the best in the production of an advanced anti-aircraft weapon which may be available in the latter stages of the war, and at the least in considerable valuable progress in research and development on jet propulsion techniques, self-guided techniques, and other technical matters of great importance to the future of ordnance.[37]

From the outset, the Bumblebee programme was assigned top priority by the BuOrd. The new undertaking led to the withdrawal of the APL from the highly important proximity fuse programme. The BuOrd specifications were also for a two-stage missile with a launch weight of 1,800kg, evenly divided between the booster and the sustainer rockets, of which about 270kg would be taken up by the warhead. The envisaged weight was relatively heavy – a slightly greater mass than the *Rheintochter* R-I, almost half the weight of the C-2/W-3 prototype and around four times heavier than the LOP/GAP. The intended operating altitude was 30,000ft and the desired maximum velocity about Mach 1.65. In concurrence with opinion within the GAP organisation, a ramjet was also considered as the best method of propulsion; American research into ramjet combustion had begun at the National Bureau of Standards and MIT in 1943. By 1945, significant progress was made with ramjet technology in the programme – in April of that year, a 6-inch (15cm) diameter ramjet test vehicle, code named 'Cobra', was successfully burned in a flight test. On 13 June 1945, the first successful supersonic test flight of the Cobra was made at a proving ground at Island Beach in New Jersey. The Cobra was boosted by four

5-inch solid propellant high-velocity aircraft rockets (HVAR), developed in 1943 by the BuOrd and nicknamed the 'Holy Moses' by American personnel who witnessed their performance, something the Nazis and scientists and engineers in the German guided missile industry probably would have laughed at. The successful telemetry of a burning ramjet in flight was also made that month. These experiments almost certainly demonstrated the viability of ramjet propulsion in a SAM application.[38]

Conclusion

By the end of World War II on 2 September 1945, the status of British and American SAM research and development can be summarised as follows. During the war, the Western Allies did not fire any anti-aircraft missiles more powerful than small solid propellant rockets in combat. When the military planners in both countries decided to authorise surface-to-air guided missile research and development programmes in 1944 and 1945, the greatest impetus was provided by the war at sea – to defend against German guided weapons and Japanese suicide aircraft. Britain was the first of the Allies to successfully develop SAMs with the 3-inch rocket and was also the first to design a surface-to-air guided missile, the LOP/GAP. Incidentally, the British were also the first to devise an anti-ballistic missile defence system, to defend against the A-4/V-2.[39] Both Allies initiated long-term SAM research and development programmes with the knowledge that the nascent technology and development of jet aircraft meant that the realisation of an operational system was still a number of years into the future. Engineers in both countries considered a LPRE the preferred propulsion system for a sustainer rocket over a SPRE due to the higher specific impulse of certain liquid propellants, yet ramjet propulsion was seen as very promising. Although Germany had a demonstrable lead in most aspects of the science and technology, in microwave research, radar techniques and proximity fuse technology the Germans lagged behind; in the two former domains, the Anglo-Americans had a lead of perhaps two or more years.[40] The American armed forces had more SAM projects underway than the British and were positioned to make more rapid progress with the technology after the war, due in part to the lack of a suitable supersonic wind tunnel in the UK, which placed the British at a relative disadvantage. Scientific and technical intelligence on the major German flak rocket projects had begun to be produced by the Western Allies in the last few weeks of the war in Europe, and by early September 1945 the German SAM development programme was considerably documented. This is covered in the next chapter.

CHAPTER THREE

Anglo-American Investigations of Intelligence Targets Linked to the German SAM Development Programme, 1944–48

The loss of the war by Germany resulted in the scientific and engineering knowledge generated in the SAM development programme passing into the hands of the Allies. Remarkably, during the war the Germans almost managed to completely conceal the existence of the programme from the Allies. Consequently, the technology transfer process began much later compared with the other German guided weapons, beginning with the A-4/V-2 in 1939 and the Hs 293, X-1 and Fi-103/V-1 in 1943. As this chapter will show, the Anglo-American operations to collect information and produce intelligence on German SAM research and development, like on other German guided weapons, were very successful. There was great interest in the German advances with SAMs, and in the scientists and engineers who invented the technology. The intelligence targets where research and development were carried out for the programme, and the Germans themselves, were heavily investigated by the numerous British and American scientific and technical missions. The seizure and examination of flak rocket technology by technical intelligence officers, interrogations of captured German specialists, and the documentation of the research and experimental establishments and firms where the technology was developed and manufactured formed the basis upon which the German knowledge was transferred and ultimately exploited by the Allies.

Anglo-American intelligence on German SAM capabilities

British and American intelligence on German anti-aircraft rocket defence prior to April 1945 mostly comprised information on the co-ordinated use of barrage rockets. The first information on the use of these weapons by *Luftwaffe* anti-aircraft units was received in 1943, when Allied bomber pilots reported attacks from what appeared to be small unguided flak rockets. Subsequently, during the last two years of the war, Squadron Leader Bernard Babington-Smith (the brother of celebrated WRAF photographic analyst Constance Babington-Smith) and another officer at the Central Interpretation Unit (CIU) at Medmenham in Buckinghamshire paid very close attention to the possible use of these weapons by the Germans. To attempt to determine if the projectiles were indeed flak rockets, the RAF conducted tests with British 3-inch rockets on the nights of 24/25 and 25/26 March 1944 at Manorbier in south-west Wales. Photographs were taken of the weapons to serve as standards of comparison against the German projectiles, and to make them more easily identifiable when salvoes were launched against British and American aircraft at night. The results of the tests were negative, however: the photographs of the 3-inch rockets, taken at 10,000ft, did not match the appearance of the projectiles that were filmed by the pilots.[1]

On 14 May 1944, however, a memorandum sent from the Assistant Director of Intelligence (Photographic Reconnaissance) (ADI (Ph)) at the British Air Ministry to the commanding officer of RAF Medmenham, 11 days after the report on the night-firing trials was produced, contradicted the conclusions from the night tests that were carried out in March. The memorandum stated that large salvoes of barrage rockets were reported to have been recently fired from 51 locations in Germany and occupied France, both during the day and at night, with the most recent attacks from Karlsruhe and from Dusseldorf and Aachen in the Ruhr region. The projectiles were described as bursting in square formations around the bomber streams, and in contrast to the tests from March, were reported as resembling the British 3-inch rocket in behaviour. Reports from so many locations would seem to confirm the earlier reports, yet definitive proof of launching sites on the ground remained elusive.[2]

British and American intelligence unknowingly came into possession of hard evidence that the Germans were developing guided SAMs after an A-4/V-2 crashed in Sweden in June 1944 (the so-called 'Swedish Incident'). The missile was launched to test the *Kehl-Strassburg* remote control equipment in the C-2 and other flak rockets. Soon after the launch, the ground operator lost control of the missile and it veered erratically off course and crashed. The British Air Attaché in Stockholm was provided with photographs of the wreckage by Swedish military intelligence and subsequently sent a report with the photographs back to British Air Intelligence in London for analysis. The photographs showed the components of the *Kehl-Strassburg* system, but the technology was not recognised as previously part of either the Fi-103/V-1 or the A-4/V-2. The British analysts logically assumed that the electronics were designed for the latter missile. Based on this evidence, the origin of the transfer of German SAM technology to the Allies can be traced back to this point, ironically through a neutral country.[3]

Prior to the landings at Normandy on 6 June 1944, German progress with rocket and guided missile technologies, along with numerous other technologies including jet aircraft and tanks, and in scientific research in fields such as nuclear physics, were considered by the Western Allies to be high priorities for capture and/or exploitation. Thus, in May 1944 the headquarters of the Assistant Chief of Staff, G-2 (Intelligence), at the Supreme Headquarters Allied Expeditionary Force (SHAEF) instructed the

Joint Intelligence Subcommittee (JIC) of the Chiefs of Staff Committee in London to establish a combined Anglo-American committee to catalogue the technical intelligence requirements of American and British government departments, and to prioritise intelligence targets in German-held territory for investigation. This committee, initially an all-British subcommittee of the British JIC called the Intelligence Priorities Committee, laid the groundwork and did much of the organisational planning for the short-lived Combined Intelligence Priorities Committee (CIPC) and its successor, the Combined Intelligence Objectives Subcommittee (CIOS) of the Combined Intelligence Committee (CIC) in Washington DC. During July, August and September 1944, the CIPC was reconstituted into CIOS under CIC control following objections from within SHAEF and the American JIC to arguments by the British JIC that the CIPC should have a British civilian as the chairman and be outside of the SHAEF command structure, the latter of which was actually in accordance with the original directive from SHAEF G-2. CIOS was officially established by an authorisation of the CIC on 21 August 1944 and was headquartered in London. The main functions of CIOS were essentially the same as those of the CIPC.[4]

The provisions for seizing German industrial (intellectual) property after the final surrender of Germany were agreed upon at a meeting of British, American and Soviet representatives on an inter-Allied body called the European Advisory Commission in London on 25 July 1944, and were contained in the text of a document entitled 'Unconditional Surrender of Germany'.[5] On the Anglo-American side, in addition to the enemy armaments that were captured on the battlefield, the governments of both countries had created legislation which enabled them to take possession and exploit intellectual property that belonged to the enemy. In the UK, this was enabled through the Patents, Designs, Copyright and Trade Marks (Emergency) Act, 1939.[6] In the US, under the War Powers Act of 1941, the Office of Alien Property Custodian, which was established on 11 March 1942, issued vesting orders seizing all patents and patent applications owned by enemy aliens residing in enemy or enemy-occupied territory, and most of the patents or patent applications owned by persons residing in countries occupied by enemy forces during the war.[7]

The 'Unconditional Surrender of Germany' document, which received the approvals of the three governments later in 1944, comprised several articles which applied to the seizure of material assets from experimental weapons programmes within the broader objectives of disarming the *Wehrmacht* and liquidating the German armaments industry. Article 5(a) stated:

> The German authorities will hold intact and in good condition at the disposal of the Allied Representatives, for such purposes and at such times and places as they may prescribe all arms, ammunition, explosives, military equipment, stores and supplies and other implements of war of all kinds and all other war material.

The material assets of the German SAM development programme would have fallen under article 5(a) as:

> All aircraft of all kinds, aviation and anti-aircraft equipment and devices ... all factories, plants, shops, research institutions, laboratories, testing facilities, technical data, patents, plans, drawings and inventions.

Under article 5(b) it was stated that:

> The German authorities will at the demand of the Allied Representatives furnish any information or records that may be required by the Allied Representatives.

And under article 8 it was added that:

> The German authorities will prevent the destruction, removal, concealment, transfer or scuttling of, or damage to, all military, naval, air, shipping, port, industrial and other like property and facilities and all records and archives, wherever they may be situated, except as may be directed by the Allied Representatives.[8]

The CIPC created a 'Black List' of 28 categories of technical items to orderly catalogue the intelligence targets for capture and investigation. Four of these categories were directly applicable to SAM technology: (1) radar; (2e (ii)) artillery and weapons (explosives and propellants); (4) rockets and rocket fuels; and (6) directed or controlled missiles. Under technical item 1, radar was a broad subject due to the many applications of the technology; applied to SAM systems it was used for target and missile tracking. Under technical item 2e (ii) were targets where solid and powder rocket propellants were manufactured. On this topic, by September 1944, the now-renamed CIOS (see below) considered the main methods of manufacture and use of these propellants and explosives by the Germans to be well known, but there were a number of subjects of special interest that were included on the Black List. One of these was research, development and production of diethylene glycol dinitrate (DGDN) – which unbeknownst to the Anglo-Americans was a common constituent in solid propellants that were used in flak rockets, such as the *Rheintochter* – and included double-base propellants that contained the material, along with their ballistics. The CIPC had questions in relation to the Germans' introduction of DGDN as a substitute for nitroglycerine – was it done because of better ballistics, less gun-barrel wear, increased safety in manufacture or purely for economic reasons? Under technical item 4, rockets and rocket fuels, information was sought on solid and liquid propellants, field rockets and anti-aircraft rockets. The Black List stated that

> anti-aircraft rocket developments are certainly of interest and any indication of the use of light alloy rocket bodies [a feature of British rocket technology], or of any other novelty of design, would invite close investigation.[9]

Under technical item 6, directed or controlled missiles, the CIPC desired information on guided missiles in four general areas. The first was on equipment in use or under development, including the principle of operation, construction and design, method of use and performance. The second was techniques embodied in missiles or used at places of development, manufacture or test, such as novel control techniques, electro-mechanical or other servo links, radio-control or homing means, testing and proving ground methods, and training devices, procedures and techniques. The third area was methods and results of research, development and manufacture, such as locations, fields of activity, facilities and archives. The fourth area – potentially the most important – was information on the personnel who were engaged in work on guided missiles, including the names of key individuals who were associated with development, the organisations by whom they were employed and their principal past achievements and present activities.[10]

Of the British missions or outfits who operated under CIOS, the most prominent was the Royal Navy/Royal Marines 30 Assault Unit (30AU), led by Commander Ian Fleming RNVR (author of the James Bond novels) of the Naval Intelligence Division (NID) in the Admiralty, who also represented the NID on the CIPC and CIOS. 30AU was created by the Admiralty in 1942 to conduct operations to capture German technical intelligence of military value. Government departments, for example the MoS and MAP, also sent investigators, as did ADI (Science) in the Air Ministry and scientific research establishments such as the RAF Physiological Laboratory (from April 1945 the RAF Institute of Aviation Medicine) at the RAE.[11] The US sent many missions into the field. The War Department sent the Alsos mission to search for evidence of German atomic weapons research. The US Army missions included the Ordnance Enemy Equipment Intelligence mission and investigators from Project Hermes. The AAF dispatched Operation *Lusty* (an acronym for *Luftwaffe* *S*ecret *T*echnolog*y*), led by Colonel Donald Putt, a former test pilot and technical intelligence officer. There was also a civilian mission from the AAF, the Army Air Forces Scientific Advisory Group (AAF SAG), which was headed by Professor Theodore von Kármán, a Hungarian Jew and immigrant from Germany, who was the director of the JPL at the CIT. The US Navy dispatched the US Naval Technical Mission in Europe (NavTechMisEu) to:

> exploit German science and technology for the benefit of the Navy Department technical bureaus and the Co-ordinator of Research and Development.[12]

The creation of NavTechMisEu was authorised by the Secretary of the Navy, James Forrestal, on 4 December 1944, and was led by Commodore Henry A. Schade, who headed the US Navy's contingent in Alsos and had been the Navy Department's representative on the CIPC.[13]

Investigative teams working on behalf of CIOS (on occasions there was only the one investigator) comprised military and civilian investigators from the British and American missions, who came from the US, the UK, Canada and British India. The composition of the teams varied, depending on the nationalities and professions of the personnel and the nature of the target(s) to be investigated. For example, CIOS team

183, which investigated German guided missile targets, comprised nine members, four American and five British. The team leader was an AAF officer, Lieutenant Colonel J. O'Mara, who was accompanied by three American civilians – two from Project Hermes and one from the AAF SAG. One of the civilians presumably also acted as an interpreter. Of the British team members, two were army officers attached to the MoS and three were RAF officers, of whom two were intelligence officers and one was attached to the MAP. Another CIOS team, number 367, also investigated German guided missile targets and also comprised nine members, of whom six were British personnel from the MAP, five civilian and one military. Three were American military officers – one from the US Navy Reserve, one US Army Signal Corps officer (who also presumably acted as an interpreter) and an AAF intelligence officer.[14]

Captured items of importance, such as documents, films and instruments, were dispatched through official CIOS channels to the CIOS Secretariat in London for examination, reproduction, translation and distribution to American and British agencies. One such agency was the joint US–UK Air Documents Research Centre in London, although not all the relevant captured items were forwarded there through the official CIOS channels. It has been documented by historians including Lasby and Bower that the intelligence-gathering effort in Europe and later in Germany under CIOS was not an orderly and systematic process. A discourse about this aspect of the history is unnecessary except to mention the reasons why there were problems with co-ordination: there were organisational flaws in the CIOS hierarchy; the scientific and technical missions in the field were very numerous; Germany collapsed more rapidly than was anticipated, which put pressure on resources; there were opportunistic seizures of material by American and British units outside of the regulations; and the sheer quantity of captured archives led to problems with storage and preservation.

By August 1944, the CIPC, through the efforts of British Air Intelligence, unknowingly came into possession of more intelligence about German SAM development. This time it concerned the *Flakversuchsstelle der Luftwaffe* at Peenemünde, where the C-2 and the *Taifun* were being developed and where all the flak rockets with LPREs were test fired. The source of the information was the logistics and organisation section in the Enemy Order of Battle department of British Air Intelligence (AI3(e)). Highly reliable information rated the highest classification (A-1) was obtained on an experimental establishment at Karlshagen, near Peenemünde, the purpose of which was only described in the CIPC Black List as 'Flak: special rockets'. No other information was included.[15]

During the next five months of the war, from September 1944 to mid-January 1945, British and American intelligence organisations did not acquire any conclusive evidence to ascertain the development or operational use of high-altitude SAMs by the *Luftwaffe*, as stated by an MI15[16] paper written on 15 January 1945 for a report to the US War Department reviewing flak prospects for 1945 as they would affect air operations in Europe:

> No intelligence has been received which would confirm the frequent reports from aircrews of high-altitude flak rockets, and it is thought that a large proportion of such reports are in fact attributable to observations of the aftermath of smoke markers. If a high-altitude rocket does exist, experience so far suggests that very considerable improvement in performance will be required before it constitutes a serious menace to bomber formations.[17]

By late February 1945, SHAEF had also compiled the numerous reports about supposed sightings of high-altitude flak rockets in several technical intelligence summaries, but there was still no interpretation data available.[18] The existence of a small, high-altitude flak rocket was finally confirmed two weeks later in a MI15 intelligence assessment, the last of the war to address the use of rockets by *Luftwaffe* flak formations, dated 12 March 1945:

> The first tangible substantiation of the numerous aircrew observations of rockets at high altitudes has now been provided by a reliable report of a HE [high explosive] rocket of about 20-centimetre calibre. It is thought that this projectile may have a warhead of about 50lbs, of which about a quarter is HE content, giving a lethal radius of burst of some 85 feet, and that its ceiling would be in the neighbourhood of 25,000 feet; there may possibly also be a multi-barrelled version. It is open to question whether many of these weapons are yet in service, but they may well have been developed experimentally in small numbers for some time, thus accounting for some of the aircrew observations reported in the past.[19]

Several 20cm parachute-and-cable type rockets were also recovered and described as a crude version of a 15.2cm (6-inch) type rocket on which MI15 already had intelligence.[20]

In the three weeks after Cologne was captured by troops from the First Army of General Omar Bradley's 12th Army Group on 6 March 1945, documents seized at the facilities of the rocket engine manufacturer *Wilhelm Schmidding* in that city provided British and American intelligence organisations with the first hard evidence of the German SAM development programme. As mentioned in Chapter 1, the firm's facilities at Bodenbach in the Sudetenland supplied the 109-553 booster rockets for the *Enzian* and the Hs 117, as well as LPREs and SPREs for other HFW guided missiles. A British Air Intelligence, Technical (AI2(g)) report, dated 29 March 1945, refers to technical data about the 109-553 and the Hs 117 project in the captured documents. But based on the information, British analysts could only confirm the existence of the air-to-air variant of the Hs 117, the Hs 117H:

> The Hs 117 appears to be essentially a large version of the Hs 298 [guided air-to-air missile], but may be fitted with a rocket motor capable of operating over a considerable period ... It seems that the Hs 117 was originally called the Hs 297 but that the name was later changed, possibly as a result of some major design alteration. The Hs 117 is a radio-controlled missile which can be launched from aircraft but may also be capable of being launched from the ground. It is being developed with a priority slightly less than that allotted to the Hs 298.[21]

Despite the discovery, as late as 1 April 1945, MI15 and the CIU had concluded that the *Luftwaffe* only envisaged plans of organised flak defence with barrage rockets. Whether the intelligence in the AI2(g) report was taken into account is unknown. The CIU assessment stated:

> a very large number of avenues of research have been explored and in all cases they have led to negative answers.

The determination was reached on the basis of information from other photographic interpretations and two additional sources – a statement by a German POW who had been interrogated and a captured German photographic interpretation manual. The POW stated that he was engaged in the production of 7.5cm flak rocket projectors, but these were only employed experimentally because of a lack of ammunition. A 7.5cm calibre projectile would have had similar dimensions to the British 3-inch rocket.[22]

The fact that British and American intelligence organisations had still not learnt about the existence of the German SAM development programme until so late in the war can be considered an intelligence failure. The German success in retaining almost complete secrecy can be explained by three main reasons. In the German armaments industry, there were elaborate security measures in place along the production lines and supply chains from manufacture to delivery, whereby each person only knew as much as was necessary for the completion of a particular task according to their position in the hierarchy. Another reason was the location of the test-launching site on an island, the Greifswalder Oie, about 10km off the coast of Peenemünde. The island was the perfect location for such a secret operation, providing a higher level of secrecy from ground and air reconnaissance than the test-launching facilities on the mainland. The third reason was that unlike what took place during the development of the A-4/V-2, forced labourers were not utilised near or within sites at Peenemünde where the C-2 and the *Taifun* were assembled and tested. The precise activities of the *Flakversuchsstelle der Luftwaffe* were not exposed to spies and agents who worked for the British-run underground intelligence networks. Nevertheless, it must be noted that by the time of the first test-launch of the C-2 on 29 February 1944, most of the forced labourers at Peenemünde had been dispersed to other locations following the RAF bombing raid against the installation on the night of 17/18 August 1943.

The successful concealment of the German SAM development programme from the Western Allies was one of the success stories of German secret weapons development. It was remarkable because the Western Allies possessed so much intelligence on the Peenemünde establishments, and by 1944 could photograph and bomb the facilities almost at will. What is staggering to the imagination (although entirely counterfactual) is the possibility that had the war continued for another six months, guided SAMs could have been introduced into service without the Allies having any intelligence on the new weapons. But the war in Europe was not going to last until October 1945, and the Americans had the atom bomb. As the German defences crumbled before the Allies' onslaught, it was only a matter of time before the German SAM development programme was discovered.

The German SAM development programme is uncovered by the Western Allies

During the final weeks of World War II in the European theatre, the Allied armies that were driving deeper into the remnants of Hitler's Third Reich discovered a variety of Germany's most highly advanced armaments and weapons systems. Prior to late March/early April 1945, the Allies already knew that Germany led the world in the design of numerous categories of armaments that first appeared during the war – ballistic missiles (A-4/V-2); cruise missiles (Fi-103/V-1); air-to-surface guided missiles (HFW Hs 293); precision-guided munitions (*Ruhrstahl* X-1); air-to-air guided missiles (*Ruhrstahl* X-4, but only in the case of France);[23] jet aircraft (*Messerschmitt* Me 262 fighter and the *Arado Flugzeug-Werke GmbH* Ar 234 *Blitz* bomber and reconnaissance aircraft) and tank designs (*Henschel und Sohn* Tiger). Besides SAMs, there were other advanced German armaments and weapons systems that were in varying stages of development and service which the Allies did not know about, such as submarine designs (Type XVIIB and Type XXI U-boats). The Anglo-Americans were by no means technologically inferior to the Germans; indeed, they led Germany in numerous fields of technology, including radar, heavy bombers, electronic warfare and aircraft carriers, all of which were of crucial significance to the victory in Europe.

The location in Germany where evidence was first found to prove the proposition in the AI2(g) report of 29 March 1945 – or found to reveal the existence of the other major flak rocket projects – is hard to determine. There were numerous locations in central Germany – for example at Göttingen, where the *Aerodynamische Versuchsanstalt* (AVA) with its supersonic wind tunnels were situated, which was captured by the US 2nd Armored Division on 8 April; the area around Nordhausen, where a number of underground aircraft and missile factories were situated, which was occupied by troops from the US 3rd Armored Division on 11 April; and Braunschweig, near the *Luftfahrtforschungsanstalt Hermann Göring* (LFA) south of Völkenrode, which was captured by the US 30th Infantry Division on 12 April. Within the first week following the occupation of these localities by American troops, captured documents, hardware and interrogations of German scientists, engineers and technicians would have revealed the astonishing progress of German SAM development to British and American investigators.

The richest source of evidence during April and early May 1945 was probably the LFA, which was Germany's – and the world's – premier aeronautical research establishment. The facilities, including its library, were captured virtually intact. British and American intelligence agencies knew of the establishment's existence during the war but had not discovered its exact location. Most of the buildings were covered in an elaborate camouflage of thick fir trees that concealed them from the air. The exposed buildings were well camouflaged to look like farm buildings, even at low altitudes, and the airfield itself was also camouflaged, sown with different coloured grasses. As a result, the establishment was not damaged by bombing and remained active until the end of German military operations in the area. Allied investigators found supersonic wind tunnels, thousands of reports and publications on all aspects of aeronautical research, complete prototypes of the *Feuerlilie* F-55 research rocket and the *Enzian*, and some of Germany's leading aeronautical specialists, several of whom had contributed valuable research to the German SAM development programme. One of those was Prof. *Dr.-Ing.* Adolf Busemann, a prominent German figure in the domain of supersonic aerodynamics. During interrogation, Busemann made what was described by American investigators as an 'amusing comment' that seemed to condemn the desperate and frantic efforts by the RLM to quickly develop an effective SAM system to defend against the Anglo-American strategic bombing campaign:

in recent years everybody was working on 'homing missiles' and nobody got anybody anywhere.[24]

The LFA was situated within the boundaries of the future British Zone of Occupation, and until the transition to the quadripartite occupation of Germany was completed, the facilities were administered by the US Strategic and Tactical Air Forces (USSTAF). USSTAF officers classified and microfilmed all the documents at the establishment. While the documents were being assessed, Allied investigators were not permitted to view any of the reports or papers but were allowed access to a card index of the collection and to interrogate the staff. Once the documents were classified and microfilmed, the USSTAF ordered the collections to remain at the establishment pending a final decision on whether the UK or the US had the right to the archives. After some lengthy negotiations, it was agreed that the MAP had custody of the documents. The policy on captured documents that was formulated between the two allies stipulated that in whichever zone of occupation the original documents were found, the nation which administered that zone could take possession. Many American, British, and French investigations would ultimately be conducted at the LFA, on various subjects relating to guided missiles and aircraft including aerodynamics, propellant and rocket engine research.[25]

Another possible location was Stassfurt, about 30km south of Magdeburg. At Stassfurt, captured by American troops on 12 April, was found the electronics firm *Stassfurter Rundfunk* and Dr. Theodor Sturm, a physicist who oversaw the development and production of all the remote control technology for guided missiles in the German industrial sector. He was found by a forward team of detachment nine from the Enemy Equipment Intelligence Service (EEIS) of the US Ninth

Army Signal Section, and with some captured equipment and documents from the firm was brought back to the Ninth Army headquarters for interrogation. Sturm subsequently provided a fairly comprehensive account of the development of the *Kehl-Strassburg* remote control system for German guided missiles, as well as a report on the latest radio technology in the Hs 117. The plant was investigated on at least one occasion by British and American investigators on behalf of CIOS, on 10 May 1945.[26]

In mid-April, British and American investigators found some personnel from the other leading electronics firm that developed remote control and radar technologies for guided missiles, *Telefunken*. They were located at sites in Thuringia in central Germany, where a part of the research and development organisation of the firm was situated. British and American military intelligence were well informed about the firm's radar technology. Intelligence on the *Würzburg* and *Würzburg Riese* radars had been produced as early as mid-1941. A specimen of the *Würzburg* was captured during the Bruneval raid by British commandos in February 1942, and information gained later that year and in 1943 enabled the Anglo-Americans to acquire a clear picture of the capability of both radars and their use in the *Luftwaffe* air defence network. By early 1945, the Anglo-Americans also had enough information on the successor of the *Würzburg Riese*, the *Mainz*, and the latest radar for flak units, the *Mannheim*, to produce reliable intelligence on each apparatus. What they did not know was that *Telefunken* modified the radars to track SAMs.[27]

The investigation of *Telefunken* began when SHAEF was notified that one of the general managers of the firm, a Dr. Engels, was being held by 12th Army Group at Wiesbaden, 25km west of Frankfurt-am-Main. A five-man CIOS team subsequently interrogated Engels and an assistant, an engineer named Urtel, on 22 April, and the two Germans agreed to travel east with the investigators into Thuringia to visit the firm's facilities in several towns and to search for personnel from the company. At Bad Blankenburg there were laboratories and workshops that were associated with development; at Bad Liebenstein there were research laboratories; at Steinbach there were laboratories and a factory; and at Erfurt a factory that had been in operation since 1937 was found intact. A number of other leading staff from the firm were also interrogated, including the head of research, Prof. *Dr.-Ing.* Hans Rukop, who was found at Bad Liebenstein. Amongst his numerous responsibilities, Rukop headed the laboratories at Berlin-Zehlendorf where radio-control, guidance and radar technology for guided missile systems was researched as part of the SAM development programme. The main purpose of the interrogations was to obtain information about the research and development organisation of *Telefunken* in order to improve and append the CIOS Black List, but also to enable the acquisition of some technical information on the current projects.[28]

Allied investigators would discover that German scientists and engineers had studied and developed two techniques for guiding SAMs towards an airborne target using radio and radar technology, namely command-link (or command-guidance) and beam-rider. Generally speaking, the command-link technique (the preferred method) had originated in the Hs 293 and Fritz X projects, which involved radio remote control and the use of optical means for missile and target tracking, as in the *Burgund* system, to be eventually replaced with the use of radars and a computer. In the beam-rider technique, which was also originally designed for air-to-surface guided weapons and the A-4/V-2, the missile would follow a control beam (*Leitstrahl*) from a radar directed towards the airborne target.

In Bavaria, an RAF officer attached to a CIOS Consolidated Advance Field Team (a CIOS reconnaissance unit), Flight Lieutenant P. R. Price from the MAP, was the first Allied investigator to reach the BMW facilities in the Munich area, where the firm's development of LPREs was centred. Price arrived at BMW on 25 April, five days before the city was captured by the US Seventh Army. During an investigation that eventually ran until 30 May 1945, it was found that air raids had severely interrupted production towards the end of the war and forced the dispersal of some the company's operations to alternative locations. The experimental production of LPREs – such as the 109-558 for the Hs 117 – was one of the effected operations. The manager for all reciprocating, jet and rocket engine development at BMW, Dr. Bruno Bruckmann, was located, but the head of the rocket engine development department, *Dipl.-Ing.* Helmut von Zborowski, an SS officer, was missing as he had apparently gone to fight shortly before the capitulation and could not be found. Von Zborowski was described by his colleagues as being more concerned with the SS and political affairs than his work for the company, and apparently left most of the technical matters to his assistants. The designer of the 109-558 and the 109-548 LPRE for the *Ruhrstahl* X-4 AAM, Hans Ziegler, was found at the manufacturing plant at Bruckmühl near Rosenheim, about 40km south-east of Munich. He was co-operative and agreed to build four samples of the 109-558, some samples of the 109-548 and three samples of the 109-718, an experimental auxiliary rocket unit for one of the firm's jet engines, for analysis by British and American agencies. Some company documents were also found, which included a report by von Zborowski on rocket development at BMW up until November 1944 and a box that contained all material specifications, drawings, calculations and specifications for rocket and jet engines.[29]

Allied investigators were also shown through the firm's facilities by Dr. Hermann Hemesath, the chief chemist in the rocket engine development department. BMW undertook a substantial amount of research into liquid propellants during the war. Hemesath claimed that around 6,000 fuel combinations were tested with nitric acid, which he felt was the most promising oxidant for LPREs. A NavTechMisEu survey of rocket engine development in Germany mentions that Hemesath's department carried out research with about 3,000 *Tonka* fuel combinations. At least two *Tonka* fuels were developed for the 109-558 LPRE: *Tonka* 250 was a mixture of triethylamine and xylidine, and *Tonka* 500 comprised 35 per cent octane mixtures, 20 per cent benzene, 12 per cent xylidine, 10 per cent optol, 10 per cent aniline, 8 per cent methylaniline and 5 per cent ethylaniline. During one CIOS investigation, several partially destroyed 109-558 engines were located near some test pits, along with several propellent tanks and a burnt-out combustion chamber in a workshop.[30]

In north-western Germany, the remnants of the *Erprobungsstelle Karlshagen* (Peenemünde-West), which had been evacuated to the *Seefliegerhorst Wesermünde* near Bremerhaven, were captured by British troops in early May. The base was secured by personnel from T-Force (a joint US Army and British Army unit) from 9 May, and on around 18 May, 39 scientific and engineering personnel from the *E-Stelle* were removed from the site by 30 Corps of the British Second Army to a nearby POW camp between Cuxhaven and Stade. When a four-man British CIOS team consisting of representatives from the MAP, the Admiralty and the Air Ministry began an investigation on 20 May, the 39 personnel could not be located. A number had undertaken the testing of ground-based remote control and guidance equipment for the SAM development programme, but like the specialists, no evidence of the programme was found at the site. Some guided weapon hardware was discovered, including about 10 *Blohm und Voss* Bv-246 glider bombs, a number of burnt and severely damaged Hs 293s and X-4s, some burnt radio transmitters and three *Askania* cinetheodolites. The only documents that were recovered were record books of the photographic section. The investigators were credulously convinced that practically all of the documentation from the *E-Stelle* had been burnt, when in fact the establishment's documents concerning guided weapon development had not been destroyed, but were buried. Almost two years would pass before British and American investigators recovered these documents.[31]

At Kiel, British troops captured the facilities of *H. Walter KG*, the company that designed the 109-502 LPRE for the *Enzian* and the 109-729 LPRE for the Hs 117. Troops from 30AU captured the firm on 5 May and soon located its director and founder, Prof. Hellmuth Walther, for interrogation. Walter told 30AU investigators that prior to the surrender of the *Wehrmacht* forces in the area at 0800 hours on 5 May, he had destroyed all the original reports and drawings in accordance with an order from *Großadmiral* Karl Dönitz, the former chief of the *Kriegsmarine* who had succeeded Adolf Hitler as the German head of state after the suicide of the *Führer* in the bunker under the Reich Chancellery on 30 April. But in a *volte-face*, on 10 May, the *Oberkommando der Marine* (OKM) at Flensburg near the Danish border in Schleswig-Holstein issued an order that instructed all naval establishments to hand over all documentation to the Allies. 30AU personnel subsequently found microfilm copies of the firm's documents buried in a coal cellar in the main office, in addition to a roll of 16mm film of new secret weapons. Whether the original company documents were actually destroyed is a matter of speculation. Also at Flensburg was the former Minister for Armaments and War Production, Albert Speer, who was being interrogated at length by investigators about the German armaments industry. Speer provided his interrogators with some basic information on SAM development. On 28 and 29 May, US Army G-2 intelligence officers from SHAEF presented Speer with lists of code names for armaments projects, and although Speer was familiar with the activities at Peenemünde, he stated that he had not heard of the *Taifun*.[32]

Within a few weeks during April and early May 1945, the British and American situation *vis-à-vis* scientific and technical intelligence on German SAM research and development quickly transformed from a dearth of evidence into a deluge. Throughout the recently captured central, northern and southern regions of Germany, the brains behind the SAM technology, the German scientists and engineers, were found along with the artefacts they designed and the remains (damaged and undamaged) of their various concerns. Intelligence was produced on the basis of discoveries pertaining to aerodynamics and hypergolic liquid propellant research at the LFA; LPRE technology and liquid propellant research at BMW (nitric acid engines) in Munich and HWK (hydrogen peroxide and nitric acid engines) at Kiel; electronic radio remote control technology for guided missiles at *Stassfurter Rundfunk* at Stassfurt; and the evacuated RLM experimental and testing establishment at Peenemünde-West, the *E-Stelle Karlshagen* (albeit without documents). The Anglo-American investigations of the prime contractors are the topics of the next two sections, along with the discovery of two of the world's first SAM training simulators. These touch on the next stages of the technology transfer process following the documentation of the intelligence targets by the officers in the field and the production of intelligence reports, which involved the physical removal of human and material assets from Germany to the US and the UK for the purpose of exploitation.

Intelligence on the C-2 and Hs 117 projects in Thuringia and Upper Bavaria

An interesting fact about the recruitment of German specialists by the Allies after World War II was that the discoveries of the Hs 117 and C-2 projects prompted some of the earliest requests to transfer German specialists to the US. It began at Oberammergau in early May, when the chief designer of guided missiles at HFW, Prof. *Dr.-Ing.* Herbert Wagner was found by counter-intelligence officers from the NavTechMisEu. Wagner was subsequently interrogated, and within the next two days escorted the officers to the Harz Mountains, where he unearthed seven enormous cases of blueprints and models of the firm's guided missile projects. Engineering drawings of the Hs 117 that were subsequently obtained by the British may have originally been recovered from this location. From there he took the officers to the *Hydra* underground factory at Woffleben near the *Mittelwerk* A-4/V-2 factory and displayed plans and prototypes of the Hs 117. In transmitting information from Germany to the US, the NavTechMisEu command gave highest priority to what was applicable to the war against Japan. Information which required rapid transmission was forwarded by dispatch or letter report. Aware of the potential value of the Hs 117 in the Pacific War, the officers cabled a request for the immediate evacuation to the US of Wagner and two of his assistants, *Dipl.-Ing.* Reinhard Lahde, a chief constructor, and Wilfried Hell, an aerodynamics specialist, both of whom had worked on the Hs 117 project. The NavTechMisEu officers afterwards examined the factory at Woffleben for almost a week with Major Robert Staver, from Ordnance Technical Intelligence Team No. 1, US Army Ordnance Technical Intelligence Branch. An almost complete Hs 117 prototype and a guidance and control unit for the Hs 298 AAM were found.[33]

In the vicinity, CIOS investigators subsequently located two other engineers from HFW who had been centrally involved with the development of the Hs 117. One was *Dr.-Ing.* Eduard Marcard, the former chief electrical engineer in *Abteilung F*, who had worked under Wagner since 1 April 1941. The other man was *Dipl.-Ing.* Julius Henrici, the former head of the mechanical construction section for the Hs 117 project. Both men could speak and write English, and they agreed to co-operate with the investigators. Over the next several weeks, they wrote reports on their wartime work that provided the investigators with a general picture of the technical details of the Hs 117. The search for records that pertained to the aerodynamic research for the project was less fruitful, revealing a pattern of behaviour that was by now common among captured German specialists – concealing knowledge of hidden documents. From May 1944 to March 1945, wind tunnel tests at speeds up to Mach 0.85 were mostly carried out in the subsonic low- and high-speed tunnels at the DVL at Adlershof in Soviet-controlled Berlin (see Appendix 3), and to a lesser degree at the AVA under the direction of Professor *Dipl.-Ing.* Otto Walchner, one of Germany's leading experts on high-velocity ballistics and swept-wing designs. Henrici asserted to the investigators that all the DVL reports had been destroyed, but the AVA reports might still be found.[34]

Subsequently, Walchner was found and interrogated by the investigators. He claimed that the reports on wind tunnel experiments for the Hs 117 project at the AVA had been destroyed, although he added that it might be possible to reconstruct the reports from the tunnel laboratory records, which he believed still existed. He agreed to try to find the records and rewrite the reports if possible. It later turned out that a complete file of reports on experiments in the Mach 3.7 wind tunnel at the AVA was recovered by Colonel Leslie E. Simon, the director of the US Army Ballistic Research Laboratory at the Aberdeen Proving Ground in Maryland. Most of the reports were passed on to CIOS, with a few forwarded to the BRL for analysis. Questions regarding the honesty of Walchner in this situation can clearly be raised, since there are numerous examples throughout this chapter where German specialists were co-operative and forthcoming, while simultaneously withholding information on the location of hidden documents. No-one can blame the Germans for engaging in this practice; usually it was done for the purpose of keeping the knowledge in German possession, but on occasions it was also used as a tool to gain leverage in future negotiations with the Allies. Marcard, Henrici and five of their colleagues were subsequently employed by the MAP in the British Zone of Occupation to write monographs on the HFW guided missile projects.[35]

At Nordhausen, US Army Ordnance Department investigators searched for senior staff from Peenemünde but did not locate any until 12 May 1945. Those found included *Dipl.-Ing.* Eberhard Rees, von Braun's deputy and former head of the production section; Walther Riedel, from the design bureau in the planning sub-section; and Karl-Otto Fleischer, the business manager of EMW. The revelation of the sophisticated and revolutionary C-2 made a considerable impression on Major Staver of the US Army Ordnance Department. Partly as a result of the information that he gathered on the C-2, in addition to fears that the best German specialists might be recruited by the Soviet Union, by 18 May, Staver had sent a letter by courier to the Ordnance Technical Division in Paris recommending that 100 specialists from Peenemünde-East be evacuated to the US within the next 30 days to continue the development of the C-2 for the war against Japan.[36]

Meanwhile, the location of the 14 tons of the Peenemünde-East archives had been disclosed to Staver by Fleischer, and a recovery operation was under way. On 22 May, Staver flew to Paris to inform Colonel Joel Holmes, the Chief of the Ordnance Technical Division, that the archives had been located and to obtain authorisation for two 10-ton trucks to transfer the precious cargo to Nordhausen. A secondary objective of Staver's mission to Paris was to follow up the letter sent several days earlier. Staver had by now learned of the detention of the specialists at Garmsich-Partenkirchen, and convinced Colonel Holmes to send a cable to Colonel Trichel, the chief of the Ordnance Rocket Branch at the Pentagon, requesting the transfer of part of the Peenemünde-East organisation to the US. Staver again used the discovery of the C-2 to support his argument. A portion of the cable read:

> Have in custody over 400 top research and development personnel of Peenemünde ... Latest development named *Wasserfall*, a 3,000kg flak rocket ... Believe this development would be important for the Pacific war. The research directors believe if their group were taken to US that after one month of adjustment and reorganisation and three months of hard work could produce complete drawings of *Wasserfall* ... Recommend that 100 of very best men of this research organisation be evacuated to US immediately for purpose of reconstructing complete drawings of *Wasserfall*. Also recommend evacuation of all material drawings and documents belonging to this group to aid in their work in the US [*sic*].

Staver's cable was well received at the Pentagon, although there was far more interest in the A4/V2. On 1 June, he was asked to submit the names of five Germans who could be employed in the US as instructors to train American personnel how to launch the A-4/V-2. On the following day, the Supreme Commander Allied Expeditionary Force (SCAEF), General Dwight D. Eisenhower, gave his support to Staver's proposal in the cable.[37]

More components from the C-2 were recovered around the area. Staver and another officer from Ordnance Technical Intelligence Team No. 1, Lieutenant Hochmuth, found a complete guidance unit in a barn; a complete gyroscope assembly was found in a shed; several complete *Mischgerät* units (an electronic mixing device that was originally designed for the A-4/V-2 to modify radio transmissions from the ground and gyroscope signals in the missile to ensure stability during flight; it was modified for use in the C-2) were discovered in another shed; a number of fire-damaged E-230 *Strassburg* receivers were found amongst some badly looted laboratory gear; and a quarter-scale aluminium model of the missile was found in a pond near Bleicherode. The gyroscope and one *Mischgerät* unit were sent to the RAE for analysis. These finds probably represented only a portion of the C-2 technology that remained in the area; von Braun, who had surrendered with some colleagues in Upper Bavaria, told one interrogator at Garmisch-Partenkirchen that there would be enough parts for several complete C-2 missiles in the vicinity of Bleicherode or in transit thereto.[38]

A large amount of technical information on the C-2, and to a lesser extent concerning the *Taifun*, along with a small amount on the *Rheintochter*, were acquired during May and June through the interrogations of the approximately 450 personnel – most of whom were from Peenemünde-East – who surrendered to US troops around Garmisch-Partenkirchen in Upper Bavaria. The foremost were *Generalmajor* Walter Dornberger, the former chief of liquid propellant rocket development for the German Army, and von Braun. A number of von Braun's subordinates at Peenemünde-East who were in the vicinity and interrogated included *Dr.-Ing.* Martin Schilling, the director of the testing department, and *Dr.-Ing.* Ernst Steinhoff, the director of the guidance and control department. Several specialists from the C-2 steering and remote control section (EW 224) were also found in the area and interrogated. They included Drs. Oswald Lange and Theodor Netzer, both of whom had headed the section; Dr. Ernst Geissler, one of the three deputy section managers, who headed the design and development of servomechanisms; Dr. Guntram Haft, who was in charge of ground apparatus development; and Dr. Helmut Weiss, who was responsible for the development of an infra-red seeker. Also found was Dr. Gerhard Heller, a chemist who researched the propellants for the C-2 LPRE.

At first, a number of the Germans were reluctant to disclose the full details about their work, and in some cases sent the investigators around in circles. The investigators sometimes had to conduct additional interrogations in order to acquire a fuller picture of the particular systems the Germans worked on. Not all the Germans were so protective, however. At least one of the specialists at Garmisch-Partenkirchen had a relative who had been killed by the Nazis. On 18 May 1945, Dr. N. J. H. David, a physicist who had been employed in EW 224 and tasked with designing an electrical model of the control system in the C-2, told his two interrogators that around the same time he was posted to Peenemünde in January 1943, his father had been arrested for expressing critical opinions (of an unspecified nature) and was eventually executed. Consequently, David was not particularly concerned about the success of the C-2 project. Overall, the German specialists provided the investigators with sufficient technical information that enabled a preliminary reckoning of the development status of the C-2 and *Taifun* prior to more thorough investigations. In addition to these personnel, five railway wagons full of radio equipment (some belonged to the C-2) were discovered at a train station at Pieting, about 25km north of Garmisch-Partenkirchen.[39]

About 30km north-east of Garmisch-Partenkirchen, the US Army took over *WVA Kochelsee GmbH*, formerly the *Aerodynamische Institut der Heeresversuchsanstalt Peenemünde*. When American troops captured the facility in early May, most of the equipment was intact except for some minor damage caused by looting. The facility's documentation and archives had been preserved, with microfilm copies of reports made just prior to its capture. Between May and September 1945, about 150 investigators from the three Western Allies visited the facility. The Mach 4.4 wind tunnel – which was the fastest in the world – and the work for the A-4/V-2 and C-2 projects, drew much attention from the scientific and technical missions. The director of the establishment, Dr. Rudolf Hermann, was keen to promote the unique role that the facility had played in the research and development for the C-2 project. In a report for British and American investigators, he recounted that representatives from the RLM, the *Forschungsführung* and the HWA all agreed that in no other supersonic wind tunnel in Germany could the extensive and new kinds of aerodynamic experiments that were demanded be carried out with the necessary versatility, deep research and rapidity.[40]

Intelligence on the *Enzian* and *Rheintochter* projects and training simulators

In southern Germany, Allied investigators found two other important intelligence targets linked to the German SAM development programme. Located at Oberammergau in Upper Bavaria was the *Messerschmitt AG Oberbayerische Forschungsanstalt* and materials concerned with the *Enzian* project. Also found in the vicinity was the designer of the *Enzian*, Dr. Hermann Wurster. During interrogations at the *Messerschmitt* facilities, Wurster was co-operative and provided information about the missile's development. When questioned about surviving hardware by four American civilian investigators on 20 May, he told them that in the final days of the war, 20 prototype *Enzian* missiles that remained from a total of 60 produced were destroyed to prevent them being captured. The search for documents was frustrated by Wurster's assertion that he did not know the whereabouts of the facility's archives or those of the client firm that built the *Enzian* prototypes, *Holzbau-Kissing KG*. Films of the test-launchings were recovered, and Wurster screened these and other films about *Messerschmitt* developments for the investigators on 19 June. The films of the *Enzian* showed problems with the stability and control of the missile.[41]

An officer from AAF Air Technical Intelligence, H. Sparrow, searched for the missing documents by following up on some information provided by Wurster to Hugh Dryden of the AAF SAG on 15 June. Wurster had told Dryden that *Luftwaffe Oberstleutnant* Dr. Friedrich Halder, the military director of the SAM development programme, and *Hauptmann* Werner Jank, from the same section, were staying at the *Haus Alpblick*, a hotel in Sonthofen, and could assist in the investigation. Wurster told Sparrow that he believed he would find Jank and the *Enzian* drawings at Sonthofen, and so on 20 June Sparrow took Wurster to the *Holzbau Kissing* plant at Sonthofen to interrogate the general manager of the company, *Herr* Kissing, and the technical manager, *Herr* Jacobsen, about the matter. In a by now familiar answer, Kissing claimed that all the blueprints related to the *Enzian* prototypes were burned two days prior to the arrival of French troops on 30 April, under the orders and supervision of a *Luftwaffe* officer, a *Leutnant* Kissel, who was staying in Sonthofen with a group of officers that included Dornberger, Halder and Jank. Kissing further stated that the group had left Sonthofen on the day the blueprints were burned with the original drawings of the production version of the *Enzian*, and headed for Pussen, Schongau or Salzburg. Kissing also said that the original drawings could be found if Jank could be located. The following day, further interrogation of Wurster, in addition to the interrogation of another man, Willy Borcheding, produced definite information that the original drawings were contained in four square metal boxes and one round metal box that were loaded into Jank's car with assistance from Borcheding, most probably on 30 April at Oberammergau. Where Jank went with the documents was apparently unknown.[42]

The second important intelligence target was the *Deutsche Forschungsanstalt für Segelflug Ernst Udet* (DFS) at Ainring near Salzburg in Austria. The CIOS report about the DFS – one of the most voluminous at over 200 pages – is a key source of information about the establishment, its activities and its staff. After American troops occupied the area, the AAF took over the establishment and investigators managed to find most of the staff around Ainring. The staff had dispersed their equipment to numerous locations in the surrounding area, including a barracks and cellars at the Ainring airfield; workshops and laboratories at Reichenhall, Teisendorf and other nearby villages; and various castles and farmhouses. The equipment and three containers of DFS reports were recovered by an AAF officer with assistance from a German, a Dr. Stamer. AAF personnel assembled the staff, equipment, and documents, and ordered the Germans to complete any unfinished research with the aid of the materials.[43]

The research work at the DFS for the SAM development programme included studies of missile trajectories, steering and theoretical studies concerning the connection of seekers to automatic control. The development work included the stabilisation and control of the *Enzian* on behalf of *Holzbau-Kissing* and an infra-red seeker for the missile in conjunction with the firm *Kepka* of Vienna (the project was transferred to *Messerschmitt* at Oberammergau in January 1945). One of the most interesting discoveries was in a laboratory and workshop at Teisendorf, where investigators found a three-dimensional flak rocket training simulator that was used to train *Luftwaffe* ground crews how to control and steer the C-2. The simulator, designed and built for the *Flakversuchsstelle der Luftwaffe* by a team led by Dr. Johann Schedling, a physicist in the DFS Institute for Aeronautical Equipment, replicated the conditions of the flight and trajectories of a target aircraft and a missile. The dimensions of the simulator were 10 metres x 3 metres x 3 metres, with a linear scale of 1:5,000. It worked by the use of motor-driven trolleys and pulleys, whereby the missile and the target were represented by spheres that were attached by string to the trolleys, which were moved by the use of a control stick. The positions of the missile and target trolleys were fed into two calculators, which computed the azimuth and elevation of the missile and target. The results were displayed as spots of light on a screen so that the data could be measured and recorded. Construction of this apparatus, referred to as a qualitative simulator because the data from the simulations was represented visually, began in May 1943, and the machine became operational in February 1944. The construction of a more sophisticated apparatus, referred to as a quantitative simulator because it was fitted with a computer and the acquired data was represented numerically, began in February 1944, but was only 50 per cent complete by May 1945. This simulator was designed to train operators to control the *Rheintochter* and *Enzian* as well as the C-2. As for the Hs 117, Wagner designed simulators at HFW for all the company's guided missile projects.[44]

The intelligence targets in Berlin were inaccessible to American, British, and French scientific and technical missions until the three countries could enter the respective sectors of Berlin on 1 July 1945 (see Appendix 3). Two of the prime contractors had their head office in Berlin – HFW was located at an airfield in Schönefeld, just outside East Berlin in the Soviet Zone of Occupation, and the headquarters of *Rheinmetall-Borsig* was at Marienfelde in the future American sector. Also located in Berlin were several firms that designed and developed electronic equipment in missile guidance systems, including *Kreiselgeräte* and *Telefunken* at Zehlendorf in the American sector, and *Askania-Werke* at Fridenau and *Siemens* at Hakenfelde in the British sector. The DVL at Adlershof, where aerodynamic research was done for the Hs 117 project, was situated in the Soviet sector.

There was at least one instance where a sample of prototype flak rocket technology in Soviet-controlled Berlin was smuggled out and ended up in the hands of the Western Allies. When the Red Army was in the process of dismantling the *Kreiselgeräte* facilities, an apparatus for calculating the parallax angle between the direction of a ground-transmitted radio-control beam and the direction of the Hs 117 missile launcher was found. Basically, how it worked was that when a Hs 117 was launched, about 50–100 metres away the radio-control beam was transmitted. When the missile intersected the radio beam, a receiver in the missile picked up the transmission and from there the Hs 117 would presumably proceed on a trajectory towards the target. *Kreiselgeräte* began the development of the apparatus at the Niedereinsiedel branch in Berlin 1944, and not until March 1945 did the main works in Berlin complete the first in a series of 10 apparatuses. The firm delivered two apparatuses to the *Luftwaffe* in the same month, apparently unaware that the Hs 117 project had been cancelled the previous month. A further two sets were left in the main works at the end of the war. One was removed by Soviet investigators and the other by German civilians, probably ex-employees of the firm, and subsequently ended up in the possession of the British Naval Gunnery Mission (BNGM) at Minden in the British Zone of Occupation (a group of ex-*Kreiselgeräte* employees, led by *Dr-Ing.* Johannes Gievers, a gyroscope expert, were briefly employed by the Soviets in Berlin prior to going over to the British side; they were sent to the BNGM for exploitation). The apparatus was reassembled at the BNGM, evaluated, and the intelligence passed on to the Americans.[45]

Many other targets in Soviet-occupied territory were off limits to American, British and French investigators. The remains of Peenemünde-East and West remained inaccessible, as were a number of the firms that were sub-contracted to manufacture and supply subsystems and components for the four prime contractors. Much of the infrastructure that was associated with the *Rheintochter* project was located in eastern Germany. The test-launching and static testing facilities were located at Leba in Pomerania prior to the evacuation to Peenemünde; *Ardelt-Werke* at Breslau manufactured the fuselages of the R-I and the *Feuerlilie* missiles; *Seyffarth* of Ebersvelde, located 30km north-east of Berlin in Mark Brandenburg (on the railway line to Peenemünde), manufactured the LPRE for the R-IIIf; and *Gerätebau Bismarck AG* at Zwickau in Saxony manufactured the SPRE for the R-IIIp. Staff from *Rheinmetall-Borsig* at Marienfelde evacuated the firm's archives westwards, evidenced by the capture of certain documents by the Air P/W Interrogation Unit (AP/WIU), RAF Second Tactical Air Force, prior to 18 June 1945. Among the documents were correspondence files, which contained communications concerning the development of the *Rheintochter* project up to 1945, and maker's handbooks on the R-I (dated April 1943) and R-III (dated September 1944), with specifications, technical details and drawings of each missile. These items were sent to the British Air Ministry ADI (K) – Assistant Directorate of Intelligence (Prisoner Interrogation) – documents centre in England for evaluation and reproduction. The documents assisted ADI (K) personnel in their interrogations of German POWs, and enabled the personnel to write a brief intelligence report about the missile prior to more detailed analysis.

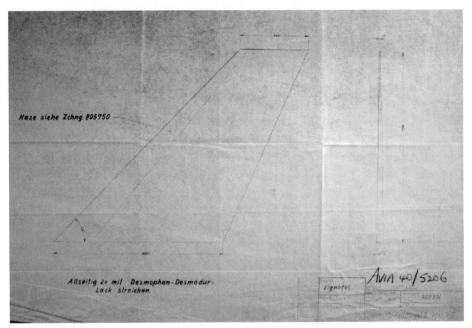

A captured engineering drawing of the *Rheintochter* R-III swept wing design, produced at the *Rheinmetall-Borsig* Berlin-Marienfelde facilities on 20 June 1944. The wings were made out of Lignofol, a highly compressed laminated wood. (Source: TNA, AVIA 40/5206)

The same RAF unit also captured a maker's handbook on the *Taifun* in Germany, which was likewise sent to the ADI (K) centre for reproduction.[46]

The sections above have discussed the key sites in Germany (and Austria) where British and American investigators captured the most important German scientists and engineers who were involved in the German SAM development programme, and also where artefacts and archives were captured, during the period when the armed forces of the Western Allies were under the combined command of SHAEF from June 1944 until the official quadripartite partition of Germany and Austria took place in July 1945. The British and American scientific and technical missions were fortunate in their activities because a number of concerns in Berlin evacuated their human and material assets westwards and southwards to prevent their capture by the Red Army. These actions significantly benefited the Anglo-Americans in particular, and the French to a limited extent (see Chapter 6), to the detriment of the Soviets.

American and British investigations after the dissolution of SHAEF

American and British intelligence organisations continued to gather information on German SAM technology after CIOS was dissolved along with SHAEF on 13 July 1945. The functions of CIOS in the new American and British Zones of Occupation were taken over by separate British and American intelligence organisations. On the American side, the War Department established the Field Information Agency, Technical (FIAT) in May 1945 for the principal purpose of aiding American civilian organisations to exploit German economic, industrial and technological intelligence. British and French arms of FIAT were also created. In June 1945, the War Department also created the Joint Intelligence Objectives Agency (JIOA) as a subcommittee of the Joint Intelligence Committee (JIC) of the Joint Chiefs of Staff, to produce intelligence and administer the US government programme to recruit German specialists for employment in the American defence sector, code named Project Overcast, which was established in July 1945. Initially, the programme was limited to recruiting specialists for military purposes. After much deliberation, it was subsequently extended to recruit specialists for both the military and civil sectors, and was assigned the new code name 'Paperclip' in March 1946. The Western Allies needed as many of the top-ranked German scientific and engineering minds as they could find for three main reasons: in the case of the US, their talents could be used in the Japanese war; after the war, their knowledge could be exploited for the benefit of the armed forces and industries of the victors; and it would deny their services to the Soviet Union and non-allied countries.[47]

To collect intelligence on German guided missile technology – SAMs being an important component – the US military set up an inter-service Joint Working Group for Guided Missiles. It drew its members from USSTAF Air Technical Intelligence; the US Army Ordnance Department, Corps of Engineers and Signal Corps; and the US Navy. Joint field teams were organised and reports were written. The scientific and technical missions from the armed services continued their operations in the field after CIOS was dissolved. One of these organisations that played an active role in producing intelligence on German SAM technology was the US Naval Technical Mission in Europe, which remained in operation until 1 November 1945. Table 6 shows the organisational structure of NavTechMisEu. Originally, NavTechMisEu had three sections that searched for information on guided missiles – ordnance, air and electronics. On 2 June 1945, a dedicated guided missiles sub-section was set up to study the German attainments in that field, which combined personnel from the three sections. Further studies were also made of launching mechanisms, control and guidance systems, and fuse developments. The subjects of technical reports that pertained to SAMs included 'General Survey of Rocket Motor Development in Germany', 'Underground Factories at Niedersachswerfen and Woffleben', 'Fusing System of the *Rheintochter* R-IIIf and R-IIIp', 'Trainers for Operators of Guided Missiles', and 'Nitroglycerin, Diethylene Glycol Dinitrate and Similar Explosive Oils – Manufacture and Development in Germany'. The bulk of the information collected by NavTechMisEu was reduced to detailed technical reports; by September 1945, hundreds of reports were written by mission investigators, which were also distributed as CIOS reports.[48]

Table 6. The organisational structure of the US Naval Technical Mission in Europe and its position within the hierarchy of the US Navy

The Navy Department
James Forrestal, Secretary of the Navy

Commander in Chief, US Fleet

Office of Naval Intelligence (ONI)	Chief of Naval Operations (CNO)
Section OP-16-R	(responsible for operational purposes)
Navy Department representative of	Section OP-16-PT
NavTechMisEu	

Commander, US Naval Forces, Europe (ComNavEu)
Admiral Harold Stark
(Responsible for administrative purposes and liaised with the Admiralty on the activities of NavTechMisEu)

US Naval Technical Mission in Europe

Commanding Officer
Commodore Henry A. Schade
Chief, NavTechMisEu and also the Deputy Head of Naval Target Sub-division (NTS), Special Sections Sub-division, G-2 SHAEF (held the latter position until 22 July 1945), in addition to continuing to serve as the senior naval member of the ALSOS mission.

NavTechMisEu officer, co-chairman,
CIOS Naval Group (Group 6)

Executive Officer
Responsible for the different sections until the position was abolished on 10 May 1945. The duties were thereafter divided between a Technical Branch and a Services Branch.

Technical Branch	Service Branch
Sections:	Sections:
(a) Ordinance*	(a) Intelligence
(b) Ships	(b) Supply
(c) Air*	(c) Operations
(d) Yards and Docks	(d) Administration
(e) Electronics*	
(f) Hydrogen Peroxide	

*Guided Missiles Sub-section (SAMs)

Source: 'Historical Data on US Naval Technical Mission in Europe, First Narrative', 1.11.1945, https:// www.fischer-tropsch.org

On the British side, the British Intelligence Objectives Subcommittee (BIOS) of the JIC succeeded CIOS. BIOS was created on 30 July 1945 with the approval of the 'BIOS Directive' by the Deputy Chiefs of Staff (DCOS) Committee. The JIC co-ordinated the British intelligence-gathering effort in Germany, while the

DCOS Committee was responsible for the exploitation of German scientific and technical knowledge for the benefit of the British armed forces. The JIC appointed the chairman and deputy chairman of BIOS, which initially comprised representatives from the Admiralty, War Office, Air Ministry, Foreign and Colonial Office, Board of Trade, MoS, MAP, Ministry of Fuel and Power, Department of Scientific and Industrial Research (DSIR) and the Government of the Dominion of Canada; at a later date also from the Control Office for Germany and Austria and the Government of the Commonwealth of Australia. Representatives from FIAT and the Control Commission for Germany (British Element) (CCG (BE)) – the organisation that administered the British Zone of Occupation – also attended. The functions of BIOS were to receive, approve and co-ordinate all requests of British government departments for military, political, industrial or economic intelligence which was available in Germany and in European countries formerly under German occupation. BIOS was not responsible for obtaining combat or technical intelligence for the armed forces, or for fulfilling any of the intelligence-gathering requirements of MI5 and MI6, but it did advise the DCOS Committee (and where necessary appropriate departments) on the steps they should take to implement the policy for the exploitation of German scientific and technical intelligence. BIOS also dealt with routine matters connected with the employment of German specialists in the UK, the US or Germany.[49]

As the CIPC and CIOS had done, BIOS drew up target lists, prepared plans for the exploitation of targets and provided expert personnel for investigations. Target lists and plans were forwarded to FIAT to ensure there was maximum transparency with the American authorities on intelligence targets, so all the information resulting from BIOS missions that corresponded with CIOS investigations was forwarded to American agencies. Targets were assigned under different group numbers. Guided missiles mostly fell under two groups: Group I (radio including television equipment, radar, infra-red, acoustics and electronic devices forming parts of weapons) and Group II (bombs and fuses, rockets and rocket fuels, and directed and controlled missiles). BIOS remained in existence until 15 May 1947.[50]

In the first six months after the final surrender, captured German guided missile specialists in the west of Germany were divided into several groups. Practically all the specialists from Peenemünde-East were assembled in the American Zone of Occupation for interrogation and possible recruitment. At first, they were located in the areas around Nordhausen, Garmisch-Partenkirchen and Kochel. Later, some were interned at Witzenhausen, 25km east of Kassel, and Eschwege, 50km south-east of Kassel, both in the province of Kurhessen. The last location prior to recruitment was Landshut, 65km north-east of Munich, where the US Army set up an accommodation barracks for the specialists and their families. While awaiting transfer to the US (to be later joined by their families), the US Army authorities agreed to lend some of the specialists at Witzenhausen to the War Office (the British government department that was responsible for the British Army) to participate in Operation *Backfire*, a British operation to assemble and launch A-4/V-2 rockets at the former *Krupp* proving ground at Altenwalde, near Cuxhaven, which ran from July–October 1945. Around 600 German personnel were employed in the operation, of whom about 70 were engineering, scientific and technical specialists. After Operation *Backfire* was completed, the MoS took over the site on 1 November 1945 and renamed it MoS Establishment Cuxhaven, with the acronym MOSEC. A group of specialists from Peenemünde-East were retained at this location to write reports and monographs, a number of whom were later recruited by the French government department *Direction des Études et Fabrications d'Armement* (see Chapter 6). MOSEC was in operation until mid-1946.[51]

Elsewhere in the British Zone of Occupation, German SAM technology continued to be documented at a number of other sites that were set up by the Admiralty, the MAP and the MoS. German specialists were also assembled and employed at these locations to write monographs and reports, in some cases to aid in their future employment in the UK. At the BNGM in Minden, 60 km south-west of Hanover in Westfalen, the Admiralty assembled German specialists with expertise of value to the Royal Navy, who included engineers and physicists specialising in electronics, gyroscopes and servomechanisms. Six sites were run as part of an operation to obtain, transfer and exploit German aeronautical knowledge from facilities, archives and human sources, called Operation *Surgeon*. Initially, the operation was jointly run by the RAF and the MAP, but after the MAP and MoS were amalgamated into two branches of the MoS in late 1945 – MoS (Air) and MoS (Munitions) – the former became responsible for the operation, although the recruitment of guided missile specialists was actually carried out by the latter.[52]

German guided missile and rocketry specialists were assembled at three of the six 'Surgeon stations.' On the outskirts of Fassburg airfield near Trauen, about 80km north by north-west of the LFA, the MoS set up an operation to exploit the knowledge of German rocket engine specialists at a former outstation of the LFA Institute for Engine Research, where rocket engines had been tested. The facility was renamed MoS Experimental Station Trauen, and a total staff of 82 German scientists, engineers and technicians were assembled to write reports on their work with LPREs. Generally speaking, the topics of the German research at Trauen related to the three oxidising agents used in German rocket engines: gaseous oxygen/fuel combinations, nitric acid (although limited to a study of the possibilities of generating pressurised gas for turbine drives) and hydrogen peroxide-based rocket systems, plus a few miscellaneous subjects such as theoretical work on combustion chamber design. The other two locations were the LFA near Völkenrode and the AVA at Göttingen. At these centres, one operation to gather information about the German SAM projects was the employment of seven former employees of HFW to write monographs about their wartime work on guided missile projects. The group eventually wrote 10 monographs that were edited by Julius Henrici, under the umbrella title of *Guided Missiles: Based on the Work of Professor Herbert Wagner*.[53]

Outside of Operation *Surgeon*, the MoS also assembled German guided missile and rocketry specialists at the former *Rheinmetall-Borsig* research and experimental establishment at Unterlüss, about 15km south-east of Trauen. The facilities, which had been badly damaged during the war, were renamed the Unterlüss Work Centre by the MoS. At this location, a group of engineers from the firm – led by *Dr.-Ing.* Heinrich Klein, the former head of the Unterlüss establishment – were employed to write monographs and reports on their wartime work for the company. Klein had previously spent some time at the 'Inkpot' detention centre at the Beltane School in Wimbledon, an unpleasant place where German specialists were sent to be interrogated and employed to write reports for a moderate sum of money. The Unterlüss Work Centre remained in operation beyond a deadline of 30 June 1946 that was agreed between the Allies for all war-related research in Germany to be concluded. The British were not the only one of the Allies to break this rule; the other three also did. From the perspective of the Western Allies, the continued employment of the Germans in the western zones was seen as preferable to their services being procured by the Soviet Union. The reports that were written by the specialists at Unterlüss which pertained to the *Rheintochter* project concerned the experimental techniques that were used in determining the paths of projectiles by both photo- and cinetheodolites, and the calculation of trajectories; the evaluation of velocity and drag from phototheodolites; one specifically about the R-III; and a general report that discussed the *Rheintochter* project with a full description of the aerodynamic characteristics of the missile and the means of guidance. Still in the progress of being written in January 1947, by early May of that year the reports were essentially completed, along with a supplementary report on aerodynamics and controls for the newly created Rocket Propulsion Department of the RAE at Westcott in Buckinghamshire. The MoS activities at Unterlüss were not terminated until mid-1948.[54]

Meanwhile, in early 1947, the American and British intelligence organisations in Germany learned that there were possibly caches of buried documents relating to German guided weapon development that were not recovered by Anglo-American investigators in 1945. An investigation had followed the arrest in February 1947 of Rudolf Brée – a former *Flieger-Oberstabsingenieur* (Wing Commander) in the engineering corps of the *Luftwaffe* who was in charge of air-to-air guided missile development in the RLM during the war – for removing classified documents from dossiers belonging to his current employers in Germany, the US Navy. In the same month, AAF Air Intelligence launched an investigation called Project Abstract with assistance from the British branch of FIAT to find any documents. The investigation discovered that a group of German specialists, led by Walter Dornberger, then in detention in South Wales, and von Braun, who had emigrated to the US in September

1945 through Project Overcast, were, along with a number of their German colleagues and contemporaries in Germany and the US, involved in a conspiracy to conceal from British and American intelligence agencies the existence of a number of buried document caches from Peenemünde-East and the *E-Stelle Karlshagen*. Their objective was to use the documents to rebuild a guided missile industry in Germany after the Allies relinquished control of the country. On the basis of information that was acquired from interrogations, four recovery operations were subsequently carried out at Wesermünde, Bad Sachsa, Oberjoch and Rechlin.[55]

The outcome of Project Abstract was mixed. Buried document collections from the *E-Stelle Karlshagen* were recovered by the investigators but represented the only tangible success of the operation. Over 1,000 pages concerned the development of the Hs 117. Most of the buried documents from Peenemünde-East were unearthed by unknown persons in the two years after the end of the war, apart from a few scattered and weather-damaged reports that were left at one burial site. In the course of the investigations, complete details were also revealed about the working group that was formed in early 1945 by the RLM and Armaments Ministry, and led by the SS, to consolidate German guided missile development and production, the *Arbeitsstab Dornberger* (see Chapter 1).[56]

Conclusion

The British and American scientific and technical intelligence on the German SAM development programme was thorough and comprehensive. It can be reasonably estimated that British and American investigators acquired at least 75 percent of the total stock of knowledge by February 1947. The documents that were recovered in Project Abstract essentially completed the recovery operations. Despite the success in gaining practically all the German secrets, the one burning question that the documents apparently did not answer concerned undocumented and unverified accounts of the use of the SAMs against Anglo-American aircraft near the end of the war. There is no evidence to point to the operational use of ground-launched anti-aircraft rockets by the *Luftwaffe* heavier than the rudimentary projectiles described earlier in this chapter, with one exception (see Chapter 6), among the hundreds of independent statements by German specialists, some of whom did not know each other. Much speculation has ensued because sightings of strange phenomena by Allied pilots in the skies over Europe were thought to be advanced German weapons, possibly AAMs or SAMs. As for any connection between the atmospheric phenomena of the so-called 'foo fighters' – the coloured balls of light that were witnessed by American pilots in the skies over Europe – and the German SAM development programme, there is also no evidence.

The collection of information about German SAM research and development, which was subsequently produced into intelligence for transmission to the personnel in the higher echelons of their governments, armed forces and industries so as to inform them about the German advances with the technology, and to guide postwar policy decisions concerning the exploitation of the knowledge, was really the first stage in the technology transfer process. In this chapter, what was essentially the beginning of the next stage of the process, the removal of materials from Germany back to the Allied countries, has been touched upon. The following chapter accounts for the transfer and general exploitation of the German intelligence by the greatest recipient and beneficiary of that knowledge, the US. The processes by which these operations were undertaken will provide a basis for comparison with the corresponding activities by the British Commonwealth and France in the proceeding chapters.

CHAPTER FOUR

The Transfer and Exploitation of German SAM Technology by the United States After 1945

Captured human and material assets from the German SAM development programme were transferred to the Western Allies in several forms. As discussed in the previous chapter, from 1945–49, American and British (and French) armed forces, and scientific and technical missions, recovered and transferred documentation from western, northern, central and southern Germany. Besides the archives of German government departments and military organisations, such as the RLM and HWA, there were the archives of three of the prime contractors and other armaments manufacturers in the German SAM development programme; archives of the aeronautical research establishments, for example, the LFA and DFS; and of experimental establishments, such as Peenemünde-East and West. The mass of documentary material that was brought back to each country comprised departmental, inter-departmental and external bureaucratic correspondence, such as letters and memoranda, along with minutes of conferences, scientific and engineering reports, records of company products, engineering drawings, schematics, films, patents, photographs, books, and journal articles.

Another form of knowledge was the captured technology. Included in this category were entire research facilities such as wind tunnels and laboratories, training simulators, complete and partial prototype missiles and ground apparatuses, electronic equipment in guidance systems, warheads, rocket engines and associated equipment, and materials and instruments such as rocket propellants and cinetheodolites. While the captured war materiel was in the custodianship of the military, intelligence or civilian organisations who were responsible for its security during storage in or transit from Germany, for storage in the three victor countries and dissemination to interested agencies, national government systems of classification were applied. During World War II and the second half of the 1940s, the US government had a three-tier system: 'Secret', 'Confidential' and 'Restricted'. The British government also had a three-tier system: 'Most Secret', which equalled the US 'Secret'; 'Secret', which equalled the US 'Confidential'; and 'Restricted', the same as the US classification. After the war, the American and British governments synchronised their classification systems. The Americans added 'Top Secret' to their system, while the British government changed 'Most Secret' to 'Top Secret', and added 'Confidential' to equal that of the US. In France, documents that were classified as sensitive were stamped *'Confidentiel', 'Secret'* or *'Très Secret'* (Very Secret).

The third and perhaps most crucial assets were the German (and Austrian) scientists, engineers and technicians themselves – collectively called specialists – who carried out research and development work at the research establishments, experimental establishments, armaments manufacturers and technology companies. As with any profession, the expertise and experience of the specialists depended on their educational background, employment experience, age and particular skills and abilities. Due to the greater size of America's armed forces over those of the British and French, and the country's superior economic and industrial resources, the US was in the best position to offer German specialists the most favourable employment opportunities after the war. There were other reasons why the US was a more attractive destination – neither the British nor French governments could afford large-scale research and development programmes, and the British were only prepared to recruit a comparatively small number of specialists. Within the construct of the Anglo-American scheme to pool German specialists and allocate them to both countries, these factors balanced the combined recruitment efforts heavily towards the side of the Americans. As a result, of the Allies, the US was the greatest beneficiary of the German SAM development programme.

These assets, considered as war booty and forms of reparations, were transferred to the Allies and subsequently exploited. The scientific and engineering knowledge was disseminated by various organisations in each of the three Western countries in hierarchical structures that differed according to each country. On the American side, the technical and intelligence services of the American armed forces (in concert with contractors in industry and the university sector) evaluated captured artefacts and documents and sponsored the recruitment and employment of the German specialists in the US. The processes on the British and French sides are addressed in the two ensuing chapters about their respective technology transfer operations.

The first part of this chapter focuses on the technology transfer by two US Navy bureaus, the Bureau of Aeronautics and the Bureau of Ordnance. Between them, these two agencies recruited between 20 and 30 German guided missile specialists, of whom several had degrees of scientific and engineering knowledge of the subsystems in SAM technology. Of particular note was the exploitation and adaption of the *Taifun* flak rocket design concept as an air-to-surface missile. The second part of the chapter is concerned with the transfer, dissemination and exploitation of the German knowledge by the AAF Air Technical Service Command, from March 1946 Air Materiel Command (AMC). All of the German guided missile technology, documentation and specialist personnel in the custody of the AAF was initially concentrated at one location, Wright Field at Dayton in Ohio. Around a dozen

German specialists brought experience with SAM technology. The third and fourth parts of this chapter are devoted to the transfer and exploitation activities by the US Army Ordnance Department Rocket Branch. That branch's acquisitions were initially concentrated at two locations: the Aberdeen Proving Ground in Maryland and Fort Bliss at El Paso in Texas, the latter not far from the White Sands Proving Ground in New Mexico. The large group of guided missile specialists whom the US Army Ordnance Department recruited – almost all from Peenemünde-East – consisted of several dozen who brought various amounts of experience from the C-2 and *Taifun* projects. Both of these German design concepts were exploited by the armed services in concert with American defence contractors.

Initially, the common interests between the Germans and their erstwhile employers did not extend very far beyond a mutual interest in science or the application of scientific knowledge in their specialised fields to guided missile technology, and a shared disdain of the communist Soviet Union. As time progressed, and after the Germans (and Austrians) assimilated themselves and their families into American society, there was a shared sense of purpose – for example, the advancement of the American ballistic missile and space rocket programmes, and to put American astronauts on the moon. In all cases, the armed services gradually relinquished custody of their German specialists for employment by the other armed services, in industry, in the university sector, by government departments or so they could return to Germany.

Due to the sheer quantity of scientific and technical intelligence that was brought back from Germany after World War II, it would be impossible to determine the quantitative and qualitative value of the German knowledge to the American armed forces, or how much time and money was saved on SAM research and development in the US after 1945. What can be determined for certain is that the intelligence from Germany played a significant role in the burgeoning American guided missile programmes of the late 1940s and early 1950s.

The US Navy Bureaus of Ordnance and Aeronautics share the spoils

The US Navy was the first of the armed services of the Western Allies to acquire physical assets from the German SAM development programme after the war. As one of the objectives of the Bureau of Ordnance (BuOrd) Bumblebee programme was to collect and assess background information from which a class of SAMs could be designed, intelligence received from Germany was therefore important. The masses of scientific and technical intelligence on German guided weapons technology that was acquired by American missions working under CIOS, such as the US Naval Technical Mission in Europe (NavTechMisEu), provided a substantial body of knowledge for exploitation by the American scientists and engineers on the Bumblebee programme and other US Navy SAM projects. However, the German SAM development programme was focused on ground-launched rather than ship-launched missiles, and thus in Germany by 1945 the application of the technology for naval purposes was still only theoretical. There were some areas where German research and development with flak rocket technology was applicable to ship-launched systems, such as SPREs, the beam-rider guidance technique (which was more suitable for ship-launched SAMs than the command-link guidance technique), and as Theodor Sturm wrote for US Army investigators in July 1945, a *Rheintochter*-type missile was considered as a ship-to-ship weapon. Ergo, the Germans could not bequeath much specific knowledge to the US Navy in that regard.[1]

Once counter-intelligence officers from NavTechMisEu recognised the areas where German technology and know-how were in advance of American, the first priority was for exploitation in the war against Japan, and they were quick to act. After the discovery of Herbert Wagner in Upper Bavaria in early May 1945, and the Hs 117 project and the underground factory at Woffleben in the Harz Mountains very soon thereafter, NavTechMisEu officers sent a signal to the Director of Naval Intelligence (DNI) in Washington DC, Rear Admiral Leo H. Thebaud, recommending that Wagner and two of his assistants, Reinhard Lahde and Wilfried Hell, be evacuated to the US. On 4 May, the DNI asked the War Department to have SHAEF arrange for the three to be immediately transferred to the US. SHAEF approved the request, and the group were subsequently evacuated to London where, in addition to another two engineers from Wagner's staff, they were interrogated by officers from ADI (K) (Prisoner Interrogation) and US Air Intelligence before Wagner, Lahde and Hell were sent to Washington DC on 19 May. These events took place before the US government formulated a policy to regulate the recruitment of former enemy aliens for work in the US. On 6 July 1945, the War Department General Staff first circulated to the relevant government and military agencies the policy principles and procedures of the recruitment programme, which was initially called Project Overcast and restricted to the US defence sector, and subsequently renamed Project Paperclip in March 1946 and widened to include the American civil sector. The British chiefs of staff agreed in principle with the American policy (recruiting and employing German specialists for exploitation after the war was actually a British idea), although a combined Anglo-American recruitment policy was not agreed upon until late September 1945. During August 1945, Wagner and his two assistants were transferred to the 160-acre former Guggenheim Estate (which ironically enough was once the property of Jews) at Sands Point on Long Island for a US Navy Bureau of Aeronautics (BuAer) undertaking called Project 77.[2]

Meanwhile in Germany, the transfer of German technology to the US Navy was well and truly under way. In the American Zone of Occupation during the second half of 1945, the BuOrd acquired the two supersonic 40cm x 40cm Mach 4.4 wind tunnels from *WVA Kochelsee* in southern Bavaria, where aerodynamic and thermodynamic research for the A-4/V-2, C-2 and *Taifun* projects was carried out. The booty also included a partially completed hypersonic Mach 7–10 wind tunnel. In the weeks after the German capitulation there was contention between the AAF and the US Navy over which service should be allocated the tunnels. The US Army Ordnance Department relinquished any claim because it possessed the supersonic tunnel at the Aberdeen Proving Ground. The dispute remained unresolved until late July 1945, when an arbitration board which consisted of representatives from the army, the navy and the NACA awarded the wind tunnels to the navy. NavTechMisEu personnel subsequently supervised the disassembling and shipment of the installation to the new Naval Ordnance Laboratory (NOL) at Silver Spring in Maryland. This operation was a large undertaking, requiring the services of six NavTechMisEu officers for more than three months, plus 30 railway cars to transport the installation to the American enclave at Bremerhaven before shipment to the US. The two wind tunnels became operational in July 1948 and were designated as Aeroballistics Tunnels 1 and 2.[3]

By March 1946, the American armed forces had recruited between approximately 140 and 150 specialists from the Peenemünde-East organisation. Since some were former members of National Socialist political organisations and more than nominal followers of the Nazi regime, the recruitment of these specialists (and others) had the effect of circumventing and undermining the denazification policy in the American-controlled areas of Germany and Austria. A grading system, whereby Germans were classed into categories depending on their level of association with the Nazi regime, determined their future role in German society and government. German (and Austrian) specialists who had associations with Nazi political organisations, or were suspected of having committed war crimes, would ordinarily have been barred from certain occupations and positions under the denazification policy, but in an ignominious and technically illegal practice, they were allowed to work in the US because their scientific or engineering knowledge was of vital interest.

The BuOrd recruited 11 specialists from the WVA, all for employment at the NOL. Table 7 lists the staff at the WVA with their destinations after the war, which serves as an example of a German organisation whose employees were dispersed to more than one of the Allies, in this case the US and France. Four of the specialists did research for the C-2 project: Drs. Hermann Kurzweg, Peter Wegener and Richard Lenhert, all physicists, and Wolfgang Zettler-Seidel, a mathematician. Kurzweg was the deputy director of the WVA and department leader in the research laboratory, and specialised in all aspects of aerodynamics, thermodynamics, experimental model testing, theory and mathematics. He authored or co-authored 21 reports pertaining to the C-2 project, on subjects including three-component measurements (lift, drag and pitch), trajectory calculations, pressure distribution measurements at sub- and supersonic velocities, vibration and manoeuvrability measurements, rudder and control vane design, and steering. Kurzweg also co-authored single reports on *Rheintochter* rudder control and the development of the *Taifun*. After starting work at the NOL, a security report revealed that Kurzweg had been a member of four NSDAP (Nazi Party) organisations, including the SS. The revelation ultimately did not prevent his continued employment in the US.[4]

Wegener specialised in basic aerodynamic research, which involved theoretical calculations and experimental proofing of aerodynamic characteristics. From December 1943, he authored or co-authored 11 reports on the C-2 and co-authored the report on the *Rheintochter* with Kurzweg. Lenhert specialised in aerodynamic measurements, including the theoretical and experimental analysis of developmental aerodynamic designs. He co-authored one report on the C-2, dated February 1944, on three-component and zero moment measurements. Zettler-Seidel specialised in trajectory calculations, which involved the application of theoretical and experimental data to external ballistics, and the development of systems for the actual calculation of missiles in free-flight. He co-authored two reports concerning trajectory calculations for the C-2 project, dated November and December 1944. The specialists were sent to the US in January 1946 on board a troopship, the *Central Falls Victory*, which also carried 30 specialists who had been recruited by the US Army Ordnance Department for employment at Fort Bliss in Texas.[5]

At the NOL, Kurzweg, Wegener, Lenhert and Zettler-Seidel were engaged on applied research with super- and hypersonic wind tunnel technology that in many respects continued the research they did at Peenemünde-East and Kochel. The NOL was not initially utilised to conduct aerodynamics research for the Bumblebee programme; wind tunnels were constructed at the APL/JHU and the Ordnance Aerophysics Laboratory near Daingerfield in Texas for that purpose. However, in 1949, a BuOrd Committee on Aeroballistics – which consisted of representatives from the Bumblebee programme, NOL, Naval Ordnance Test Station (NOTS) at China Lake in California, Project Meteor (an air-to-air guided missile research programme at MIT that was sponsored by the BuOrd) and the Naval Proving Ground at Dahlgren in Virginia – decided to test models of the first missile to be produced by the programme, the 'Terrier', in the wind tunnels at the NOL.[6]

Aside from exploiting German knowledge in wind tunnel technology, the BuOrd also exploited the expertise of German chemists and the design of the *Taifun* barrage rocket for the purpose of developing an unguided air-to-surface missile for aircraft carrier-borne aircraft. The story began in late 1949 when the BuOrd asked the NOTS to submit a proposal for a liquid propellant rocket with performance similar to the 5-inch High Performance Air-to-Ground (HPAG) Rocket, a solid propellant projectile with a light aluminium body and fins which replaced the World War II HVAR, but with a markedly shorter burning time. Commander Levering Smith, the acting head of the NOTS Rockets and Explosives Department, became interested in the *Taifun* design, particularly in how the hypergolic propellants in the tanks were pressurised by a solid propellant gas generator in the propellant feed system, after being given the task of reviewing captured documents from Peenemünde-East.

Table 7. The staff of the WVA who were recruited by the US Navy and AAF, and also by the French DEFA. An asterisk indicates experience and expertise appertaining to flak rockets

Administration		
Dr. Phil. Rudolf Hermann*	Director	AAF
Dr. Phil. Herbert Graf	Head of department	France (DEFA)
Günther Herrmann	Assistant	France (DEFA)
Research Laboratory		
Dr. Phil. Hermann Kurzweg*	Head of department	US Navy BuOrd
Dr. Phil. Richard Lenhert*	Main group leader (aerodynamic measurement)	US Navy BuOrd
Dr. Phil. Peter Wegener*	Main group leader (basic aerodynamic research)	US Navy BuOrd
Dipl.-Ing. Hans-Ullrich Eckert*	Group leader (dynamic investigations)	AAF
Dr. Phil. Willi Heybey	Main group leader (mathematical office)	US Navy BuOrd
Elsbeth Hermann	Group leader (test evaluation)	France (DEFA)
Feodor Schubert	Group leader (theoretical gas dynamics)	France (DEFA)
Wolfgang Zettler-Seidel*	Group leader (trajectory calculations)	US Navy BuOrd
Dr. Phil. Werner Kraus	Group leader (thermodynamics)	France (DEFA)
Dr. Phil. Ernst-Hans Winkler	Group leader (optics)	US Navy BuOrd
Dr. Eva Winkler	Physicist (wife of Dr. Ernst Winkler)	US Navy BuOrd
Erich Rott	Experimental/test technician	France (DEFA)
Electrical Laboratory		
Dipl.-Ing. Heinrich Ramm	Head of department	AAF
Dipl.-Ing. Siegfried Hoh	Main group leader (electrical measurements)	AAF
Dipl.-Ing. Gottfried Arnold	Group leader	AAF
Dipl.-Ing. Max Peuker	Main group leader (electrical drive)	US Navy BuOrd
Ing. Emil Walk	Test engineer/workshop manager	AAF
Ing. Fritz Vollmer	Test engineer	France (DEFA)
Technical Development Department		
Obering. Hans Gessner	Head of department	France (DEFA)
Ing. Josef Kuckertz	Main group leader (design bureau)	France (DEFA)
Ing. Edmund Stollenwerk	Main group leader (manufacturing)	US Navy BuOrd
Ing. Hans-Ludwig Kleinekuhle	Workshop manager	France (DEFA)
Ing. Florian Geineder	Constructor	France (DEFA)
Ing. Kurt Seyfurt	Constructor	France (DEFA)
Wilhelm Buckesfeld	Constructor	France (DEFA)
Gerhard Hentsch	Constructor	France (DEFA)
Windkanal Süd (Projekt A) (Mach 7–10 wind tunnel)		
Dr.-Ing. Gerhard Eber	Head of department	US Navy BuOrd
Dr. Phil. Karl-Heinz Grünewald	Main group leader (scientific questions)	US Navy BuOrd
Dipl.-Ing. Günther Dellmeier	Main group leader (measuring section)	AAF

Also to France (DEFA): *Dipl.-Ing.* Rolf Trotz (engineer), Martha Knoop (mathematician), *Ing.* Martin Fiechter and *Ing.* Eckart Finger (test engineers), Heinz Klein, Gerhard Ebert and Willy Schmidt (precision mechanics), Georg Reetz (interpreter for special scientific and technical problems), Gretel Ruf (technical secretary), Josef Held (fitter) and Ernst Vogt (foreman).

Source: F1 Kochel (Personal), Inventaires – List de personnels, 12.9.1945 – SHD, Châtellerault, AA 910 6F1 11

Subsequently, he accepted some resources that were offered to the BuOrd in order to exploit the weapon's design, which included a solid propellant continuous extrusion press and three German specialists. All were sent to the NOTS.[7]

Two of the German specialists can be linked to the *Taifun* project through their professions and place of employment in Germany during the war. They were Dr. Wolfgang Noeggerath (who was originally recruited by the AAF) and Dr. Hans Haussmann, both chemists and former colleagues at the LFA. Each had done research with hypergolic rocket propellants at the LFA, one group of which were fuels called *Visols* that were found to be suitable with mixed acid (around 90 per cent nitric acid with 10 per cent sulphuric acid added to prevent the interior of the tanks from corroding). *Visol* and nitric acid propellant combinations were used in the LPREs that powered the C-2, *Rheintochter* R-IIIf and *Taifun*. It appears that these specialists were selected for the BuOrd project on the basis of the decision by the project engineers to use hypergolic propellants in the rocket's engine. By early 1950, work on the projectile, called the Liquid Aircraft Rocket (LAR), was within two years to produce a rocket that could withstand a 40ft drop onto steel or concrete without leaking or malfunctioning. The first LAR was successfully flight tested in early 1952 and continued to be used as a test vehicle over the next several years with the intention of introducing the rocket into service. Despite defence budget cuts by the American government in 1953 as a response to public pressure to reduce expenditure for the war in Korea, the LAR programme continued until 1958, but the weapon never became operational. By this time, the navy had realised that liquid propellant rockets were not a viable alternative to solid propellant rockets due to the unacceptable fire risk they posed aboard a ship.[8]

The other US Navy agency that had the prerogative to develop guided missiles, the Bureau of Aeronautics (BuAer), also recruited guided missile specialists with experience and expertise from the German SAM development programme. Consistent with American recruitment practices, all held doctorates and each was a leader in his particular scientific or engineering field in Germany. In addition to Wagner from HFW, the chief designer of all the firm's guided missiles, there was Theodore Sturm, the physicist from *Stassfurter-Rundfunk* who had overseen the development of all radio remote control technology for guided missiles in the German industrial sector. There was also Edgar W. Kutzscher, the former director of infra-red research and development at *Electroacoustic AG* (ELAC) at Kiel. Kutzscher's research focused on the design of infra-red seekers for guided missiles that used lead sulphide cells to detect the infra-red radiation. One of his projects was an infra-red seeker for the C-2 that was being developed in conjunction with Dr. Helmut Weiss at Peenemünde-East. Another prominent specialist was world-leading aerodynamicist and swept-wing expert Prof. *Dr.-Ing.* Adolf Busemann, who co-designed the *Feuerlilie* series of research missiles with *Dr.-Ing.* Gerhard Braun at the LFA. Busemann was recruited by the British MoS in mid-1946 for six months' work at RAE Farnborough (see the following chapter), but returned to Germany in early 1947, after which he was recruited by the BuAer through Project Paperclip. Busemann later went into the American university sector, while Wagner, Sturm, Kutzscher and other Germans were sent to work at the Naval Air Missile Test Center at Point Mugu, California, in 1947.

The Naval Air Missile Test Center had a central role in the testing and evaluation of captured German LPRE technology. In the US, the exploitation of the technology was only one part of many programmes by the armed services to develop LPREs for experimental and applied purposes, including rocket-powered aircraft, ATO units, sounding rockets and guided missiles. By 1 March 1947, 75 per cent of the development work was divided into four categories according to the oxidising agent: nitric acid, gaseous and liquid oxygen, hydrogen peroxide and nitromethane (a colourless, oily liquid used as a fuel in racing car engines) systems. The BuAer set up a project at the Naval Air Missile Test Center to test captured German LPREs for educational and familiarisation purposes. A test stand was constructed to test a number of different engines. They included three BMW LPREs that utilised nitric acid, which were more than likely the samples that were acquired from the BMW facilities around Munich by the Naval Technical Mission in 1945. The three engines were the 109-548 for the *Ruhrstahl* X-4 AAM, the 109-558 for the Hs 117 and the 109-718 auxiliary propulsion system which was to be mounted on two BMW 109-003R gas turbine jet engines under the wings of the *Messerschmitt* Me 262 jet fighter. Two HWK hydrogen peroxide LPREs were also tested at the new facility – the 109-501 ATO unit and the 109-509A for the *Messerschmitt* Me 163 rocket-powered interceptor. The firm Reaction Motors also evaluated HWK hydrogen peroxide engine technology for the BuAer at other sites under separate contracts.[9]

In conclusion, the two US Navy bureaus had access to not only all the scientific and technical intelligence that the Naval Technical Mission brought back to the US, but also that obtained by the other American missions and the British. The intelligence from the German SAM and other guided weapon programmes, and the knowledge of the specialist personnel, were evaluated and exploited at naval facilities all over the US – concerning aerodynamics and wind tunnel technology at the NOL in Silver Spring, Maryland; air-launched missile technology at the Naval Ordnance Testing Station at China Lake in California; LPRE and guided missile technology at the Naval Air Test Missile Center; and the Bumblebee programme. While the intelligence would have provided many lessons to the American scientists and engineers about the various aspects and principles of SAM design and development, what could be exploited in the near-term was sometimes limited. Hardly any research and development of ship-launched guided anti-aircraft missiles was carried out in Germany during the war. Germany lagged behind the US (and the UK) in microwave radio research; the preferred beam-rider guidance technique for ship-launched SAMs, while studied to a certain degree in Germany, was not extensively developed; and a single-stage liquid propellant SAM like the C-2, which weighed 4 tons at launch, was far too heavy for a ship-launched system. Furthermore, subsonic SAMs of the Hs 117 and *Enzian* types had become obsolete by the end of 1945; and liquid rocket propellants were very dangerous to store on board naval vessels. Solid propellant SAMs of the *Rheintochter* types – radio-beam guided with American microwave radio technology and of reduced weight with more powerful propellants – offered the most promising prospects to the US Navy over the long term in conjunction with ramjet propulsion.

The AAF concentration of German resources at Wright Field

During late 1945 and 1946, the AAF transferred its share of captured German technical documentation, guided missile technology and specialists to one location – Wright Field at Dayton in Ohio. In June 1945, the copious quantities of captured technical documentation on aeronautical subjects had led to the formation of a combined Anglo-American Air Documents Research Centre (ADRC). The ADRC was based in London and was organised by representatives from the armed forces of both countries. Its purpose was to screen, assess, index, reproduce and distribute all captured German documents that related to aerial warfare. The sorting of the captured documents was undertaken on a five-year approach, whereby during the first year, documents that could be utilised for the war against Japan received top priority; during the second and third years, all documents were processed; and during the fourth and fifth years, the documents were assembled for historical purposes. According to the agreement, all captured documents were to be channelled to the ADRC, but this rule was not always adhered to by both allies. One example of the circumvention of the agreement by American personnel concerned the removal from Germany to the US of the archives of the *Messerschmitt AG Oberbayerische Forschungsanstalt* at Oberammergau in Upper Bavaria, which contained technical documentation concerning the *Enzian* project. When a British investigator from the RAE, aircraft designer Morien Morgan, travelled to Upper Bavaria in the autumn of 1945 to locate German aeronautical experts, he was informed at the end of his tour by Colonel Gifford, the leader of the American mission at *Messerschmitt*, that 17 sealed crates of the company's blueprints had been sent to the ADRC, when in fact the lot was already on its way to Wright Field. It seems apparent that the AAF officers jumped the gun and determined that the scientific and technical knowledge had to be exploited as expeditiously as possible.[10]

The German documents on guided missiles were evaluated at Wright Field and then listed in bibliographies for distribution. The process whereby the documents were translated, abstracted and disseminated by AAF intelligence (T-2) is quoted in the first such bibliography:

This bibliography lists German documents on Guided Missiles which are now available at the Air Documents Division of Intelligence T-2, Air Material Command at Wright Field. This bibliography is only preliminary and, when further documents on this subject can be released, revisions and supplements will be prepared and mailed to interested parties. In its final form, the bibliography will give both English and German titles and contain brief abstracts of the documents listed. All documents included in this bibliography are available on microfilm which can be obtained upon written request. These requests should provide complete information as given for each item and should state whether a copy of the original document or translation is desired. (The number of documents of which complete translations are available is still limited … In a few instances, translations only are available.)[11]

The documents concerning the *Rheintochter* that were available for dissemination provide an insight into how successful the Anglo-American intelligence-gathering effort was in relation to the German SAM development programme, and also the type of technical intelligence on the weapon's development that was available for exploitation in the US. Captured documents from *Rheinmetall-Borsig* included reports from 1944 on the development of the SPRE for the R-I and the LPRE for the R-IIIf; a report on the testing of the electrical installation in the R-IIIf and R-IIIp dated November 1944; ballistics and test-launching evaluations from 1943 and 1944; reports pertaining to the guidance system (remote control, steering and gyroscopes); schematics and photographs; information about the contracts made between *Rheinmetall-Borsig* and German industry for the manufacture of components; records of conferences that were held at the firm's premises at Berlin-Marienfelde and Leba; records of correspondence between the firm and the RLM; and records of conferences held at the RLM in Berlin by GL/Flak-E. and TLR/Flak-E. Also listed were reports and technical drawings from the second and third series of wind tunnel tests at the AIA and reports on wind tunnel tests at the AVA and the WVA.[12]

The AAF also acquired samples of German flak rocket technology, including prototypes of the C-2 and the Hs 117, amongst a consignment of captured guided missile technology that was sent to the US via the UK during July 1945. After the hardware reached the US, it was sent by train to Freeman Field in Indiana and then forwarded to the Air Technical Service Command (ATSC) at Wright Field. The AAF also acquired the second flak rocket training simulator (quantitative) from the DFS at Ainring, which was designed to train ground operators how to control and steer the C-2. The incomplete apparatus was shipped to the US during the autumn of 1945. The American general public had an opportunity to view the captured flak rockets and aircraft from Germany, and numerous other countries, at the AAF Air Fair at Wright Field – the purpose of which was to display advances in aviation technology during World War II – that ran from 15–21 October 1945.[13]

The AAF and its successor, the US Air Force (USAF; formed in 1947 after the AAF became an independent branch of the American armed forces), recruited more than 700 German scientists and engineers under Projects Overcast and Paperclip by

1952, somewhat more than the army and more than twice as many as the navy.[14] Most of the German specialists at Wright Field were recruited for their experience and expertise on subjects related to aircraft, such as jet engine propulsion and parachutes, rather than guided missile technology. The Germans at Wright Field remained in the custody of the Intelligence Department of the ATSC and the AMC for about a year, where they existed in a relative state of inertia while they wrote more reports about their work in Germany, representatives from the armaments industry and the military interviewed them, and while the AAF/USAF assessed the German technical documentation for its own projects. Thereafter they were transferred to the Engineering Division of the AMC, also at Wright Field, whose staff were interested in using the specialists on research projects. The AMC did most of the research and development activities for the AAF/USAF, including for guided missile projects.

Several of the German specialists brought research and development expertise with guided weapons, including SAMs. From the LFA aerodynamics institute, where wind tunnel experiments were conducted for the *Enzian* and *Feuerlilie* projects, there was *Dr.-Ing.* Gerhard Braun, a former *Abteilungsleiter* who co-designed the *Feuerlilie* series of research missiles with Adolf Busemann; and *Dr.-Ing.* Theodore Zobel and Prof. Dr. Rudolph Edse. Also from the LFA was Dr. Wolfram Kerris, a physicist who specialised with instrumentation. From the DVL at Berlin-Adlershof there was aerodynamics expert Dr. Bernhard Göthert, who conducted wind tunnel research in the low-speed tunnel facilities at the establishment for the Hs 117 project. There was also swept-wing and external ballistics expert Prof. *Dipl.-Ing.* Otto Walchner from the AVA at Göttingen, who had carried out aerodynamics research for the *Rheintochter* project in the establishment's Mach 3.7 supersonic wind tunnel. In the fields of proximity fuses and seekers there was physicist Dr. Werner Rambauske, formerly of the *Institut für Physikalische Forschung* at Neu Drossenfeld, who during the war was developing a television seeker for the *Enzian*. Another physicist, Dr. Walter Wessel, who was formerly employed in the C-2 steering and control section (EW 224) at Peenemünde-East in a sub-section under Dr. Karl Slevogt, brought expertise with high-frequency radio technology.[15]

Of the seven specialists who were recruited from the WVA, two had done aerodynamic research for the C-2 and *Taifun* projects. The most notable was physicist Dr. Rudolf Hermann, the former head of the institute. Hermann joined Peenemünde in 1937 from the AIA, where he designed a supersonic 10cm x 10cm Mach 3.3 wind tunnel. The two 40cm x 40cm supersonic wind tunnels at Peenemünde-East were based on Hermann's original design. Together with Kurzweg, Hermann co-authored 19 reports that concerned the C-2, in addition to the single reports on the rudder for the *Rheintochter* and the development of the *Taifun*. He also authored a report on the aerodynamic development of the C-2. The other specialist was physicist *Dipl.-Ing.* Hans-Ulrich Eckert, who co-authored four reports on the C-2 beginning in February 1944. Three reports concerned studies of rudder hinge torque, rudder developments at sub- and supersonic speeds and pressure distribution. The other report concerned the vane hinge torque on the external control vanes at the base of the missile.

The specialist recruited by the AAF/USAF who perhaps had the most significant role in the development of flak rockets by the prime contractors was engineering physicist *Dr.-Ing.* Werner Fricke, the former technical director of the *Rheintochter* project at *Rheinmetall-Borsig.* Fricke was educated at the TH Munich and the University of Jena. From 1934–45, he was employed by *Rheinmetall-Borsig,* as section chief of ballistics instrumentation from 1935–40, then as section chief of experimental ballistics from 1940–43, and from 1943–45 as project manager for missile development. With Fricke's recruitment, the Americans succeeded in obtaining the services of three of the four technical directors of the guided flak rocket projects, the others being Wagner (Hs 117) and von Braun (C-2). In May 1945, Fricke was amongst the 450–500 guided missile specialists captured in Upper Bavaria. After working for the USAF, he went into the private sector, as section chief of environmental studies at Bell Aerosystems Company, a division of Bell Aerospace Corporation, from 1954–68, then from 1968–71 as chief engineer. Walter Dornberger, the ex-head of liquid propellant rocket development in the HWA, was employed by the same company after working for the USAF as a guided missile consultant from July 1947 to May 1950. In the field of LPREs, the AAF also recruited *Dipl.-Ing.* Heinz Müller from BMW, who from 1943–45 was the chief of the technical office for rocket engine development at the firm and deputy chief of the rocket engine development department at Munich-Allach under von Zborowski.[16]

A comprehensive overall analysis of German technological developments in rocket engine design was undertaken on behalf of the AMC Intelligence Department at the now-renamed Wright-Patterson Air Force Base (which was created on 13 January 1948 following the merger with nearby Patterson Field). The AMC contracted the American Power Jet Company of Montclair, New Jersey, to analyse and study the German LPREs that were developed up to 1945 as part of the company's foreign liquid rocket analysis programme. Information was acquired from the approximately 55,000 captured documents relating to German rocket engine research and was supplemented with interrogations of German specialists in the US. By early 1952, the results were published in a series of 14 volumes under the umbrella title of 'Analysis and Evaluation of German Attainments and Research in the Liquid Rocket Engine Field'. All the power plants that were fitted in the guided flak rockets were analysed as part of the study. These reports clearly show that the US acquired practically the entire reservoir of German knowledge on LPREs, which constituted more than two decades of German research and development.[17]

The AAF/USAF had access to the same scientific and technical intelligence on German guided missile development as the US Navy bureaus and the US Army Ordnance Department for the purpose of exploitation. Like the US Navy bureaus, the AAF ATSC/AMC aimed its recruitment efforts at the best scientists and engineers from the German aeronautical research complex, who had research expertise and experience with cutting-edge design concepts and technologies, primarily for advanced aircraft but also for guided missiles. The AAF/USAF pursued a far more ambitious postwar guided missile programme than the US Navy and US Army. The AAF/USAF did not have the prerogative to develop land-based SAMs – that was the responsibility of the US Army – and thus studied 'pilotless aircraft' for tactical and strategic area defence, which were SAMs in all but name. By the end of April 1946, the AMC had awarded contracts for 28 missile projects across the four categories of guided missiles, composed of three relating to SAMs—the Boeing GAPA study that commenced in 1945 and two theoretical studies. The projected thinking within the AAF in regard to SAM technology was already well beyond that in Germany by 1945, and far in excess of anything the British and French were conceiving. The performance of the A-4/V-2 rocket, rather than the performances of the German flak rockets, provided the stimulus for investigations into the feasibility of long-range strategic SAMs that could combat jet aircraft and ballistic missiles at extreme ranges and altitudes. Astonishingly, the two theoretical studies, 'Wizard' at the University of Michigan and 'Thumper' by General Electric, envisaged an anti-ballistic missile system with a range of 880km and an altitude of 500,000ft, or just over 152km. By comparison, the C-2 was intended to be effective up to a maximum altitude of 58,000ft. These studies indicate that from the perspective of the AAF/USAF, the wartime German SAM technology was already starting to become obsolete. These three AAF/USAF projects eventually led to the development of the ramjet-powered BOMARC (Boeing Michigan Aeronautical Research Centre) SAM system, capable of being armed with a nuclear warhead. The BOMARC was a strategic weapon designed for the area defence of the US and Canada, which along with the US Army's Nike-Hercules, the successor of the Nike-Ajax, which could also be armed with a nuclear warhead, was one of the most formidable SAM systems to appear in the late 1950s.[18]

The US Army Ordnance Department and the Peenemünde group

The US Army Ordnance Department Rocket Branch exploited the German knowledge from the C-2 and *Taifun* projects more than any other American military or civilian organisation. This would not have been possible had the specialists and archives from Peenemünde-East been captured by France or the Soviet Union. The complete development histories of the A-4/V-2, C-2 and *Taifun* projects were contained within the approximately 3,500 reports and 510,010 engineering drawings (which weighed about 14 tons) that were recovered from the mine at Dörnten in the Harz Mountains in May 1945.[19] The precious cargo was transported by road from Nordhausen to France, and then shipped to the Foreign Documents Evaluation Centre at the Aberdeen Proving Ground (APG) in Maryland for sorting, cataloguing and for abstracts of the reports to be written. At Peenemünde-East, the reports which dealt with subjects that pertained to flak rocket research and development were catalogued in the same numerical referencing system which was used for other reports. For example, reports that concerned the ballistics and control of the C-2 were in *Archiv 86* series from 86/113 to 86/150.[20] The task of writing the abstracts was a time-consuming activity because it required the employment of translators.

Other tasks that were undertaken at the APG included the testing of models of the C-2 (and the A-4/V-2) in the supersonic wind tunnel (the 'Bomb Tunnel') at the Ballistic Research Laboratory on behalf of GE as part of Project Hermes. The purpose of these tests was twofold – to obtain aerodynamics data for comparison with the German results that were obtained in the Mach 4.4 wind tunnels at Peenemünde-East and the WVA, and to contribute to American guided missile research, for internal distribution within GE and external distribution to the relevant agencies in the US for further exploitation. These tests would have served as a continuation of phase two of Project Hermes, which was the dispatching of a scientific mission to Germany to survey German developments and obtain samples of weapons and components. However, the Bomb Tunnel had a maximum Mach number of 1.72, and thus the US Army and GE scientists and engineers were restricted by this limitation until wind tunnels capable of generating greater wind speeds were constructed or installed in the US, including the two Mach 4.4 wind tunnels.[21]

The captured technical documentation and hardware was accompanied by a group of German specialists who were recruited through Project Overcast. As has been well-documented since the mid-1960s by authors and historians including Huzel in 1962, von Braun in 1963,[22] McGovern in 1964, Ordway and Sharpe in 1979 and Neufeld in 2007, initially around 115–120 specialists from Peenemünde-East were offered Ordnance Department contracts, which was later increased to 127. The contracts were for six months for employment at Fort Bliss at El Paso, Texas, and the White Sands Proving Ground across the state border in New Mexico. A first group of seven departed from Witzenhausen in September 1945, followed by three more groups that left Landshut between November 1945 and January 1946. By February 1946, almost all of the eventual total number had arrived at Fort Bliss. The vast majority were recruited on the basis of their qualifications and their contribution to the development and production of the A-4/V-2. Of a lesser importance to the Rocket Branch – but by no means insignificant – was the knowledge that several of them brought from their work on the C-2 and *Taifun* projects (see Table 8). The group comprised electrical and mechanical engineers, physicists and chemists who were employed at Peenemünde-East or at other establishments as outside contractors in various capacities such as designing, engineering, testing and manufacturing.[23]

Amongst the outside contractors, the work histories of three provide an indication of the specialised knowledge of SAM technology that they brought to the US. There was Prof. *Dr.-Ing.* Eduard Fischel, the former chief of the Institute for Flight Equipment at the DFS, who had designed and built guided weapon simulators, initially to train bombardiers how to control the Hs 293 and the X-1, then from mid-1943 the qualitative and quantitative simulators to train *Luftwaffe* personnel how to steer the C-2 and other flak rockets. From the TH Darmstadt were Prof. Dr. Theodor Buchhold and Dr. Walter Haussermann, both of whom were contracted to work on stability problems for the A-4/V-2 and C-2 projects. Haussermann was employed at Peenemünde-East from December 1939 until May 1942, where he worked on stability calculations. One of his projects at Darmstadt was the construction of a model to simulate flight conditions in order to test the control equipment in the C-2. By 1960, almost half of the 127 – which included Haussermann – were employed at the NASA Marshall Space Flight Center at Huntsville in Alabama.[24]

Concerning German influence on the development of the US Army's first guided SAM system, Project Nike (from the 1950s Nike-Ajax, named after the ancient Greek goddess of victory and the Greek hero), there are several factors which must be considered. With regard to the design of the booster rocket and sustainer missile, the general philosophy was to adhere to established techniques as much as possible without departing from the original AAGM Report concept of July 1945 unless forced by necessities. In the early stages of the research and development processes during late 1945 and 1946, the scientific and technical intelligence from the German SAM development programme, and German guided missile and rocket developments more generally, required time to be fully assessed. The available information mostly comprised intelligence reports that were based on statements made by German specialists under interrogation and cursory examinations of captured documents. Information from these sources had to be verified through the testing of captured hardware, consultation with the German specialists and closer studies of the documentation recovered in Germany. In addition, time was needed to recruit those German specialists who were of interest to the Rocket Branch. A full assessment of the German attainments would have taken many more months, well into 1946.[25]

Once the Germans had been processed into the US and sent to Fort Bliss, their knowledge was made available to representatives from the armed forces, defence contractors and organisations with an interest in guided missile and rocket technology, who were permitted to interrogate the specialists. The group also assisted in the reconstruction, testing and launching of captured A-4/V-2 rockets at WSPG as part of Project Hermes. Von Braun – who in the absence of Dornberger was the designated leader and representative of the German group – assigned a group of 36 specialists to travel daily to WSPG to assist personnel from GE in this task. The German group consulted on various projects, including 'Bumper', which involved the mounting of a WAC Corporal onto the nose of an A-4/V-2 in order to carry out scientific studies of the upper atmosphere. Another project was the Hermes II, which had as its objective the development of a two-stage research test vehicle for a tactical SSM, with a ramjet-powered winged missile as the second stage. From 1950, the German group was occupied with the Hermes C-1, an SSM intended to have a range of 500 nautical miles that evolved into the Redstone ballistic missile, a significantly improved design of the A-4/V-2.[26]

A clue that points to the German influence on Project Nike is found in *History of German Guided Missiles Development* in 1957. On page six, one of the editors, Theodor Benecke, a former *Luftwaffe* engineering officer in the RLM during World War II, observed:

> the experience gained in Germany with the *Wasserfall* development had its outcome in the well-known Nike development in the United States.[27]

Table 8. The heads of departments and sections in EMW (Peenemünde-East), and some other personnel, who brought experience and expertise from the C-2 and *Taifun* projects to the US Army Ordnance Department Rocket Branch after World War II. The specialists with an asterisk were recruited by the Americans

Management department (EW 1)	Director Paul Storch and Karl-Otto Fleischer, commercial manager*
Development department (EW 2)	Prof. Dr. Wernher von Braun*
Planning section (EW 21)	*Dipl.-Ing.* Konrad Dannenberg*
• Statics:	*Dipl.-Ing.* Emil Hellebrand*
• Stability:	*Dr.-Ing.* Friedrich Bornscheuer (France)
• A-4/V-2 design:	August Schulze*
• C-2 design:	Kurt Patt*
• Taifun design:	(probably *Dipl.-Ing.* Klaus Scheufelen)*
• Propellants:	*Dipl.-Chem.* Gerhard Heller (C-2)*
• Materials research:	Dr. Wolfgang Steurer*
• Graphics:	Gerd de Beek (C-2)*

Specialists from this section who also brought expertise from the C-2 and *Taifun* projects were Dr. Helmut Zoike, *Dipl.-Ing.* Ludwig Roth, *Dipl.-Ing.* William Mzarek and *Dipl.-Ing.* Werner Dahm.

Guidance and control section (EW 22)	*Dr.-Ing.* Ernst Steinhoff*
• EW 220 technical department (*Technischestelle*)	*Dipl.-Ing.* Helmut Gröttrup (Soviet Union)
• EW 221 onboard and ground electrical wiring systems:	Dr. Hans Wierer (France, DEFA)
• EW 222 steering and remote control technology:	Helmut Hoelzer*
• EW 223 measuring and shading technology:	Dr. Friedrich Kirschstein
• EW 224 C-2 steering and remote control:	Dr. Theodor Netzer
• EW 225 workshop and supply:	*Herr.* Brützel
• EW 226 design, data and instrument testing:	*Dipl.-Ing.* Josef Böhm*
• EW 227 proof stands and test firing:	Dr. Kurt Debus*

(Continued)

Table 8. (Continued)

- EW 228 ground installation construction and testing: Werner Gengelbach (II)*
- EW 229 preparation of flight equipment manufacture: Dr. Erich Neubert*

From EW 224, Dr. Ernst Geissler and *Dipl.-Ing.* Johann Klein brought expertise from the C-2 project.

Ground equipment section (EW 23)   *Dipl.-Ing.* Hans Hüter*
Hans Lührsen, chief architect of EMW, and Bernard Tessmann, the designer of test stands (including four test stands for the C-2), brought expertise from the C-2 project.

Production department (EW 3)   *Dipl.-Ing.* Eberhard Rees*
- EW 326 final assembly of the C-2.

Testing department (EW 4)   *Dr.-Ing.* Martin Schilling (C-2 and
                                                     Taifun propulsion).* Also Theodore
                                                     Poppel and Karl Heimburg.*

Administration department (EW 5)
Accounts department (EW 6)

Sources: *The Story of Peenemünde, 79–83; Stokes, 'Interrogation of Dr. Netzer, EW 224, re Wasserfall', 24.5.1945, in The Story of Peenemünde, 387*

However, there is no evidence in the authoritative published primary and secondary sources to indicate that the German influence on Project Nike went beyond a consultative role on scientific and engineering questions. The official US Army history of the development of the Nike, *Development, Production and Deployment of the Nike-Ajax Guided Missile System 1945–1959*, that was published in 1959, does not even mention any German contributions to the missile system's development. The only reference to the transfer of knowledge from Germany is a footnote on page five, where the author mentions that when the AAGM Report was being written in mid-1945, the technical information on four of the guided flak rockets had not yet been fully assessed as part of a survey of the current state of the art. McGovern, in *Crossbow and Overcast*, first published in 1964, also makes no reference to German work in relation to Project Nike. Bullard's 1965 official US Army history of the Redstone ballistic missile project, which was under the leadership of von Braun, likewise does not mention any prior German work on Project Nike. Most importantly, Neufeld's 2007 biography of von Braun, which in part was based on research done in American archives, also does not link the Germans to Project Nike.

Consistent with the observation in *History of German Guided Missiles Development*, there were aspects of the Nike development where German advances up until 1945 were undoubtedly influential. Radar technology was not one of them. The Germans obviously possessed more technical know-how about the beam-rider and command-link guidance techniques than the Americans. But by the end of the war,

Germany was far behind the US and UK in microwave radar research and had not reached a stage whereby the technology could be soon considered for the SAM development programme. By contrast, the US was the world leader in anti-aircraft gun-laying radar technology, exemplified in the SCR-584 microwave radar by Bell Telephone Laboratories (BTL), which was superior to the *Telefunken Würzburg Riese*, *Mainz* and *Mannheim* gun-laying radars. In December 1945, W. H. Pickering, a New Zealander in the Electrical Engineering Department of the CIT, who was a member of the AAF SAG, summed up the value of German radar research to the US in the *Towards New Horizons* series of reports for the AAF:

> it can be stated unequivocally that German radar research has nothing to contribute to our radar techniques, present or future. Some minor developments are of course of interest as presenting new ideas on certain problems.[28]

The engineers at BTL utilised and improved upon its radar designs for Project Nike. The test-launchings of the first Nike prototype, the Nike-46, which took place from 24 September 1946 to 28 January 1947, were beacon-tracked with a BTL SCR-545 gun-laying radar modified for operation in the X-band (microwave frequency bands between 8 and 12 GHz). The second prototype, the Nike-47, was also beacon-tracked, but this time with the SCR-584 that was also modified for operation in the X-band. As was done in Germany, the beacon was fitted in the rear of the sustainer rocket in these trials.[29]

With regard to the equipment in the German command-link guidance systems, which were being designed to operate with or without radar, Pickering saw promise in the *Burgund* system, although it was originally designed to track subsonic winged missiles of the Hs 117 type. In December 1945, Pickering recommended to the AAF that:

> because of the importance of this problem [tracking and correction of the trajectory of a guided missile], and because of the early stage of development in each country, complete information on *Burgund*, including the actual equipment, be made available to US workers. As far as other [German and American] systems are concerned, I believe that none is sufficiently far advanced to warrant much interest.[30]

The eventual design of the Nike command-link guidance system was the realisation of the pioneering research and development that was done in Germany on the *Rheinland* guidance system programme for the C-2. In summary, like the German systems in development, the Nike guidance system had two radars, one for tracking the target aircraft and another for tracking the missile, with a ground computer to calculate the information on the position of the target and the missile. About five seconds after launch, once the sustainer rocket had separated from the booster, the computer issued a command to the sustainer rocket to turn on a general trajectory towards the predicted point of intercept, which was the purpose of the *Einlenkrechner* line-of-sight trajectory computer that was designed to turn the C-2 towards the target in the first 200 metres of flight. From that point the ground computer steered the missile on its trajectory towards the target.[31]

What may be surprising, considering the German lead in rocket engine development during the war, was that the experimental propulsion system for the early Nike prototypes was largely the product of American research and development. To begin with, the Project Nike engineers preferred a solid propellant booster/liquid propellant sustainer rocket design in a tandem configuration – similar to the British LOP/GAP and American WAC Corporal designs from 1945. The Germans did not develop a missile with such a configuration during the war. In addition, the chemists and engineers at the two concerns which were contracted to supply the propulsion system for the Nike, Aerojet and the JPL/CIT, had made significant advances with liquid propellant research and both liquid and solid propellant rocket engines during the war. The confidence in American progress was summed up in August 1945 by Dr. Theodore von Kármán, the director of the JPL/CIT:

> our early perfection of long-duration solid propellant rockets, and the promising results obtained with nitric acid/aniline … should be further exploited.[32]

The most promising propellant combination at the time was red fuming nitric acid (RFNA) as an oxidising agent with an 80 per cent aniline and 20 per cent furfuryl alcohol mixture as the fuel, a combination that was researched in the JPL and selected for use in the LPRE for the WAC Corporal. The booster for the WAC Corporal was also not based on German developments but was rather a modified 'Tiny Tim' aircraft rocket. All of these American technologies successfully functioned together in the first series of test-launches of the WAC Corporal, which took place between 26 September and 25 October 1945 at WSPG (just as the advance group of seven German specialists and the archives from Peenemünde-East had arrived at the Aberdeen Proving Ground). During the series of test-launches, the projectile impressively reached an estimated altitude of 230,000ft (around 70,100 metres) vertically and an estimated top speed of Mach 3. It was on the basis of the experience gained from the test-launches of the WAC Corporal that Aerojet undertook design studies of cooled and uncooled LPREs for the Nike. The development of the LPRE for the Nike commenced in late 1945.[33]

For Project Nike, the engineers at Aerojet chose to continue with the propellant combination used in the WAC Corporal LPRE, but with a ratio of aniline and furfuryl alcohol of 65/35 per cent instead of 80/20 per cent. This engine, designated X21AL-2600, produced 2,600lbs of thrust (1,180kg) with a burning time of 21 seconds. The first prototype engine was static-fired during May 1946, and subsequently underwent flight tests in the Nike-46 prototype. Like the LPRE in the WAC Corporal, the LPRE in the Nike also proved itself. In the first of eight launches of powered Nike-46 rounds, which took place on 8 October 1946, the missile achieved an estimated altitude of 140,000ft (about 42,700 metres), which probably surpassed the maximum altitude that the C-2, with the most powerful engine, attained during the test-launching programme at the *Flakversuchsstelle*. Two further rounds reached altitudes of over 100,000ft.[34]

The booster rockets that were used for the development of the Nike, and the propellants in these rockets, were also not selected on the basis of German research and development, but rather on proven American experience with the technology. Initially, the booster rockets for the Nike were supplied by the Monsanto Chemical Corporation, in a configuration of eight T-10E1 10-inch rockets that each produced 5 tons of thrust. In March 1946, the design of that booster was discarded and Aerojet was contracted to supply the booster rockets. Since 1942, Aerojet had acquired considerable expertise in the design of SPREs, especially in collaboration with the Guggenheim Aeronautical Laboratory (GAL) at the CIT. Aerojet built engines for numerous applications, including ATO units, for the US Navy's first surface-to-air and air-to-surface guided missiles, and the sustainer rockets in the Ordnance Department's Private A and F experimental rockets.[35]

On this point, the Nike launcher system should be briefly mentioned, because the changes in the design of the booster rocket necessitated changes to the launcher design. The first mechanical launcher for the Nike-46 comprised four parallel steel rails of hollow rectangular cross-section welded to a pivoted root frame, which could be tilted to a horizontal position for loading and raised to almost vertically for launching. When the cluster booster was replaced with a single long booster in 1948, a series of single-rail launchers were built that eventually led to a lightweight design that was mounted on the platform of an M-2 40mm anti-aircraft gun carriage, reverting back to previous design concepts. As previously stated, the German flak rockets were fired from a launcher mounted on the platform of an 88mm flak gun, while the first LOP/GAP launcher was mounted on a standard British 3.7-inch (94mm) heavy anti-aircraft gun.[36]

It was in the fields of aerodynamic and structural properties – particularly ideas and principles that were established during the development of the C-2 and *Rheintochter* – where German influences on Project Nike can be more clearly seen. On the question of whether to use fins or wings for SAMs, in either a one- or two-plane arrangement, there were two schools of thought in Germany during the war. The specialists at the LFA favoured wings or fins in a one-plane arrangement, like in the *Enzian* design, whereas at Peenemünde-East two planes at 90° to each other were preferred, as in the C-2. The aerodynamics specialists at the LFA also preferred wings or fins with much sweep-back to raise the critical Mach number (when the velocity of air flow over wings or fins is greater than the velocity of the missile), and a preference for the control surfaces to be located at the nose for missiles operating at supersonic velocities, to get greater moment arm about the centre of gravity, as in the designs of the *Rheintochter* R-I and R-III. From the start, the Nike sustainer rocket was designed with a two-plane arrangement, like the *Rheintochter* and the C-2. The 1945 design of the Nike sustainer rocket had parallelogram-shaped fins as the control surfaces that were positioned at the centre section of the fuselage, with rear fins that stretched from where the trailing edge of the control surfaces were attached to the fuselage to the rear of the rocket. During the preliminary design studies that were carried out by DAC in late 1945 and early 1946, the control surfaces were moved close to the nose of the missile for greater leverage, they were reduced in area and their shape was altered to a 23° semi-vertex angle delta for lower drag and smaller centre of pressure shift. Models of this design were tested in the Mach 1.72 wind tunnel at the APG, and subsequently the design was frozen in mid-February 1946. By 1947, the leading edges of the rear fins and the control surfaces were at the same angle, which was also a characteristic of the C-2 design. American missile designers concurred with the Germans in regard to greatly swept-back fins; however, as a result of further research and development in the US, it was determined that trapezoidal or delta-shaped fins were more suitable than swept wings for SAMs that travelled at supersonic velocities.[37]

The preference at the LFA for spoiler control of two different types instead of conventional ailerons for sub- and transonic velocities was also rejected, as was the preference for booster and sustainer rockets in a parallel arrangement (the Hs 117, *Enzian* and R-III) as opposed to a tandem arrangement (the R-I). The specialists at the LFA had believed that the parallel arrangement was better for aerodynamic stability. The engineers at Peenemünde-East bequeathed to the designers of the Nike at DAC one very important innovation that was a hallmark of the design of German guided missiles – monocoque construction (a technique pioneered by *Dr.-Ing.* Herbert Wagner), whereby the propellant tanks were constructed as an integral load-bearing part of the fuselage. This feature was incorporated into the Nike-46 prototype that was frozen in mid-February 1946. The exploitation of German knowledge in Project Nike was essentially over by 1948, the progress having exceeded the limit of German research and development in 1945.[38]

The US Army Ordnance Department exploited the knowledge of German specialists from Peenemünde-East primarily for the purpose of missile developments other than SAMs – assisting the GE personnel to assemble and launch A-4/V-2 rockets; improving the design of the A-4/V-2 to gather scientific data from the upper atmosphere as the preliminary stage of space exploration; designing test vehicles; and, from 1950, work on the development of the Redstone short-range ballistic missile, and thereafter more powerful ballistic missiles and space rockets. Nevertheless, the realisation of the Nike SAM system embodied some of the ground-breaking scientific and engineering attainments that were established by German research and development for the supersonic C-2 and *Rheintochter* projects and their ancillary ground apparatuses and equipment. The Project Nike contractors and the US Army deserve the credit for bringing to completion and into service the world's first operational surface-to-air guided missile system by successfully integrating the different subsystems and technologies – as originally engineered by the Germans – into a functional weapons system that fulfilled the US Army's requirement. In 1947, two other Ordnance Department missile projects saw the further exploitation of German scientific and technical intelligence from Peenemünde-East, from the C-2 project but also from the *Taifun*.

The Nike SAM. (Source: The Nike Historical Society, https://www.nikemissile.org/Ajax.shtml)

The Loki barrage rocket and Hermes A-1 test vehicle: Ordnance Department missile designs based on German ideas

In the late 1940s and early 1950s, American industrial firms built and tested derivatives of two flak rockets that were developed at Peenemünde-East. While the Nike was under development, the Ordnance Department sought to develop a tactical SAM to temporarily fill the gap between anti-aircraft artillery and guided missiles in the US Army's ground-based air defence capabilities. It appeared that a weapon designed along the lines of the *Taifun* would fulfil the requirement. Consequently, *Dipl.-Ing.* Klaus Scheufelen, the designer of the *Taifun*, assisted by three other ex-employees of Peenemünde-East at the Fort Bliss Rocket Sub-office of the Ordnance Research and Development Division – Friedrich Dohm, an electrical engineer with experience in high-frequency radio research, Herbert Dobrick and Kurt Neuhoffer – were employed to prepare a feasibility study of a liquid propellant barrage rocket of the *Taifun* type and to propose a development programme.[39]

The Department of the Army subsequently approved the proposal. The desired specifications of the missile were a range of around 81,000ft (24,700 metres) and

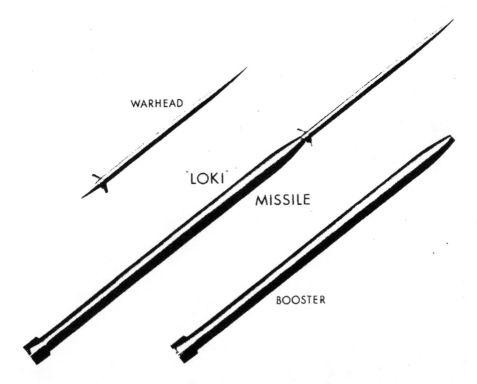

The Loki barrage rocket. (Source: Cagle, *Loki Antiaircraft Free Flight Rocket*, December 1947–November 1955)

a ceiling of around 71,800ft (21,900 metres) to engage supersonic aircraft with speeds of up to 1,600 km/h (about Mach 1.3). The programme also called for the

A Hermes A-1 C-2 derivative at the White Sands Proving Ground in New Mexico in 1951. (Source: 'Missile, Surface-to-Surface, Hermes A-1, Experimental,' *Smithsonian NASM*, https://www.si.edu/object/missile-surface-surface-hermes-1-experimental%3Anasm_A19800214000)

development of a suitable launcher and mount, high-explosive warhead, fuse and fire control system. The OCO (Office, Chief of Ordnance) selected the Eclipse-Pioneer Division of the Bendix Aviation Corporation as the prime contractor. The contract was signed on 28 November 1947, and under the agreed terms the missile system was to be delivered by 31 December 1954. The Ordnance Department made the four German engineers available to Bendix on a full-time basis and transferred them from Fort Bliss to Teterboro in New Jersey to live in apartments near the Bendix plant. According to Ordway and Sharpe, the four were the first Germans brought to the US through Projects Overcast and Paperclip to live in the local economy (outside military accommodation). Although the relaxation of supervision was another significant step in the recruitment programme, after the specialists' families began to arrive in the US from Landshut in Bavaria in December 1946, the Germans were still not trusted – the four remained under surveillance by intelligence personnel from the US First Army for an unspecified length of time.[40]

The weapon was dubbed 'Loki', after the mythological Norse god of trouble and mischief. The design of the Loki incorporated two major changes to the *Taifun*. Firstly, the missile had two stages – a booster with a warhead fitted to the front that detached prior to impact – and secondly, the *Visol* fuel was replaced by aniline. The use of liquid propellants for propulsion was soon brought into question as a result of a study by the JPL and the APG. From 1948–50, studies were conducted to determine the feasibility of a solid propellant version. As a result of these studies, Redstone Arsenal was assigned the task of supervising the development of a solid propellant version concurrently with the evolution of the liquid propellant version. In March 1951, the JPL was awarded a contract to develop the solid propellant missile. Meanwhile, trial firings of the liquid propellant version had commenced at the APG. Later that year, the Loki was the first projectile test-fired at the newly opened 16km x 1.5km Small Missile Range at WSPG. By early 1952, the potential of the solid propellant version led the Ordnance Department to place greater emphasis on it over the liquid propellant design. Subsequently, the solid propellant version was considered for an undertaking far greater than an anti-aircraft weapon. In mid-1954, at a meeting in Washington DC between von Braun and a group of officers and civilians from the Office of Naval Research about the development of a launch vehicle for a space satellite, the German (now an American citizen) proposed a Redstone rocket with two upper stages powered with clusters of solid propellant Loki rockets. Returning to the SAM system, in addition to the aforementioned changes, the project was delayed by numerous problems. The contractors and sub-contractors were often behind schedule in the supply of materials and in the design and manufacturing of subsystems. Furthermore, the launcher needed to be redesigned to accommodate the solid propellant version. These delays, in addition to the progress made with the Nike and also the Hawk SAM by Raytheon, convinced the Army Chief of Staff to terminate the project in November 1955.[41]

The second German flak rocket design that was exploited by the Rocket Branch was the C-2. A derivative of the missile was utilised as an experimental prototype for the research and development of a tactical SSM as part of Project Hermes. The missile was designated the Hermes A-1, which as we have seen had been relegated to a test vehicle after the contract for Project Nike was awarded to Western Electric in early 1945. The A-1 also received the US Army designation of Control Test Vehicle, Army, fifth missile type (CTV-G-5) in April 1948. The project engineers retained many of the original features of the C-2 design, an indication of how well the Americans viewed the usefulness of the German technology. The dimensions and aerodynamic shape of the A-1 were identical to the C-2/W-3 prototype – 7.83 metres in length, with a span of 2.51 metres at the gas vanes, and a fuselage diameter of 88cm. Steering was by the same gyroscope-controlled, electro-hydraulic method of operating the four control surfaces at the tail of the missile. The A-1 was launched vertically.[42]

Nevertheless, the German technology was considerably modified. The *Visol*/nitric acid LPRE that was designed at Peenemünde-East was replaced with a new engine supplied by General Electric. By March 1947, GE was engaged on a design study which had as its objective a LPRE which utilised a propellant combination of LOX and alcohol, pressure-fed from the tanks into the combustion chamber, to produce a thrust of around 7,250kg and a burning duration of 50 seconds. GE initially built smaller-scale prototypes of the engine, which were designed and tested at the GE Malta Test Station, 35km north of the main GE plant at Schenectady in New York state, where the engines were fabricated. In the full-scale engine, the propellant feed system used pressurised helium gas instead of pressurised nitrogen gas, and the resulting thrust was slightly higher than the German LPRE, at approximately 8,600kg at sea level fully fuelled. The take-off weight was about a quarter lighter at around 3 tons.[43]

In place of the ground-based *Einlenkrechner* (line-of-sight computer) that was designed to steer the missile onto its trajectory after six seconds, a device was fitted in the missile that was programmed to automatically turn the missile after five seconds. Guidance was by the command-link technique, whereby a modified SCR-584 radar tracked the missile throughout its flight. During the upward part of the trajectory, a pre-programmed flight control system automatically maintained the missile on course by correcting deviations in azimuth. Once the missile reached the desired velocity after around 45 seconds, the radar transmitted a signal to the missile to cut off the engine. The tracking radar then guided the missile towards the target point, around 64km away.[44]

Based on the successful performance of the CTV-G-5, the Rocket Branch considered the design as a possible tactical SSM. The missile was redesignated XSSM-A-15 (indicating that it was in an experimental or developmental stage), but by September 1951, it was again relegated to a test vehicle with the designation RV-A-5 (Research and Test Vehicle, Army, design number five). The RV-A-5 was one of two missiles used as test vehicles for the Hermes A-2, the other being the RV-A-10, which eventually became the Sergeant SSM. The Hermes A-2 was originally envisioned by GE to be a wingless SSM based on the original SAM concept of the A-1, but the A-2 was cancelled during the planning stage. In 1949, the A-2 was revived as a low-cost SSM to carry a 680kg warhead over a distance of 120km, fitted with a LPRE that was jointly developed by GE and the Thiokol Chemical Corporation. By 1951, the A-2 had received the designation XSSM-A-13, but after September 1951 the project was again cancelled.[45]

Table 9. The type designations and popular names of the guided missiles under the cognisance of the Rocket Branch, Ordnance Research and Development Division, 1948–51

Popular name	1948	1951	
WAC Corporal	RTV-G-1	RV-A-1	Development concluded 29.7.1950
Corporal E	RTV-G-2	RV-A-2	Development concluded 11.10.1951
Hermes II	RTV-G-3	RV-A-3	Cancelled in September 1953
Bumper	RTV-G-4	RV-A-4	Programme concluded 29.7.1950
Hermes A-1	CTV-G-5	RV-A-5	Formerly XSSM-A-15. Became a test vehicle for the Hermes A-2
Hermes B-1	RTV-G-6	RV-A-6	Cancelled in 1954
Nike I	XSAM-G-7	XSAM-A-7	Entered service in December 1953
Hermes A-3	XSSM-G-8	RV-A-8	As the A-3A, became a test vehicle for the XSSM-A-16. Cancelled in 1954
Hermes B-2	XSSM-G-9	XSSM-A-9	Cancelled by September 1951
Sergeant	-	RV-A-10	Became a test vehicle for the Hermes A-2. Cancelled in 1951 then revived in 1953. Entered service in 1961
Lacrosse	-	XSSM-A-12	Originally a US Navy BuOrd project. Entered service in 1959
Hermes A-2	-	XSSM-A-13	Revived in 1949, cancelled after 1951
Hermes C-1	-	XSSM-A-14	Became the Redstone SSM, which entered service in June 1958
		XSSM-A-15	Ex-CTV-G-5. Became RV-A-5
Hermes A-3B	-	XSSM-A-16	Cancelled in 1954
Corporal I and II	-	XSSM-A-17	Entered service in February 1955
Hawk	-	XSAM-A-18	Entered service in 1958
Plato	-	XSAM-A-19	Anti-missile missile. Cancelled in 1959

* RTV-G = Research Test Vehicle, Army
RV-A = Research and Test Vehicle, Army
XSAM-G and XSAM-A = Experimental Surface-to-Air Missile, Army
XSSM-G and XSSM-A= Experimental Surface-to-Surface Missile, Army

Sources: Memorandum from Toftoy to Secretary, Ordnance Technical Committee, 29.4.1948 and 25.9.1951, in Cagle, Development, Production and Deployment of the Nike-Ajax Guided Missile System 1945–1959, 246–254; Bullard, History of the Redstone Missile System, 8–13 and 42

Overall, the exploitation of the German knowledge in the Hermes A-1 test vehicle was successful. In 1960, Ordway and Wakeford summarised the results from the construction of six A-1 missiles at WSPG. By the spring of 1949, the first test vehicle was ready for static tests, but a fire in the tail section damaged the missile. They recorded the dates of five of the test-launches: on 19 May 1950, 14 September 1950, 8 February 1951, 15 March 1951 and 26 April 1951. The design of the missile and the various subsystems in these launchings performed well. The LPRE functioned up to full thrust and enabled the missile to attain a velocity of Mach 2.4. The command-link guidance system and gyroscopic control functioned satisfactorily. The missile was fitted with a Doppler transponder (a combined transmitter and receiver that sent out a signal automatically upon receiving a predetermined trigger signal) and a radar beacon for guidance and tracking. Satisfactory stability was achieved from the launch through the subsonic, transonic and supersonic regions without aerodynamic flutter, and the structural integrity of the airframe was preserved.[46]

The development of these two projects, that were both largely built upon German technology, did not culminate in any operational missile systems. They ultimately served to generate data on the sub-systems fundamental to both types of weapons systems – aerodynamics, launching, flight behaviour, liquid and solid rocket propulsion, and in the case of the Hermes A-1, the control and guidance of a supersonic SSM, an outcome no-one at Peenemünde would have anticipated in 1942.

Conclusion

Although the majority of the German guided missile and rocketry specialists who were employed in the US after World War II were involved with aircraft, ballistic missile and eventually space rocket programmes, the transfer of knowledge about SAM technology from Germany played a significant part in forming the foundation of tactical guided missile development for the American armed forces in the late 1940s. The fundamental principles of SAM design, and the scientific and engineering problems that required solving, had been identified by German specialists several years beforehand. For example, research data on the structural and aerodynamic properties and behaviour in-flight of a supersonic, liquid propellant SAM brought important information to the engineers and scientists on Project Nike. A German concept that was not developed any further in the US was the subsonic or low transonic SAM; the Hs 117 and *Enzian* very soon became obsolete. All American SAMs were planned to be supersonic in order to intercept jet aircraft with ever-increasing speeds of the types that were currently being developed in the US and elsewhere.

American scientists and engineers were quick to improve upon some aspects of the German technology – radio technology for missile guidance and tracking was one area where American ideas very soon surpassed the limit of German advances by 1945. The ideas in the ambitious Wizard and Thumper theoretical studies by the

AAF are two further examples. In other fields, such as the development of propulsion systems, the evaluation and exploitation took more time, but by the early 1950s the principles that were established in Germany with LPREs also became obsolete. German rocket engineers and chemists had previously established that LPREs were more suitable for SAMs than solid propellant systems. The design of the LPRE in the sustainer rocket of the Nike SAM system which entered service in December 1953 confirmed the original German view about the superior performance of liquid propellants. By this time, however, American (and British and French) research into solid propellants indicated that SPREs would be more suitable for tactical guided missiles than LPREs. With the exception of ramjet propulsion, the change from liquid to solid propellants was characteristic of all other SAM developments in the US, the UK, France, and the Soviet Union during the 1950s.

The development of other technologies also took time to perfect, even for the scientifically capable and technologically adroit Americans. The problems that the Germans faced in designing a seeker were probably the most complicated and were not solved by the Americans until well into the 1950s. The American scientists and engineers who were involved with research and development for Project Nike were unable to successfully design a seeker for the missile. The proximity of the missile to the target and the point at which the warhead was detonated were calculated by a ground computer on the basis of information that was received by the acquisition and tracking radars.[47] The first American SAMs to be fitted with seekers did not enter service until 1958. These were the Hawk, with semi-active homing technology, and the BOMARC, which had an active seeker and a proximity fuse that detonated the warhead – a true realisation of unsuccessful German research and development from the war.[48]

The Transfer and Exploitation of German SAM Technology by the United Kingdom After 1945

Britain's role in the conquest and occupation of Nazi Germany resulted in a great amount of scientific and technical intelligence that pertained to flak rockets – in intelligence reports and monographs, captured artefacts and documentation, and the minds of German specialists – either transferred directly to the UK from Germany or indirectly from the US. The postwar exploitation of the recent German scientific and engineering advances in guided weapons, rocketry and aerodynamics were just three examples of where the erstwhile enemy's intellectual achievements benefited the British armed forces. Other domains where Britain profited from German knowledge included aircraft designs, aircraft instrumentation, gas turbine engines, armoured fighting vehicles, metallurgy, fire control, ballistics, gyroscopes, servomotors, small arms, ceramics, hydrodynamics, and hydrogen peroxide propulsion for submarines. Although British engineers considered utilising certain aspects of the *Wasserfall* design in a test vehicle, conducted experiments with captured *Taifun* rounds, and possessed interest in German progress with nitric acid and hydrogen peroxide LPREs, German SAM technology was not exploited by the British to the degree that it was in the US. Only 162 German specialists were recruited through the British government recruitment programme for the defence sector, the Deputy Chiefs of Staff (DCOS) scheme, of whom guided weapon and rocketry specialists constituted a significant proportion. The work of the German specialists who brought their expertise and experience with SAMs to the UK was directed towards research and development of tactical guided weapons and test vehicles, where a small number of the specialists made some important contributions to postwar British SAM development. Former political affiliations with the Nazi party did not prevent some German specialists and their families from assimilating into British society, an outcome which mirrored the American experience.

Captured German flak rocket technology is transferred to the United Kingdom

German flak rocket technology was transferred to British research and experimental establishments for evaluation within weeks of the conclusion of the war in the

European theatre. The booty from the *Enzian* project included two prototype missiles, one of which was sent to Australia in 1947 with an A-4/V-2 and a X-4 AAM; one HWK 109-502 LPRE, which was sent to the MoS ADD facilities at Fort Halstead near Sevenoaks in Kent; and the only known existing prototype of the *Madrid* infrared seeker that was under development by the firm *Kepka* of Vienna in conjunction with the DFS, which was sent with the accompanying documentation to ADI (Sc) at the Air Ministry. The booty from the *Rheintochter* project included complete prototypes of the R-I and R-IIIf, which were acquired by the MoS, and two samples of the 41-inch solid propellant sticks in the R-I booster rocket and the four-stage *Rheinbote* SSM, which were sent to the Projectile Development Establishment (PDE) at Aberporth in South Wales for chemical analysis. From the Hs 117 project, one partially complete prototype was recovered, in addition to a sample of the BMW 109-558 LPRE which was also sent to Woolwich Arsenal. From the C-2 project, the haul included a centre section of the fuselage from the W-3 prototype; a sample of the rocket engine; a gyroscope assembly accompanied by a mixing device called a *Mischgerät*, which were sent to RAE Farnborough; and by 9 October 1945, a captured electric servomechanism that was analysed and tested at the Admiralty Gunnery Establishment (AGE) at Teddington in London. Although the majority of the documentary records from the German SAM development programme were captured by the American armed forces and scientific and technical missions, agencies in the US produced microfilm copies for British organisations. The British also reciprocated. For instance, there were documents in British possession that were firstly microfilmed by the Directorate of Microgram Services in the Air Ministry and then dispatched to the Air Documents Division at Wright Field. Scientific and technical intelligence from the German SAM development programme was also distributed to the Dominions of the British Commonwealth.[1]

After BIOS was created in mid-1945, German equipment that pertained to armaments and associated topics which was seized by BIOS investigators was transferred to one location in the UK – the Halstead Exploitation Centre (HEC) at Fort Halstead in Kent. The HEC was established in 1945 under the auspices of BIOS Group II as a repository and clearing house. Documents received by the HEC were in two different categories – (a) unsorted and (b) sorted. Under category (a) were documents that arrived with no lists as to the details or contents; category (b) included documents that were scrutinised in the field and carefully selected as worthy of retention and were accompanied by lists with identification details and contents. At the HEC, documents were screened and the titles of those considered to be of technical interest were translated and entered onto accession lists which were then circulated to British, American and Dominions organisations. Documents that were subsequently applied for by those organisations were translated by German POWs borrowed from the War Office. In November 1947, the staff at the HEC comprised six military officers, 17 civilian translators and 86 German POWs. Documents were categorised under an alphabetical system, from 'A' to 'Y'. The documents concerning explosives, propellants and pyrotechnics were under 'E'; guided projectiles were under 'G'; and rockets, including rocket-assisted projectiles, were under 'R'. The HEC was used extensively by the research and development interests in the MoS, but by the end of 1947 the active exploitation of intelligence from Germany was passing. The HEC was closed down on 31 March 1948, and the documents were transferred to the Technical Personnel Administration/Technical Intelligence Branch (TPA/TIB) in the MoS, which became the responsible department for all requests for classification of translated documents.[2]

At Fort Halstead, ADD engineers carried out an overall evaluation of at least two flak rockets, the C-2 and the *Taifun*. They relied upon various sources of information, including British and American intelligence reports, examinations of captured technical documentation and prototypes which had been brought back to the UK, and interrogations of specialists from Peenemünde. In the case of the C-2, information was also obtained from other British establishments where captured artefacts were sent for evaluation. These included RAE Farnborough, the Radar Research and Development Establishment (RRDE) at Great Malvern in Worcestershire, the AGE, the recently created Guided Projectiles Aerodynamics Section in the Armaments Research Department (also a MoS department) and the company Messrs. Firth Vickers. The evaluation of the supersonic C-2, the most technologically advanced of the German flak rockets, was a somewhat challenging task because without a complete prototype, a full examination of the missile could not be made. Also, the author of the ADD technical report cited the difficulty in assessing the numerous prototypes and changes to the overall design, which was common in all experimental German guided missiles. The C-2/W-3 prototype, which was built in the greatest quantity, was therefore evaluated, along with the equipment in its guidance system. A metallurgical analysis of the missile was also carried out. Prior to the completion of the report, intelligence from Germany had already revealed that the C-2 and other German flak rockets were more advanced than the LOP/GAP. The ADD technical report therefore served to document the German technology for future exploitation.[3]

As for the *Taifun*, the MoS acquired at least 150 of the approximately 600 rounds of the liquid propellant version of the rocket (*Taifun F*) that were manufactured in the *Mittelwerk* by April 1945. British engineers found that the performance of the *Taifun F* was superior to the cordite-fuelled British 3-inch rocket. As in the US, there was interest in the design of the *Taifun*; by October 1946, the ADD had commenced the design of a similar projectile for use as a test vehicle. The ADD version was originally planned to be fuelled with nitric acid and aniline, but the fuel mixture was later changed to a 70/30 per cent by volume mixture of furfuryl alcohol and aniline (the same propellants in the fuel used in the Nike LPRE). In test-launches of the British derivative, by October 1948 the projectile was considered to have

Table 10. A comparison of the specifications and performance of the British 3-inch rocket and the *Taifun F*

	3-inch/10.2kg. warhead	*3-inch/0.9kg. warhead*	*Taifun F*
Total weight	24kg	14.5kg	21kg
Empty weight with warhead	18kg	8.7kg	10kg
Ratio of fuel/total weight	24%	40%	52.5%
Overall length	1,397mm	1,397mm	1,920mm
Thrust obtained	(approximate) 680kg	680kg	840kg
Burning time	1.6 seconds	1.6 seconds	2.5 seconds
All burnt velocity obtained	1,580ft/sec	–	3,000ft/sec
Operational height	–	–	50,000ft
Velocity at operational height	–	–	975ft/sec

Source: TNA, DEFE 15/217

performed satisfactorily. This development enabled the ADD to gain experience and knowledge in the use of nitric acid as an oxidiser in LPREs, although the rocket engines that burned from 25–40 seconds presented British rocket designers with a more complicated challenge and different sets of problems.[4]

The ADD technical reports on the evaluations of the C-2 and the *Taifun* do not offer any opinions about how advanced the technology was (or was not) compared to British developments, or on the potential implications for the war had the missiles entered service. Such comments and opinions are found in industry publications such as *Flight* magazine. In these publications, the British (and American) public got glimpses of the German flak rocket technology during the summer of 1945. A closer view of German missiles and avionics was possible at the first annual air show at RAE Farnborough since the end of the war, which began on 29 October 1945. Besides the latest and most advanced aircraft designs of the Society of British Aircraft Constructors that were on display for domestic and foreign audiences and potential customers, the captured German technology on show was also impressive, as a journalist of *Flight* magazine attested:

> In fact, the entire [German] display at Farnborough was a monument to German ingenuity and energy, while, by its very diversity, demonstrating a state of panic at a time when they must have known that the end was near.[5]

It is appropriate to conclude this section by referring to the reactions at the time to the progress of German flak rocket development. The British author Rowland Pocock, analysing the reactions in his 1967 book *German Guided Missiles of the Second World War*, described a general feeling of surprise amongst many who saw how advanced the German SAM technology was. He mentioned that the revelation of the extent of German progress by the end of the war caused a sensation almost verging on a minor panic, with the impression in the popular and technical press summed up as that the war finished just in time. Pocock urged caution when approaching these initial assessments. His analysis convincingly showed that with the damage to German industry, the time it would have taken to transition from AA artillery to guided missiles (2,170 missile batteries were required to defend German airspace) and with the associated high manpower requirements, German missiles would not have had any appreciable impact on the outcome of the aerial war, even if the conflict had lasted another six months.[6]

A captured HWK 109-502 hydrogen peroxide LPRE that was transferred to the ADD at Fort Halstead for analysis in mid-1945. (Source: TNA, DEFE 15/194)

The postwar consolidation of British guided weapons research and development, and the early influence of German flak rocket technology on British missile projects

With the end of World War II and the demobilisation of the wartime armed forces, the British government (from July 1945 led by Prime Minister Clement Attlee, after

The centre section of a C-2/W-3 prototype that was transferred from Germany to the United Kingdom for analysis. The liquid high-explosive warhead, pressurised nitrogen gas tank, fuel tank and oxidiser tank (in order) were housed in this section. (Source: TNA, DEFE 15/216)

A captured LPRE for the *Rheintochter* R-IIIf that was designed by *Dr.-Ing.* Hans-Joachim Conrad of the TH Berlin/*Vierjahresplan-Institut für Kraftfahrzeuge*. (Source: TNA, AIR 40/2532)

Labour's victory over the Conservatives in the first general election since the war began) consolidated the guided weapon research and development organisation and infrastructure in the UK. In 1993, Twigge addressed these historical events in some detail,[7] and thus only a summary of the changes as they impacted British SAM research and development is necessary. In June 1945, the British government decided that all guided weapons work undertaken to meet Admiralty and War Office requirements (that is to say except AAMs) was to be brought under the control of Sir Alwyn Crow, the Director of Guided Projectiles (DGP) in the MoS.[8] Further consolidation took place when, as we saw in Chapter 3, the government amalgamated the MAP and the MoS in late 1945 and early 1946 to create MoS (Air) and MoS (Munitions), with guided weapon research and development the responsibility of the former branch. The continued consolidation by the Labour government was not indicative of a lack of political interest in guided weapon development; the Attlee administration placed great value on scientific research and the design and development of new and advanced weapons. The decision to secretly invest many millions of pounds in the atomic bomb programme is a prime example.[9]

A new Guided Projectiles Establishment (GPE) was created at Westcott in Buckinghamshire in February 1946, which was planned as the central establishment for all rocket and guided weapon research and development in the UK. In late 1946, the GPE was absorbed into the RAE and renamed the Rocket Propulsion Department (RAE/RPD). Captured German LPREs were tested in purpose-built test beds at the new facility. There were other changes: the RAE Controlled Weapons Department became the RAE Guided Weapons (GW) Department, based at an outstation at nearby Bramshot; and the testing range at Ynyslas in Wales was closed in favour of the range at the PDE at Aberporth, which became the Trials Wing of the GW Department. By early 1947, the RAE was responsible for the development of all guided weapons for the British armed forces. Also created during this period, by the British and Australian governments, was the Anglo-Australian Joint Project to develop guided missiles in South Australia.[10]

Meanwhile, the first British SAM, the basic solid propellant Brakemine, had become a test vehicle. But the project was soon terminated, in late October 1946, by the Chief Superintendent of the GPE, William Cook. The decision to end the project was influenced by several factors. The main one was the announcement by the British government that the experimental establishment at Walton-on-the-Naze in Essex would be closed by the end of the year. Other factors were that the current design of the projectile was determined to be completely unsuitable as a test vehicle for guidance systems; it was learned that the aerodynamic shape of the projectile created excessively high drag, which provided only a matter of seconds to study the behaviour of the missile at supersonic velocities; and the A. C. Cossor control system, called 'twist and steer', was considered unsatisfactory (the system was subsequently redesigned for use in the LOP/GAP). In December 1946, further development was ceased, and the remaining projectiles were sent to the PDE to be used in a restricted firing programme for training purposes and to provide data for a new servomechanism simulator. By May 1947, the project was fully terminated.[11]

In order to analyse the general German influence on British SAM development in the immediate postwar years up until 1947, a very good source is 'History of the Development of RTV 1' by J. Clemow of RAE Farnborough, published in January 1950. The reorganisation and consolidation of the British military infrastructure saw the co-ordination of the LOP/GAP programme transferred at first from the ADD to the GPE, then from the GPE to RAE Farnborough. As a result, the programme endured a number of delays in 1946, although several factors were connected to the transfer of German knowledge to the UK. The captured German scientific and technical documentation needed to be sifted and read; the German specialists who were employed in the British Zone of Occupation had to complete monographs and reports on their work; and BIOS investigators had to be sent to Europe to obtain information on German developments. Other factors included the transferral of staff to the GPE and several setbacks in the manufacture of components. Consequently, there was an 11-month hiatus in test firings of the LOP/GAP from January–December 1946.[12]

The scientific and technical intelligence on German SAM development initially prompted interest from British engineers on the LOP/GAP project but had little influence. In 1952, Kenneth Gatland, in *Development of the Guided Missile*, was critical of the then-apparent decision by the MoS not to exploit the design of the *Wasserfall* as a test vehicle (there was a high level of security surrounding British guided weapons projects):

> It is in many ways surprising that production of the *Wasserfall* was not taken up in Britain when the guided missile programme was first laid down in 1945, if only to give our technicians 'something to play with' in the shape of a successful missile. With such a primary vehicle, they could have carried out all their basic research on stability, inherent control, booster techniques, remote guidance, etc., at an early stage, besides obtaining experience in the handling and servicing of a practical supersonic rocket – a rocket, moreover, that could not otherwise be had but for years of labour and vast expenditure. Failure to evaluate and exploit the German research effort in this practical sense has undoubtedly made the British missile programme both laborious and uneconomic.[13]

The GAP organisation did consider exploiting certain aspects of the design of the *Wasserfall*. In a paper entitled 'Vertical Launching of GAP' by the Chief Engineer, Armaments Design (CEAD) of the ADD, dated 22 July 1945 for Technical Panel A (Radio, Radar, Stabilisation and Servos) of the Guided Projectiles (GP) Working Committee (succeeding the GAP Working Committee, which was dissolved in late June/July 1945), CEAD discussed the possibility of launching the LOP/GAP vertically, including control during boost period, control subsequent to separation, abandonment of boost, and addition of vanes in the gas jet, the latter a feature of the *Wasserfall* design. A note was added on the possible performance of a new design of projectile similar to the *Wasserfall*, designed from the start for vertical launching. By 14 August 1945, the date of the 12th report of the GP Working Committee, the possibility of launching LOP/GAP vertically was being considered.[14]

Just over four months later, in a GP Working Committee paper entitled 'Summary of the present position of the design, production and testing of Liquid Oxygen/ Petrol GAP' dated 5 December 1945, CEAD reviewed the changes in the design of the LOP/GAP during the previous eight months in light of the experimental results. Up until this time, 33 dummy rounds had been fired and the LOP/GAP was becoming an experimental vehicle rather than a potential service weapon. CEAD considered that, for long ranges, a vertically launched projectile controlled by vanes in the gas jet would be the ultimate solution, as it would avoid the use of both a booster and trainable launcher. In spite of this, however, it was considered that no radical changes should be made to LOP/GAP since a suitable vehicle was urgently needed to carry out control system tests – the present horizontally projected design was being carried on allowing early testing of controls and improvement in control design. This decision seems to have been confirmed by Technical Panel A of the GP Working Committee in a paper dated 22 July 1946, on the subject of a MoS report, 'Ministry of Supply Establishment at Cuxhaven Technical Reports', which concluded that the technical reports written by the German specialists at MOSEC, mainly on the control of the A-4/V-2 and *Wasserfall*, were (strangely enough) of little use to GAPs.[15]

It would seem that beyond the same general problems that the Germans identified – for example, British engineers also recognised that vibration was caused by the rocket engine and the aerodynamic forces on the missile – the avenues of research that were pursued by the Germans in relation to their designs up until early 1945 may not have been of much value to the LOP/GAP project. The propulsion system in the British missile is a salient example. It appears that the British engineers who were responsible for developing the booster stage soon decided that the intelligence on German solid propellant rockets did not indicate that British solid propellant rockets were inferior to the type that were used to boost the *Rheintochter*, Hs 117 and *Enzian*. In early 1946, the ADD, whose personnel were in the process of concomitantly evaluating each German flak rocket, was charged with improving the design of the booster. A cluster of seven British 5-inch ATO rockets was designed, constructed out of light alloy tubing and fuelled with a recently developed plastic solid propellant, initially sodium nitrate, then ammonium perchlorate (the latter produced a higher specific impulse).[16]

With regard to the LPRE in the sustainer rocket, British engineers retained the use of liquid oxygen (LOX) as the oxidiser, which the Germans did not use in any flak rocket LPREs, even though the chemical produces a higher specific impulse than nitric acid and hydrogen peroxide. It is rather curious to observe that British rocket engineers chose to design a LPRE that used LOX first with petrol, then from 1948 with a 60/40 per cent alcohol/water mixture, the latter similar to the propellant combination in the A-4/V-2 LPRE. The British experience with LOX soon found that the chemical was unsuitable for land- and sea-launched SAMs anyway. If the filled weapon stood for too long, failure of some of the valves occurred, which prevented firing at short notice, thus defeating the main purpose of the weapon. At sea, the storage of LOX on board a warship was considered feasible, but the possible presence of high concentrations of oxygen gas made the fire risk very high, and more so in combat because steel burns readily in LOX and a fire could cause the loss of the ship. On land, it was considered very wasteful to fill small tanks with LOX due to evaporation, which was about 2 per cent per day. Nitric acid was still the preferred oxidiser for LPREs in SAMs, as it had been in Germany during the war and currently was in the US. Nitric acid was preferred to LOX and hydrogen peroxide because of its simpler handling and suitability for storage for extended periods in sealed tanks, which met the requirement for ready-to-use missile systems at sea and on land.[17]

From 1945–47, the dimensions and aerodynamic shape of the LOP/GAP were not altered very much, apart from a slightly increased length in order to fit new radio-control equipment and a reduction of the aspect ratio of the control surfaces and main wings. The guidance system used the beam-rider technique and was built from American and British technology. The ground equipment, developed at the RRDE, consisted of a modified American BTL SCR-584 microwave gun-laying radar, which was similar, if not identical, to the technology that was first used by the Americans in Project Nike. The receivers in the missile were British-designed and developed. The development of the control system was the responsibility of the AGE, with assistance from the Guidance and Control Division of the RAE GW Department from early 1947. Within a year, German specialists would be at work within the RAE GW Department.[18]

Despite Britain being behind Germany in guided weapons development during the war, the British attitude of confidence in the country's scientific and technical proficiency was manifested in the LOP/GAP, which was renamed Rocket Test Vehicle 1 in August 1948.[19] Indeed there were aspects of the technology where the German scientific and technical intelligence was of limited value, but as shown, the Germans were so much more advanced in the development of certain subsystems, notably rocket propulsion and supersonic aerodynamics. Also, the evidence indicates that the German knowledge could not be promptly exploited by the scientists and engineers in the MoS due to a lack of resources and the reorganisation and consolidation of the British defence establishment, which slowed progress.

The recruitment of German guided weapon and rocketry specialists through the Deputy Chiefs of Staff scheme

German specialists who were selected for employment in the British defence sector were recruited through a government-run programme, the DCOS scheme. Through this scheme, German specialists with expertise and experience in fields where Germany had a lead over Britain – such as in ballistics, metallurgy, armoured fighting vehicles, submarine designs, guided weapons, rocketry, gas turbines, photogrammetry, gyroscopes, and aircraft instrumentation – were employed on classified defence projects. The initial intention of the scheme was to exploit the knowledge of the German specialists in the UK for a short period and then send them back to Germany, but concerns over the security of British military secrets rendered the plan impracticable. Those specialists who were considered for employment in the UK were assembled at the MoS Experimental Station Trauen. The starting contract was for six months, which was extended for another six or 12 months based on the specialist's performance. Not surprisingly, security assessments of the Germans revealed that some had been members of Nazi political organisations. The policy of the British government towards the recruitment of Germans with connections to the NSDAP was unambiguous and clearly defined:

> Nobody whose record indicates that he was a convinced Nazi should be brought to the United Kingdom to work, however high his scientific qualifications.[20]

This policy, in addition to the denazification schemes in the British and American Zones of Occupation, were creating obstacles that barred some very talented guided weapon and rocketry specialists from being employed in the UK. On 3 September 1946, William Cook at the GPE wrote a memorandum to Sir Alwyn Crow, the DGP in the MoS, proposing that the strict conditions be relaxed:

> Under the denazification scheme key scientists are being refused entry to this country. A case in point is Dr. [Johannes] Schmidt, now at Trauen who is required at Westcott. He was, in common with many other key men, connected with the Nazi party. Entry to this country has been refused so far and I am afraid that he and his team may join the Russians. It is submitted as a point which should be taken up urgently that unless the connection with the Nazi party was deep, key scientists should be permitted to enter this country in those cases where specialist knowledge will be of benefit to current work of national importance.[21]

Subsequent to this memorandum, the DCOS scheme recruitment policy was re-examined. Pragmatism and the acceptance of the intense competition with the Soviet Union for the brains behind Germany's most advanced weapons programmes were two factors that prioritised the exploitation of German specialists with unique credentials and experience – such as Schmidt, who entered the UK around two months later – before the ethical and moral problems of recruiting Germans with dubious political backgrounds. Like in the US, the common interests between the

Germans and their erstwhile enemies did not extend very far beyond a mutual interest in science, the application of scientific and engineering knowledge in their specialised fields to the British armaments industry and a disdain of the communist Soviet Union.

Three MoS research and experimental establishments sponsored the employment of German guided weapon and rocketry specialists: RAE Farnborough (including the outstation at Bramshot), the GPE (later RAE/RPD) at Westcott and the Chemical Research and Development Department (CRDD) near Waltham Abbey in Essex, on the site of the former Royal Gunpowder Factory, which was initially a daughter establishment of the ARD at Fort Halstead in Kent. In 1948, the CRDD became an independent establishment and was renamed the Explosives Research and Development Establishment (ERDE). At first, no German specialists were recruited for employment in the Anglo-Australian Joint Project in South Australia, because the Australian government had ruled in May 1947 that former enemy aliens could not be employed in Australia on defence projects classified 'Secret' or above. However, one of the Germans employed in the RAE Aerodynamics Department, aeronautical engineer *Dr.-Ing.* Heinz Gorges, formerly of the LFA, was transferred to the Aeronautical Research Laboratories (ARL) in Melbourne in 1949 and worked on the construction of a wind tunnel at the new ARL establishment in Salisbury, South Australia, in connection with the Long Range Weapons Establishment.[22]

Like the technical branches of the American armed forces, the Admiralty and MoS tended to recruit groups of German specialists from the one company or establishment. The practice was particularly useful when the German specialists had been involved in the research and development of key technologies in which German knowledge was superior, such as in hydrogen peroxide propulsion systems, rocket engine technology, and electronics in remote control and guidance systems. The groups of specialists had worked either together or separately on projects for various applications. The members of a group of 13 specialists from HWK who were recruited by the MoS had worked on the design and development of LPREs and propellants for projects including the HFW Hs 293 guided missile, the *Messerschmitt* Me 163 rocket-powered interceptor aircraft, and ATO units for aircraft and flak rockets, namely the *Enzian* and the Hs 117. Five of the six scientists and engineers who were recruited from Peenemünde-East were clearly sought-after by the MoS for how their expertise and experience from the research and development of the C-2 could be applied to British SAM projects. Furthermore, a group of eight specialists who had formerly worked for *Telefunken* had all been engaged on either research or development of remote control and guidance technology for flak rockets. Nevertheless, there were instances where a single specialist from an organisation was recruited on the basis of particular specialised knowledge or skills.

One advantage the British had over the Americans when recruiting specialists was the geographical proximity to Germany. This enabled British government organisations to easily remove German specialists to the UK for temporary exploitation. However, the importation of specialists did not really begin in earnest until early 1946, several months after the US War Department began processing the transfers of groups of six or more specialists at a time through Project Overcast. The first specialists brought to the UK through the DCOS scheme were a group of eight from the most fertile recruitment ground for rocket engine specialists, HWK at Wik in Kiel, in January 1946, by the Royal Naval Scientific Service (RNSS) in the Admiralty. The RNSS was formed in 1944 as a civilian scientific research department in the Admiralty, and had its roots in a similar department that was created by the Admiralty soon after World War I.[23] Within the RNSS, the Department of Research Programmes and Planning was responsible for the co-ordination of matters relating to the employment of German specialists on defence projects.[24] Six of the eight specialists, led by the company's former director, Prof. Hellmuth Walter, were employed as consultants to British personnel at the Admiralty Development Establishment in the Vickers-Armstrong works at Barrow-in-Furness in Cumbria, where the Admiralty was exploiting the advances that were made by the firm with hydrogen peroxide propulsion systems for submarines, which promised much faster speeds than diesel-electric engines when submerged.[25]

The MoS recruited 13 former employees of HWK during 1946 and 1947 – 12 for employment at the RAE/RPD at Westcott and one at the CRDD at Waltham Abbey. Nine of them specialised with hydrogen peroxide propulsion systems for guided missiles and aircraft, although not for SAMs, as German studies during the war had determined the chemical to be unsuitable for flak rocket propulsion systems. Postwar research at the RAE would, however, demonstrate that the notion of using the propellant in a SAM LPRE was by no means abandoned. At least two specialists from HWK brought experience with the use of nitric acid as an oxidiser in LPREs for guided missiles, an application which German research and development had proven was more suitable than hydrogen peroxide for flak rockets.

The de facto leader of the HWK group was *Dr.-Ing.* Johannes Schmidt, an experienced aircraft and rocket propulsion specialist. From 1940–45, he was the director of the rocket engine development department at the company, and later was also the technical director at the firm's main engineering plant at Beersberg in Silesia. Schmidt was best known for his work in the design and development of the 109-509 hydrogen peroxide LPRE for the *Messerschmitt* Me-163b *Komet* interceptor, but also for the 109-502 hydrogen peroxide ATO unit, used in the *Enzian*, and the 109-729 petrol/mixed acid LPRE that was built for the Hs 117. Schmidt arrived in the UK on or around 12 November 1946 after eight colleagues arrived from Trauen on the evening of 5 November, and became the technical director of the German team at Westcott. In the group of eight was Dr. Jurgen Diederichsen, a 32-year-old Danish-born chemist who was a former assistant of chemist Hermann von Döhren, one of the HWK group who were recruited by the Admiralty. Diederichsen invented the hydrogen peroxide catalytic method employed in the firm's 'cold type' rocket engines (the 109-502 ATO was of the cold type). The others were *Dipl.- Phys.* Hermann Treutler, the chief of the physics laboratory in the rocket engine department from 1937–45; *Dipl.-Ing.* Johannes Frauenberger, who from 1942–45 was the chief designer of the group that was responsible for the development of the 109-507 LPRE in the Hs 293; Gustav Fiedler, a 46-year-old engineer whose latest work was on the development of a hydrogen peroxide steam-driven turbine engine for torpedoes; engineers Heinz Walder, Friedrich Jessen and Karl Meier, who had expertise with ATO units for *Messerschmitt* aircraft such as the Me-163 and Me-262; Walter Müller, a 56-year-old engineer, the former chief of test stands; and foreman/ fitter Walter Koltermann, a test engineer with welding skills apparently rarely found in the UK. Two more former employees from HWK completed the complement. Willi Kretschmer, a 32-year-old engineer, worked at the Kiel and Beersberg plants as a section leader under Schmidt from 1940–45, where he specialised in the design of combustion chambers. Originally recruited for work in the Supersonics Division of the RAE Aerodynamics Department, by 30 August 1947 Kretschmer had been transferred to the RPD. Lastly, there was engineer Werner Schonheit, who had worked under Schmidt from 1942–45. He was sent to Braunschweig on 17 September 1947 and signed a DCOS scheme contract on 3 October that year. Apart from the specialists from HWK who were recruited by the Admiralty, this group of Germans was arguably the most qualified group of experts on hydrogen peroxide rocket engines in the world.[26]

In what appears to have been a public relations exercise, journalists from *Flight* magazine were permitted to interview the team of Germans at the RAE/RPD in August 1947. The article succinctly described the conditions under which the Germans were employed:

> The German scientists have 12-month contracts and are housed in Nissen huts off the site. They are given a free hand to work on their own special subjects but are not permitted access to all information on current developments.[27]

The austere accommodation provided for the Germans – a Nissen hut is a semi-cy-lindrical corrugated iron structure – was in stark contrast to what the Admiralty provided for their colleagues up at Barrow. In January 1946, the Germans there were billeted in a grand old house called Rocklea, which aroused disapproval from local citizens who questioned why their erstwhile enemies should live in more comfortable conditions than most people in Barrow. Perhaps those complaints were the reason why the MoS decided on austere accommodation at Westcott. The *Flight* article also mentioned that the guided missiles captured in Germany – including one each of the *Feuerlilie* F-55, *Rheintochter* (R-I or R-III), C-2, Hs 117 and *Enzian* E-1 – were housed in a hangar at the base that had been converted into a museum. The Germans provided the journalists with a demonstration of the properties of some rocket propellants and static-tested two HWK hydrogen peroxide LPREs. Three months later, the static-testing of a HWK 109-509 LPRE in an emplacement at Westcott left a lasting reminder of the dangerous nature of rocket engineering. On 14 November 1947, Schmidt and two British technicians were killed, and 11 injured, including Walder, when the armoured glass windows in the wall between the control room and firing bay provided inadequate protection after the engine exploded. In 2019, Hall referred to a deep sense of loss within the local community for Schmidt as well as for the two deceased Britons, which he suggested was due to a greater degree of integration into British society by the German group at Westcott than was previously reported.[28]

Table 11. The German scientists, engineers, and technicians with various levels of research and/or development experience and expertise related to the development of guided weapons and rockets who were recruited by the MoS through the DCOS scheme

Royal Aircraft Establishment			
Aerodynamics Department			
Prof. *Dr.-Ing.* Adolf Busemann	Aeronautical engineer	LFA	July 1946
Heinrich Voepel	Aeronautical engineer	HFW	7 March 1948
Guided Weapons Department			
Dr.-Ing. Josef Linke	Physicist	*Telefunken*	24 February 1947
Dr.-Ing. Karl-Eduard Büchs	Engineer	*Telefunken*	24 February 1947
Dipl.-Ing. Hans Prost	Engineer	*Telefunken*	24 February 1947
Dipl.-Ing. Fritz Rockstuhl	Physicist	*Telefunken*	24 February 1947
Dipl.-Ing. Karl Wilhelm	Engineer	*Telefunken*	24 February 1947
Paul-Gerhard Rothe	Physicist	*Telefunken*	24 February 1947
Dr. Wilhelm Elfers	Physicist	P-E	3 March 1947
Dr. Oswald Lange	Physicist	P-E	4 March 1947
Dr. Karl-Heinz Schirrmacher	Physicist	*Telefunken*	4 March 1947
Hans Hasse	Engineer	*Siemens*	5 March 1947
Dipl.-Ing. Heinrich Katz	Engineer	*Siemens*	5 March 1947
Dr.-Ing. Theodor Schmidt	Engineer	LFA	17 March 1947
Dr. Martin Eichler	Physicist	P-E	29 April 1947
Dipl.-Ing. Siegfried Entres	Engineer	P-E	27 August 1947
Dipl.-Ing. Günter Pieper	Engineer	*Telefunken*	29 August 1947
Hans Rohr	Engineer	*Siemens*	22 September 1947
Instrumentation Department			
Dipl.-Ing. Gerald Klein	Engineer	LGW	15 October 1946
Rocket Propulsion Department			
Dr. Jurgen Diederichsen	Chemist	HWK	6 November 1946
Gustav Fiedler	Engineer	HWK	6 November 1946

(Continued)

Table 11 (Continued)

Dipl.-Ing. Johannes Frauenberger	Engineer	HWK	6 November 1946
Heinz Walder	Engineer	HWK	6 November 1946
Friedrich Jessen	Engineer	HWK	6 November 1946
Karl Meier	Engineer	HWK	6 November 1946
Dipl.-Phys. Hermann Treutler	Physicist	HWK	6 November 1946
Walter Müller	Engineer	HWK	6 November 1946
Walter Koltermann	Welder	HWK	6 November 1946
Willi Kretschmer	Engineer	HWK	6 November 1946
Dr.-Ing. Johannes Schmidt	Engineer	HWK	12 November 1946
Walter Riedel	Engineer	P-E	2 March 1947
Dipl.-Ing. Hermann Zumpe	Engineer	BMW	10 August 1947
Dr.-Ing. Ulrich Barske	Engineer	HFW	21 September 1947
Werner Schonheit	Engineer	HWK	3 October 1947
Dr.-Ing. Hugo Reichert	Engineer	LFA	1947
Dr.-Ing. Hans-Joachim Conrad	Engineer	VfK/TH Berlin	1948
Armaments Research Department CRDD/ERDE			
Gerhard Müller	Fitter	LFA	7 September 1946
Franz Neunzig	Engineer	HWK	19 November 1946
Dipl.-Ing. Hans Ziebland	Engineer	LFA	19 November 1946
Dr. Botho Demant	Chemist	P-W	17 June 1947
Dipl.-Ing. Norbert Luft	Chemist	P-E	18 June 1947

Sources: TNA, AVIA 54/1295, 'Scientists employed under DCOS scheme by MoS. 30.8.1947' (information added to document postdate); DEFE 43/3, 'BUSEMANN, Adolf'; DEFE 43/17, 'VOEPEL, Heinrich Menke Andreas'

There were instances where German specialists soon decided to leave. The case of *Dr.-Ing.* Adolf Busemann is an example. Busemann, an aeronautical engineer and the former head of the supersonic wind tunnel section at the LFA, was a swept-wing expert and co-designer of the *Feuerlilie* research missiles with *Dr.-Ing.* Gerhard Braun. In July 1946, Busemann began a six-month contract at the RAE Aerodynamics Department, but by December 1946 he had become disappointed with the contract and decided not renew it.[29] On 1 February 1947, the British Joint Staff Mission in Washington DC notified the JIOA that Busemann was available for exploitation in the US, and subsequently he signed a contract with the US Navy BuAer through Project Paperclip.[30] Another German specialist with experience from flak rocket research and development who was employed in the RAE Aerodynamics Department was Heinrich Voepel, formerly of HFW, who contributed to the Hs 117 project.[31] One of Voepel's other projects at HFW, from 1944, was the study of a tailless, winged, supersonic SAM, the *Zitterrochen*, models of which were tested at the AVA and provided aerodynamic data on a missile with control surfaces in the trailing edges of wings with low aspect ratios.[32]

Five specialists with expertise in rocket engines and liquid propellants were recruited for employment at the CRDD, two of whom were also former employees of the LFA. One was 32-year-old *Dipl.-Ing.* Hans Ziebland, who had worked at the LFA from 1939–45, eventually as the chief engineer in charge of proof stands, workshops and the design office for rocket research. From 1945–46 he was employed by the MoS at Trauen as the chief experimental engineer. The other was Gerhard Müller, a 26-year-old fitter, who was employed at the LFA from 1940–45. Two chemists were recruited from the inter-service installation at Peenemünde, both of whom were also employed at Trauen. They were Dr. Botho Demant, aged 39, who was formerly the chief chemist at the *E-Stelle Karlshagen*, and 27-year-old *Dipl.-Ing.* Norbert Luft, who appears to have been employed at Peenemünde-East from 1943 onwards. At Trauen, Luft was the most prolific in the writing of reports, authoring 24 and co-authoring six, on subjects relating to the thermodynamics of liquid rocket propellants and gases (thermodynamics being the science concerned with the relations between heat and mechanical energy or work, and the conversion of one into the other), the design of components for LPREs, including for the catalytic decomposition of hydrogen peroxide, and on certain liquid propellant fuels and oxidisers, including a report on the 'Optoline' rocket fuels for the LPRE in the C-2. The fifth specialist was Franz Neunzig, a former employee of HWK from 1942–45, who was an experimental engineer in rocket engine tests. From 1945–46 he also worked briefly for the British, firstly in Kiel and then for the MoS at Trauen. The work of these specialists with rocket propellants and in thermodynamics research was closely associated with the work of the German group at Westcott in rocket engine designs.[33]

During late 1946 and early 1947, a group of six former *Telefunken* employees with expertise in radio-control for guided missiles were recruited in Berlin under the conditions of an operation run by the Foreign Office code named 'Matchbox'. The purpose of Matchbox was to deny the services of certain 'war potential' German scientists, engineers, and technicians to other powers both hostile and neutral, primarily the Soviet Union. There were concerns within the British government, armed forces, and research and experimental establishments about the possibility that German specialists employed in sensitive areas in the UK may seek to leave the country to work in non-allied countries. In addition, there were many specialists currently residing in the British zone and sector of Berlin who had qualifications which the British government also wished to deny to non-allied countries. The operation restricted the movement of German specialists until March 1947, when the restrictions were relaxed, the same month when British intelligence, according to Dorril in 2002, began 'dumping' German specialists on Commonwealth countries, including Canada and Australia.[34]

The group of six were *Dr.-Ing.* Josef Linke, *Dr.-Ing.* Karl-Eduard Büchs, *Dipl.-Ing.* Hans Prost, *Dipl.-Ing.* Fritz Rockstuhl, *Dipl.-Ing.* Karl Wilhelm and Paul-Gerhard Rothe, all of whom had previously worked in either research or development capacities at *Telefunken*. Linke was employed by the firm from 1936–45, eventually as a laboratory chief in the low-frequency section under a Dr. Hugo Lichte. Rothe, a 36-year-old physicist, was employed by the firm from 1938–45 as a group leader working on short-wave transmitters. Prost was from 1934–45 employed as a development engineer, specialising in high-power short-wave transmitters. Rockstuhl, a 34-year-old physicist, was from 1937–45 employed as a development engineer in charge of a group working on various radio and radar techniques. Wilhelm was at *Telefunken* from 1931–45, eventually as a laboratory chief, while Büchs, a 36-year-old engineer, was at the firm from 1935–45, becoming the leader of a group researching electro-mechanical aspects of radio.[35]

Following the German capitulation, the six men were employed by the Soviet authorities at the '*Institut Berlin*' in the former GEMA (*Gesellschaft für Elektroakustiche und Mechanische Apparate GmbH*) plant in Berlin-Köpenick in the Soviet sector of Berlin, where the communists were attempting to reconstruct prototypes of the Hs 117, C-2 and *Rheintochter* with the ancillary control equipment to continue the experimental development of the technology (see Appendix 3). By November 1946, the six specialists – in addition to two others who resided in the British sector of Berlin, Karl Spies and Wolfgang Kodantke – were in contact with RAF Air Intelligence in Berlin, probably covertly.[36]

In early December 1946, Linke and Wilhelm left the Soviet sector to travel to the UK to discuss employment in the MoS. The trip was organised by the DGP, Sir Alwyn Crow, who contacted RAF Air Intelligence after the meeting and stated that the group could be employed, which was confirmed in a signal dated 21 January 1947. Linke and Wilhelm returned to Berlin, and by 28 January, the six had relinquished their jobs and were evacuated to the British sector. There was, however, a disruption to these plans when MoS (Air) telephoned RAF Air Intelligence on 5 February, indicating that only Linke and Wilhelm could be offered definite employment, while jobs for the other six were under consideration. By this point, it was impossible to return four of the specialists to the Soviet sector, and subsequently the six were removed to Alswede near Minden in the British zone. Büchs was transferred on 24 February, while Prost, Linke, Rockstuhl and Rothe were transferred the following day; Wilhelm was the last, on 29 February. Their DCOS scheme contracts officially commenced on 24 February, but an official decision still had to be made on their definite employment. On 13 March, D. E. Evans, the director of the Scientific and Technical Intelligence Branch (STIB), part of the British intelligence organisation in Germany, sent a stern signal to the Control Commission for Germany (British Element) (CCG (BE)), requesting that the promise of employment be honoured. Part of the signal contained the following statement:

> Failure to implement our promises will undoubtedly have severe repercussions on British prestige among German scientists and technicians. Operation 'Match Box' will be seriously prejudiced and the integrity of our intelligence organisations operating in Berlin will be brought into disrepute.

The JIC (Germany) approved Evans's signal at a meeting on 24 March. The six were subsequently allocated positions at RAE Farnborough, where they were employed on guided weapons work, but Spies and Kodantke were not. The group were transferred to Vlotho on 28 April and arrived in the UK on 29 April 1947. The families of the men were also transferred but were among the last of the family members of the specialists to arrive in Britain in what was a gradual process, on or around 23 December 1948.[37]

The six former *Telefunken* specialists were soon joined by two of their former colleagues. The first was Dr. Karl-Heinz Schirrmacher, a 33-year-old physicist who was employed at the firm's Berlin-Zehlendorf establishment until 1945, eventually as a senior group leader. Schirrmacher's work in Germany was particularly important in relation to SAMs because he brought significant expertise and experience from the development of remote control and command-link guidance technology for SAMs. He helped to develop the *Kogge* transmitter in the *Kogge-Brigg* remote control system for AAMs and SAMs and had a senior role in the development and testing of the *Rheinland* guidance system for the supersonic flak rockets, the C-2 and the *Rheintochter*. Prior to signing a DCOS scheme contract on 4 March 1947, Schirrmacher was moved around frequently by the British and American authorities. He was interned at the 'Dustbin' detention centre at Kransberg Castle near Frankfurt in the American zone, then was sent to the Inkpot detention centre in London for interrogation on 6 September 1945, before being employed at MOSEC and subsequently transferred to the BNGM at Minden. After MOSEC was closed down in June 1946, he was, along with a number of specialists employed there, in negotiations with the DEFA to work in France, 'under heavy pressure' according to one source. The other specialist was *Dipl.-Ing.* Günter Pieper, employed by *Telefunken* from 1940–45, later as the leader of a group working on control systems for flak rockets. He was contracted on 29 August 1947.[38]

The imbalance between the numbers of guided weapon and rocketry specialists who were recruited by the Americans through Projects Overcast and Paperclip, and by the British through the DCOS scheme, is represented in the numbers recruited from the Peenemünde-East organisation and its associated institutions – between 140 and 150 were recruited by the American armed forces, while only six were recruited by the MoS, consisting of one chemist, one propulsion specialist, and four guidance specialists. The propulsion specialist was Walter Riedel, the former head of the design bureau at Peenemünde-East. Like Kretschmer, Riedel was recruited to work in the Supersonics Division of the Aerodynamics Department but was soon transferred to the RPD. Among the rocket engine specialists recruited by the MoS, he had the most experience in the use of LOX for rocket propulsion; it was likely that he worked with the LOP/GAP and RTV 1 projects in some capacity.[39]

The four guidance specialists had all worked in, or with, the C-2 steering and remote control section (EW 224) at Peenemünde-East. They were clearly recruited on the basis of their expertise and experience with the C-2, although two had also been involved in the development of the A-4/V-2 guidance system beforehand. Three of them were physicists. Dr. Oswald Lange, aged 34, had been an employee at Peenemünde-East since 1940 and the *Abteilungsleiter* of EW 224 up until April 1944. Like Schirrmacher, Lange was in negotiations with the DEFA before signing a contract with the MoS. He subsequently became the leader of the German team in the RAE GW Department. Dr. Martin Eichler, who was 35, was employed at the TH Darmstadt from 1941–43 as an outside contractor to Peenemünde-East on the A-4/V-2 project. During the next two years he worked on the design of the *Einlenkrechner* line-of-sight computer for the C-2. Dr. Wilhelm Elfers, aged 46, served in the *Wehrmacht* from 1940 prior to being posted to Peenemünde-East in 1943 during the great influx of scientists and engineers to the base that year. He was also involved in the design of the *Einlenkrechner*. Following the capitulation, Elfers lectured at the Physics Institute of the University of Hamburg before being recruited. The fourth man was *Dipl.-Ing.* Siegfried Entres, who was also posted to Peenemünde-East in 1943. All four were employed in the RAE GW Department.[40]

The other German guided missile and rocketry specialists who were recruited by the MoS brought experience and expertise from flak rocket projects at various other concerns. A specialist of note was *Dr.-Ing.* Hans-Joachim Conrad, a 35-year-old engineer who was employed at the TH of Berlin from 1937–45 and oversaw the development of LPREs for the *Enzian* and *Rheintochter* R-IIIf at the *Vierjahresplan-Institut für Kraftfahrzeug* (VfK, Four-Year-Plan Institute for Motor Vehicles), also in Berlin. Initially allocated to the US for exploitation, Conrad was made available to the MoS in late 1948. Conrad's expertise with nitric acid LPREs would have been valuable at Westcott, where there was a shortage of rocket engineers with such a specialisation. Two more specialists were recruited from the LFA: *Dr.-Ing.* Theodore Schmidt, who was employed in the RAE GW Department, and engineer and scientific assistant *Dr.-Ing.* Hugo Reichert, who was employed at the RAE/RPD. Also recruited for work at the RAE were three former employees of *Siemens*. They were 45-year-old Hans Hasse, employed by *Siemens* firms from 1929–45, where he eventually became a laboratory chief; *Dipl.-Ing.* Heinrich Katz, who worked for *Siemens* in Berlin from 1935–45 as a laboratory engineer developing radio-control systems; and Hans Rohr, a 31-year-old who was also employed by *Siemens* in Berlin as a laboratory engineer, developing electrical control systems. Five former employees of *Luftfahrtgerätewerk Hakenfelde GmbH* (LGW) of Berlin-Spandau (a subsidiary of *Siemens*), the firm that supplied servomechanisms and gyroscopes for the C-2 project, were also recruited, all for employment in the RAE Instrumentation Department at

Farnborough to undertake research on aircraft instruments. The leader of the group, *Dipl.-Ing.* Gerald Klein, who was head of the department for gyroscopic devices and automatic control of aircraft and guided weapons at the company, can definitely be linked to German SAM development.[41]

The story of the recruitment of *Dipl.-Ing.* Hermann Zumpe, a rocket engineer, is perhaps the most intriguing. From 1936–39, Zumpe was employed by BMW at Spandau in Berlin on research into petrol aircraft engines, and after BMW expanded into the rocket engine field in 1939, he worked on the design and development of LPREs. When Zumpe was recruited by the British, it appears that he neglected to inform his new employers about his actual position at BMW. According to MoS documents, Zumpe stated that he was employed at BMW only until 1941, and thereafter ran his own business with funding from the RLM. But according to a paper which was presented at a NATO AGARD conference in Munich in 1956 by *Dipl.-Ing.* Helmut von Zborowski, the former head of rocket engine development at BMW, by the end of the war Zumpe was a flight construction engineer and von Zborowski's first assistant. If Zumpe did withhold the truth about his work history, the most likely reason was because the close association with von Zborowski may have revealed a link with the SS that would have been detrimental for his future employment prospects in the UK. Fear of being kidnapped by the Soviets may have been another motive.[42]

According to Zumpe, prior to being recruited by the British, he had escaped from Soviet custody on at least two occasions. On 5 September 1945, he was arrested by the NKGB (the Soviet intelligence service) through what he later claimed was a trick. He was imprisoned for three weeks and interrogated several times; at the end of September, the Soviet authorities ordered him to carry out theoretical calculations for a rocket engine that could generate 60 tons of thrust for 70 seconds. In November 1945, the calculations were completed and he was permitted to return to his home in the British sector of Berlin, with instructions to report to the NKGB every 14 days. In February 1946, Zumpe approached the authorities in the British sector seeking permission to move to the British zone (unapproved travel between Berlin and the western zones of Germany was forbidden to Germans). He provided the British authorities with two portfolios that contained details of his work on rocket propulsion systems for fighter aircraft, which were passed on to the Air Ministry but were not considered to be of outstanding interest. In May 1946, the Soviet authorities again contacted Zumpe and ordered him to report to Oberschönewelde in the Soviet sector of Berlin to complete a set of incomplete C-2 drawings. On 15 August that year, the Soviet authorities employed him at the *Institut Berlin* (see Appendix 3), where he was tasked with making improvements to the C-2 with an engine that incorporated his own principles of design. On 22 October, the date of Operation *Osoaviakhim*, the mass deportation of German specialists to the Soviet Union, Zumpe again sought asylum from the British for himself and his family, this time at the FIAT (Forward) headquarters in the British sector. Zumpe was subsequently interrogated by officers from the Intelligence Department, Air Division, CCG(BE). On this occasion the British interrogators showed more interest in Zumpe's ideas, perhaps because he disclosed information about his work for the Soviets. Zumpe was thereafter evacuated by air and lodged in the Dustbin detention centre in the American zone. Zumpe spent some time at Dusseldorf before eventually being offered a contract to work at the RAE/RPD. His escape from the Soviet sector of Berlin with current information on Soviet guided missile designs led the MoS to conclude that he may have been a 'marked man', interpreted as probably in danger of being kidnapped.[43]

By April 1949, the recruitment of German specialists through the DCOS scheme was completed. In total, 41 specialists with varying degrees of research and/or development expertise and experience related to guided weapons and rocketry (for aircraft and guided missiles) were brought to the UK through the DCOS scheme. The majority were classified as engineers in the broadest sense, 28 in total. Sixteen had degrees, and therefore in the German system were entitled to practice as an engineer – seven held a doctorate (*Doktor-Ingenieur*) and nine held a diploma (*Diplom-Ingenieur*), while three had attended university but were without a formal qualification. The remainder had passed an 'engineer exam' that qualified them to work in industry as designers, draughtsmen, or test engineers. Eleven specialists were scientists, eight of whom were physicists and three were chemists; seven held doctorates, three held a diploma, and one had attended university but also did not have a degree. There were also two specialists with trade qualifications, one a fitter and the other a welder and foreman. The specialists' families gradually arrived, beginning with the first three families on 14 October 1946; by December 1948, all the families had arrived in the UK.[44]

The MoS and the Admiralty did not recruit any specialists who had developed SAMs at *Rheinmetall-Borsig* and *Messerschmitt*. Some of the Germans from these firms went to work in non-Allied countries. For example, at least three of the HFW group who wrote monographs about the firm's guided missile projects for the MoS in 1945 and 1946 – J. Henrici, G. Mandel and C. Diederich – went to Argentina to develop a missile based on the Hs 293.[45]

German experience, naturalisations, and postwar British SAM development

For the German specialists, the process from recruitment to employment, and eventually to naturalisation, was not a smooth one. In December 1947, the Treasury approved the recruitment of Germans for continued employment in the UK, and in early February 1948 the RAE Establishment Division 4 (the security division) considered a Treasury proposal to offer contracts to selected German specialists of up to five years, which it was felt should be submitted to the JIC and MI5 for a security assessment. The subject was discussed at a meeting of the JIC on 1 April that year. The JIC saw no objection on security grounds to extending the contracts offered to German specialists from one to five years and decided to seek the views of the Defence Research Policy Committee (DRPC), which had taken over the direction of the DCOS scheme, on general policy questions in regard to employing aliens before submitting the recommendations to the chiefs of staff. The MoS was also invited to review the work of the German specialists every three months (which was already standard procedure). On 20 April, the DRPC considered the draft report prepared by the JIC and determined that the employment of selected alien specialists on a long-term basis to supplement British specialists should go ahead in the national interest. The JIC's decision was not conditional on that of the DRPC, so the determination by the JIC cleared the MoS to offer the new contracts.[46]

On 2 June 1948, the Treasury and the JIC approved the offer of contracts to the Germans for up to five years. To be attached to the contracts was a condition that for continued employment in the UK, the Germans would have to become naturalised British subjects. However, a legal problem with the new policy was soon found, because under the current legislation concerning the employment of aliens, it was impossible to offer five-year contracts. Defence Regulation 60D, which permitted the employment of aliens by the government – which superseded the Aliens Restriction (Amendment) Act of 1919 that forbid the government from employing aliens, and continued under the Supplies and Services (Transitional Powers) Act of 1945 – was due to expire in December 1950. There were four possible solutions to this problem:

1. To continue to employ the Germans on short-term contracts until December 1950, then discharge them;
2. to discharge all the German specialists once their current contracts expired;
3. to obtain a decision from the PM to extend the Defence Regulation beyond December 1950;
4. to introduce amending legislation to permit the employment of aliens on a long-term basis.

Subsequently, at the 11th meeting of the Lord President's Committee on 13 May 1949, it was decided to extend beyond December 1950 the emergency legislation enabling government departments to employ German specialists.[47]

In 1949, the actual process to retain German specialists for naturalisation commenced. The MoS establishments had to nominate which specialists they wished to retain by 31 January 1950. Those specialists who were not sought for retention, or who had made the decision to resign for reasons not related to the naturalisation process, did not have their contracts renewed when they expired in 1950. Those who were sought for retention were offered appointments in the British civil service and the opportunity to become naturalised British subjects. If they did not wish to remain in the UK, their contracts were also not renewed. The MoS helped to arrange new employment in NATO countries for those specialists who were not retained beyond 1950.

The knowledge and experience that the Germans brought to the UK was supplemented with captured technical documentation and hardware, for exploitation by the MoS for the benefit of the British armed forces and also for science. The first liquid propellant missiles that were ever fired in Australia, at the Long Range Weapons Establishment in South Australia, were some of the 150 *Taifun* barrage rockets acquired by the MoS from Germany. These projectiles were renamed '4-inch LPAA' and were filled with WAF 1 fuel, a mixture of aniline and furfuryl alcohol, instead of *Visol*, with nitric acid as the oxidizing agent (3-inch rockets were the first solid propellant missiles launched). The first series of firings commenced in late August 1949 and had several objectives. They were to familiarise and train the Australian personnel in the handling of liquid propellants and use of range instrumentation, to test the British logistics operation and to obtain ballistic data, all in preparation for the test-launchings of the RTV 1 in Australia, which commenced on 11 December

A prototype *Enzian* with booster rockets attached, sent to Australia by the British Ministry of Supply in 1947. (Source: NAA: D874, NG592)

A Thunderbird SAM on its launcher at a firing range at Woomera in South Australia. (Source: NAA: D897, K61/95)

1949. Tracking of the missile trajectories was recorded with another piece of captured German hardware, *Askania* cinetheodolites.[48]

A number of other British rocket and guided weapon projects benefited from German knowledge. These included the RTV 2 and the General Purpose Test Vehicle (GPV), the Black Knight high-altitude research and development rocket and the planned Blue Streak intermediate-range ballistic missile. The three first-generation British SAMs – the Bristol/Ferranti 'Red Duster', which entered service with the RAF as the 'Bloodhound', the English Electric 'Red Shoes', which entered service with the British Army as the 'Thunderbird', and the Armstrong Whitworth 'Sea Slug' for the Royal Navy – also benefitted, in the fields of rocket propulsion, guidance and control, and computer simulators.

In the realm of rocket propulsion, the German influence is recorded in the words of one of their British superiors. At the RAE/RPD, the German expertise with LPRE design – particularly engines that used hydrogen peroxide – was important, as a February 1949 report by A. Baxter of the RAE/RPD summarised:

> Our knowledge of liquid propellant rockets has extended rapidly since 1945. It has been largely built on the foundation of German work prior to the end of the war and supported by American results ... Developments in this country have been influenced by the German scientific teams brought here. Considerable work was done in Germany on nitric acid systems, but these were not so far forward as the HTP [High Test Peroxide, 85–98 per cent concentrated solution of hydrogen peroxide] units. Furthermore, fewer good scientists with nitric acid experience were available for carrying out work in this country. This is reflected in the German staff at Westcott, where only two out of a total of fifteen were engaged primarily for their experience of nitric acid and nine for their work on hydrogen peroxide. In consequence, the progress on HTP systems has been made greater and, therefore, the design of motors operating on HTP can be undertaken with a fair degree of confidence.[49]

The influence of German knowledge in liquid propellant rocket propulsion on British SAM development, before SPREs and ramjets became the preferred methods of propulsion in the 1950s, can be seen in the Sea Slug and Red Shoes. The Sea Slug project originated in the LOP/GAP soon after the war, but the German influence can be seen in the tentative propulsion system, a nitric acid/petrol LPRE, rather than a LPRE using LOX. *H. Walter KG* (HWK) designed engines using nitric acid and petrol for the *Enzian* and the Hs 117. The tentative propulsion system in the Red Shoes similarly bore the hallmarks of German influence, also from HWK. Prior to the development of the Red Shoes, which began in April 1950, English Electric signed a contract with the MoS to undertake a general study and theoretical research of a guided SAM system in collaboration with the RAE. The tentative propulsion system in the sustainer rocket was an LPRE using hydrogen peroxide (HTP), now considered suitable as an oxidising agent in engines for SAMs, and kerosene, ignited with 'C-stoff' (15 per cent hydrazine hydrate, 57 per cent methyl alcohol and 13 per cent water), a propellant mixture that was developed in Germany during the war, which found applications in the hydrogen peroxide LPREs that HWK designed for the Me 163 interceptor.[50]

Individual contributions to the Red Shoes project were made by Willi Kretschmer, formerly of HWK, who designed combustion chambers for hydrogen peroxide LPREs, including for the Beta and KP series of engines, in collaboration with fellow Germans including Friedrich Jessen and Werner Schonheit. When the Germans were being assessed for their suitability for naturalisation in 1952, the Director of the RAE, Arnold Hall, made reference to the importance of Kretschmer's work in regard to the Red Shoes, in a letter to the TPA at the MoS on 22 September of that year:

[Kretschmer] has been responsible, in collaboration with some other staff, for the development of the RTV 2 motor, which is to be flight tested shortly. Recently he has developed an alternative motor based on the 'thermal decomposition' principle [gaseous injection system] for GPV. This motor appears to be so successful that it has been adopted and copied by the firm developing the Red Shoes weapon missile in place of the design based previously on the RTV 2 motor. Clearly, Kretschmer has a record of accomplishment and his loss at this stage would be most serious both for the test vehicles RTV 2, GPV and the weapon Red Shoes. The motors are in their development phase where close collaboration between Kretschmer, GW Department Trials Division and the firm is essential.[51]

Kretschmer was naturalised with his wife and two children on 18 March 1953. In November 1955, he resigned to take a position in Canada as a combustion test engineer supervisor with Orenda-Nobel Engines, a division of Avro (the firm that designed and built the Lancaster and Vulcan bombers for the RAF).[52]

Contributions were also made to the propulsion system in the Red Shoes by Gustav Fiedler, also from HWK, who for two years contributed to the successful development of an expulsion system for the HTP/kerosene LPRE before the decision was made in 1954 to use an SPRE. Much of Fiedler's other work was related to nitric acid LPREs. He designed numerous valves for rocket engines, including a solenoid-operated main fuel valve; an oxidant/fuel mixture control valve that was submitted for patents action; a stop valve incorporating stainless steel bellows; and a motorised remotely operated throttle valve with a piston indicator. In 1952, Fiedler was put in charge of a new hydraulic and valve development laboratory which provided a service for hydraulically calibrating components and servicing valves; by that time he was also developing flow meters for corrosive liquids for general use. Fiedler was assessed in 1952 as having 'many original ideas and the standard of his work is high'. The RAE Promotion Board considered his ability to have been seriously underestimated compared with other German staff. Fiedler became naturalised on 26 March 1953.[53]

Of their colleagues, Jurgen Diederichsen conducted research connected with rocket engine ignition and fuel combustion. He became naturalised on 26 March 1953. Ulrich Barske and Walter Müller specialised in the design of turbopumps that were used to force propellants into the combustion chamber. Barske carried out fundamental research with the technology; the RAE considered him 'their most outstanding man' and he was regarded as an authority in his field in both government and industry. Müller became naturalised on 16 April 1953 and Barske on 19 December of that year.[54]

Walter Riedel was employed on calculations and associated design work of improved types of LPREs that used LOX, mainly for aircraft use. He was assessed as not outstanding or irreplaceable, but his loss would have slowed down important work. He became naturalised on 18 November 1953. Hermann Treutler was involved in the assessment of general projects, such as the augmentation of rocket thrust in guided rockets. Heinz Walder conducted research on various aspects of LPREs, including ignition, injection and heat transfer. He contributed to the development of the Gamma LPRE in the Black Knight and the Black Arrow satellite launch vehicle. Hermann Zumpe also worked on calculations and associated design of improved types of rocket motors. He became naturalised on 25 August 1953. Although neither an engineer nor a scientist, Walter Koltermann worked with the team of German specialists at the RAE/RPD as a foreman and technician, doing welding and associated work on test installations. Werner Schonheit, Friedrich Jessen, Karl Meier and Johannes Frauenberger were all engaged in experimental development work with LPREs. Schonheit worked on the design of combustion chambers for hydrogen peroxide propulsion systems; Jessen studied propellant injection and gas generation matters; Frauenberger and Meier did design work on combustion chambers. Schonheit became naturalised on 18 April 1953, Meier on 19 March 1953, followed by Jessen and Koltermann four days later. By early 1954, Frauenberger had changed his focus to SPREs, but was the only former HWK employee at the RPD who did not become naturalised, along with about half a dozen other Germans at Westcott. He resigned on 31 December 1955 to take up a position in the Federal Republic of Germany with the firm *Passavant-Werke* at Michelbach in Nassau.[55]

In the GW Department at Farnborough, the Germans were recognised as important to the most crucial research being undertaken at the establishment. At a MoS meeting held on 4 March 1949 to consider the long-term employment of German specialists, Dr. W. Cawood from the RAE explained the factors behind why the RAE sought to retain the German specialists there for the main reason of security:

> The German scientists at RAE, because of their specialised knowledge and experience, and the time they had been employed, have now become an essential part of the scientific complement, and their loss would be detrimental to the work being carried out. This was particularly so in the case of the Guided Weapons Department where they made major contributions. It had been found to be impracticable to keep from the Germans knowledge of other secret work going on in the establishment, and for this reason it was essential to do everything possible to stop them from going back to Germany, even on leave.[56]

The Germans in the GW Department worked with British engineers and scientists to design and develop the UK's first guided weapons simulator, the TRIDAC (Tri-Dimensional Analog Computer). TRIDAC was based on an American system developed at the Massachusetts Institute of Technology. The MoS contracted the firm of Elliott Brothers in London to design and build the computer in conjunction with the RAE. Work on the computer started around 1950, and in 1954, when TRIDAC became operational, it was the largest computing machine in the UK. TRIDAC was designed to simulate a number of functions, such as testing sub-systems and aerodynamics. In 2011, Lavington remarked that simulations using TRIDAC contributed generally to British guided missile technology and saved a great deal of money by reducing the need for live test firings. A simplified version of TRIDAC, the AGWAC (Australian Guided Weapons Analog Computer), was installed at Salisbury in South Australia as part of the Anglo-Australian Joint Project.[57]

However, most of the German specialists in the GW Department whose services were retained by the MoS beyond 1950 chose not to become naturalised and left the UK. There were numerous reasons for their departure, such as family commitments and better career opportunities and salaries in the US. Hans Hasse and Günter Pieper were the first to resign, between 1 April and 30 September 1953. Hasse worked on a number of projects – mechanical and electrical engineering problems associated with guided weapon control systems, designing and building electromechanical simulators used as test beds for control equipment, and by 1953 was working on the control and stabilisation of guided test vehicles. Pieper, meanwhile, worked on the development of electronic computing units and TRIDAC. Oswald Lange designed electronic circuits and also worked on TRIDAC. He resigned on 7 January 1954 to join his former colleagues from Peenemünde at Redstone Arsenal in Alabama, and later worked for NASA. Hans Rohr designed and developed electromagnetic devices for guided weapon control systems, in addition to components for an electromechanical computer. After resigning in January 1954, Rohr returned to Germany and moved to the American sector of Berlin. Heinrich Katz was engaged in electrical engineering development in connection with the testing and tuning of guided weapon control systems, and later worked on gyroscopes. He resigned on 28 February 1954. Wilhelm Elfers also worked in a number of areas, some involving TRIDAC. These included work on electronic computing units to investigate control problems; the construction of simulators based on techniques previously used in Germany; theoretical studies relating to simulators; and a continued interest in command-link guidance systems for guided weapons. Elfers applied for naturalisation, but since 1951 had expressed disappointment about the remuneration attached to naturalisation in correspondence to the Director of the RAE, Arnold Hall. He eventually decided to accept a contract to work for the American aircraft manufacturer Glenn L. Martin and Co. of Baltimore and resigned on 31 July 1954.[58]

Five members of the 'Linke Team' from *Telefunken* eventually left the UK. Josef Linke was transferred to the Darwin Panel scheme for the civil sector around June 1948 but returned to Germany soon after. Hans Prost worked on the development of UHF radio transmitters of wide frequency coverage for air and ground applications. He resigned on 15 June 1951 and returned to Germany, and afterwards went to work in Italy. Karl Wilhelm worked on the development of radio frequency control and 2,000-channel HF communications equipment. He left by August 1952. Paul Rothe carried out research on aircraft aerials but returned to Germany to look after his family. He resigned on 31 March 1953 and took up a job with *Telefunken* at Heidelberg. Fritz Rockstuhl worked on VHF/UHF receivers between 200 and 400 MHz; he resigned on 31 March 1954.[59]

Only four of the Germans in the GW Department became naturalised British subjects. Three were also involved with the design and construction of guided weapon simulators: Karl-Heinz Schirrmacher, Hugo Ulrich and Siegfried Entres. By 1950,

Schirrmacher was working on the accuracy of beam-rider guidance systems and assisting in the design of TRIDAC. He was assessed by the MoS as an outstanding man on high-priority work. He became naturalised on 21 October 1953. In 1956 he presented a paper at the NATO AGARD seminar in Munich about the wartime German developments with radar and guidance technology for the German SAM development programme, titled 'Guidance of Surface-to-Air Missiles by means of Radar'. Ulrich did not bring any experience from the German SAM development programme but did work on the detailed mechanical design of guided weapons simulators at the RAE. His superiors at the RAE considered him to be an outstanding and talented specialist. He became naturalised on 28 October 1953. Entres was involved with the general supervision of design and construction of electromechanical simulators with Ulrich, and later was responsible for developing high-accuracy computing methods for computers. He became naturalised on 20 April 1954. Later he transferred to the newly formed RAE Space Department in 1962 and became a leading figure in the British space community. The fourth to be naturalised was Karl-Eduard Büchs, who worked on the development of radio altimeters. He was the only member of the Linke Team to become naturalised, on 27 April 1953.[60]

Three of the five specialists employed at the ERDE became naturalised. Botho Demant departed in 1948 and afterwards worked briefly for the Admiralty, and fitter Gerhard Müller had left by 1952. Norbert Luft, a theoretical chemist who specialised with combustion, worked on general theoretical calculations on thermo-chemistry, kinetic problems connected with decomposers, catalyst stones for HTP systems and on the production of hydrazine. He resigned from the ERDE on 9 May 1952, became naturalised and took up a position with the firm of Simon-Carvers in Stockport. Hans Ziebland was a research engineer concerned with heat transfer and worked in three areas: heat transfer and its application to high pressure and high temperature systems; determination of thermal conductivity and viscosity of rocket propellants for heat transfer calculations; and general design consultant work for the group at the ERDE who worked with liquid propellants. Ziebland became naturalised on 19 March 1953. He remained at the ERDE and published extensively in government reports and industry literature on the subject of heat transfer. Franz Neunzig supervised the firing of LPREs and SPREs on proof stands, and was in charge of various engineering facilities, such as the supply of high-pressure gases, the maintenance machine shop, and the design of equipment. Neunzig also did some work for the Admiralty – by 1950 he was working on the conversion of a torpedo motor to operate on HTP. He also designed a mesh plate type of engine to work on HTP and conducted investigations on the working of hot and cold thrust units. His last role at the ERDE concerned development work on the application of concentrated hydrogen peroxide. He resigned on 27 February 1953, became naturalised a month later and afterwards emigrated to Australia to work for British Commonwealth Pacific Airlines.[61]

Finally, Heinrich Voepel, an aerodynamics specialist who had worked on the Hs 117 at HFW during the war, continued to be employed in the RAE Aerodynamics Department. MoS documents record that in November 1950 and September 1952 he was working on theoretical analysis of longitudinal stability of proposed ground-launched rocket models, possibly in relation to delta wing designs.[62] He was one of the last Germans to be naturalised, during the second half of 1954.[63]

Conclusion

This chapter has demonstrated where and how the German scientific and technical intelligence concerning SAMs – in samples of captured German technology and archives – and the knowledge in the minds of German specialists who were recruited after the war, was transferred to the British Commonwealth. How the knowledge was subsequently exploited for the benefit of the British armed forces differed from the American experience. German ideas were less prevalent in experimental British postwar guided weapon and rocket designs than those developed in the US. Nevertheless, a number of German specialists who were recruited for employment at MoS research and experimental establishments were specifically sought because of their expertise and experience with SAM technology.

When assessing the Germans' work in the UK, one must remember that their number and achievements were modest compared to those who went to the US. However, the contributions of some of the Germans were important to the development of tactical guided weapons technology in the UK during the late 1940s and the 1950s. Rocket propulsion, thermodynamics, computers and guided weapon simulators were all domains where German scientific and engineering knowledge was important. The main cultural difference between the American and British importation programmes was that British government departments did not recruit any Germans who were suspected of having committed, or were implicated in, war crimes or crimes against humanity. The six Germans from Peenemünde-East who were recruited by the MoS have not been subject to the same level of scrutiny as the group led by Wernher von Braun who were recruited by the US Army.

The British SAMs which appeared in the late 1950s and early 1960s functioned according to many of the principles that the Germans established in World War II. Like their American contemporaries, British scientists and engineers took over 10 years to successfully perfect the design of a functional seeker for a SAM. In all other aspects of SAM design, such as propulsion systems (with the exception of ramjet technology), electronic guidance and control technology, and missile construction, the British technology had considerably evolved from the experimental German and British missiles from World War II.

CHAPTER SIX

The Transfer and Exploitation of German SAM Technology by France After 1945

The history of the French activities to acquire, transfer and exploit scientific and technical intelligence on World War II German SAM development after 1945 requires a separate chapter, because the narrative of the Western Allies' liberation of France and the occupation of Germany was characterised by a dualism. Although the French played a role, albeit limited, in the liberation of their country and the defeat of Nazi Germany, there was no comprehensive intelligence-sharing agreement with the US and the UK. French scientific and technical missions were, therefore, initially at a considerable disadvantage when searching for German military secrets. The almost complete absence of guided missile and rocketry development in France during the Occupation also placed France at a disadvantage, in contrast to the early progress that was made in the US and the UK between 1943 and 1945.

There was some research into anti-aircraft rocket weapons in France prior to 1945, although not to the degree that was undertaken in the US and the UK. During the 1930s, the French Army (*Armée de Terre*) engineer and pioneering rocketry figure, Captain Jean Jacque Barré, designed a 16kg anti-aircraft rocket projectile that was fuelled with a propellant combination of nitrogen peroxide and possibly a benzene derivative. During 1937 and 1938 at Versailles, Barré successfully tested a LPRE for the SAM which produced 700kg of thrust, but the project was terminated prior to France's defeat in June 1940. During the Occupation, German electronics companies that supplied equipment for guided missiles, including *Telefunken*, *Opta* and *Siemens*, took over French technology firms and oversaw the production of equipment for the *Wehrmacht* that included electronic components for radar, radio and television systems. These production orders were for components not considered of great sensitivity, for the *Wehrmacht* generally excluded French industry from highly classified component contracts, including electronic components for guided missiles. When the occasional attempts to pass on research problems to French industry were made, they were met with extreme reluctance. The only significant instance was by *Des Compteurs* at Montrouge in Paris, which carried out some exemplary radio research for *Telefunken*. Until the Allies fully liberated France, the only knowledge that the French armed forces possessed about liquid propellant rocket and guided missile technology outside of Barré's work concerned the Fi-103/V-1, A-4/V-2, Hs 293 and the *Ruhrstahl* X-4 AAM that was acquired by the French resistance and intelligence networks. The first French liquid propellant rocket to be test-launched was designed by Barré, the EA-41, which took place on 15 March 1945 at the *Établissement Technique de la Renardière* near Toulon.[1]

French authorities eventually made reciprocal agreements with the Anglo-Americans that enabled French investigators to inspect intelligence targets in the American and British Zones of Occupation, but the outcome was that France acquired only a small fraction of the scientific and technical intelligence from the German SAM development programme. The relative lack of intelligence from artefacts and documents was offset by the recruitment of guided missile specialists, of whom French government organisations recruited a number comparable to the Americans, and more than twice the number by the British. The development of the first experimental SAM system for the *Armée de Terre* was assisted by German specialists who had helped to develop the A-4/V-2 and C-2. The French government agencies which were responsible for research, development and procurement for the French Air Force (*Armée de l'Air*) and French Navy (*Marine Nationale*), and French industrial contractors, also exploited scientific and technical intelligence on German advances with SAM technology, in France termed *fusées de DCA* or *projectiles de DCA* (*Défense Contre Avion*) (anti-aircraft rockets or projectiles), or given the more general terms used to describe guided missiles, *engins spéciaux* (special devices) or *engins autopropulsés* (self-propelled devices).

The *Première Armée Française* in southern Germany and Austria

The formations of the *Première Armée Française* (French First Army), under the command of *Général* Jean de Lattre de Tassigny, landed on French soil in August 1944, two months after American, British and Canadian forces landed at Normandy to free France and Europe from the Nazi overlord. The French forces comprised the 2nd Armoured Division commanded by *Général* Jacques Leclerc, which landed at Normandy on 5 August, and seven divisions that participated in Operation *Dragoon*, which landed on the shores of Provence on 15 August. Each branch of the French armed forces sent scientific and technical investigators to examine and document German war materiel during the march through France and Germany. The *Armée de Terre* general staff sent a *Section T*, and the *Bureau Scientifique de l'Armée* sent the *Mission d'Information Technique*. The *Marine Nationale* later sent the *Mission Technique Permanente* to Germany as its scientific and technical mission. The *Armée de l'Air* set up the *Service Central d'Information Technique* (SCIT) in September 1944 to search for German aeronautical research and hardware behind the French front lines and in Germany. Indeed, one of the first German guided missiles captured on

French territory was a HFW Hs 293, seized by Captain Michel Decker of the SCIT on 15 September 1944. To review and discuss the discoveries of German advanced weapons, the *Ministère de la Guerre* set up an inter-service committee called the *Commission d'Études des Armes Secrètes Allemandes* (Commission for the Study of German Secret Weapons).[2]

While the Western Allies were liberating France, the *Gouvernement Provisoire de la République Française*, headed by *Général* Charles de Gaulle, sought co-operation with the Anglo-Americans in the search for German military secrets on French territory. In early October 1944, the French military headquarters made a verbal request to SHAEF offering French military officers to assist in the examination of captured war materiel in the operational areas of the Anglo-American armies. On 10 October, SHAEF agreed to the request, although on a limited basis. The British 21st Army Group, commanded by Field Marshal Bernard Montgomery, accepted four French officers; the US 6th and 12th Army Groups declined the offer; while the USSTAF and the British Air Ministry each accepted six officers. There were already a number of French liaison officers attached to the three army groups in Western Europe. The functions performed by the various categories of liaison officers and responsibility for discipline and control were stipulated in a SHAEF signal dated 25 October 1944. The liaison officers who participated in the search for German military secrets on French territory belonged in two categories: liaison intelligence and counter-intelligence personnel, who co-operated with their British and American counterparts, and military liaison personnel, who liaised tactically with the Anglo-American commanders in the field.[3]

By the second half of February 1945, the *Première Armée Française*, in conjunction with the US Seventh Army, had reconquered Alsace-Lorraine and driven the German forces back across the Rhine. Later that month, *Général* Alphonse Juin, the *Chef d'État-Major Général de la Défense Nationale* (Chief of the General Staff for National Defence), wrote a letter to General Eisenhower requesting the inclusion of French investigators with teams under the aegis of CIOS, who were poised to follow in the wake of the Allied armies in the planned invasion of Germany. Up until that time, closer intelligence co-operation between France and the Anglo-Americans had been complicated for two main reasons. Firstly, CIOS investigators sought access to the industrial and commercial secrets of French firms which had collaborated with the Germans. Understandably, the provisional French government did not approve of these inquiries, and disagreed with the British and American practice of treating French companies like the enemy and taking their intellectual property, even though some French firms had chosen to freely collaborate with the Germans. In the letter, Juin asked that CIOS teams be ordered not to seek confidential information from French firms. The other main reason was that the Anglo-American military leadership was reluctant to allow equal French participation in the investigation of German intelligence targets for the purpose of exploitation. Eisenhower acceded to Juin's request, although again on a limited basis. As an example of this co-operation, on 5 March 1945, French and British intelligence services made requests to SHAEF for one officer from each country to be attached to T-Force, 12th Army Group, for the purpose of exploiting targets around Cologne in anticipation of that city's imminent capture. SHAEF approved the request the following day. Whether the French officer was privy to the seizure of documents at the *Wilhelm Schmidding* facilities in the city on or around the abovementioned date, as was discussed in Chapter 3, is not known to the author. The same two officers were also permitted to be attached to T-Force, 6th Army Group, from after 13 March to exploit targets in the Karlsruhe and Stuttgart areas, also in anticipation of the planned capture of the two cities, this time by the US Seventh Army. However, these plans were negated on 28 March when the *Première Armée Française* was assigned the prerogative of capturing both cities in addition to Pforzheim. Karlsruhe was captured on 4 April, followed by Stuttgart on 22 April.[4]

Although France was officially outside the CIOS apparatus, British and American government departments nevertheless shared important intelligence on advanced German armaments with the French government. For example, the French *Ministère de l'Air* was on the ADI (K) (Assistant Directorate of Intelligence, Prisoner Interrogation, British Air Ministry) distribution list. For example, an ADI (K) report on the interrogation of a German POW who had worked in an experimental section of *Rheinmetall Borsig* at Sommerda and Breslau on air-launched rocket development by officers from ADI (K) and US Air Intelligence, dated 20 April 1945, was passed on to the *Ministère de l'Air*, as was a report on the interrogation in England of *Dr.-Ing.* Max Kramer, the designer of the X series of weapons, dated 4 May 1945. It appears, though, that the arrangement was not continued after the war in Europe had concluded. The *Ministère de l'Air* was not on the distribution list of a report on the interrogations of Herbert Wagner and four engineers from his staff in England by personnel from ADI (K) and US Air Intelligence, dated 18 May 1945.[5]

For the most part, the *Première Armée Française* did not operate in areas of Germany where targets that were linked to the German SAM development programme were situated. Apparently, the French military command and intelligence agencies were just as unaware about the programme prior to the invasion of Germany as their British and American allies were. It was likely that regular French troops or military intelligence (*Deuxième Bureau*, Second Bureau) officers first encountered evidence of the programme during the advance through southern Germany towards the Swiss border in late April and early May 1945. In these regions were two companies that were connected to the German SAM development programme. One was *Luftschiffbau Zeppelin GmbH* (Zeppelin Airship Company) at Friedrichshafen on Lake Constance. The company was contracted by the HWA to manufacture parts and carry out aerodynamics research in its wind tunnels for the A-4/V-2 project. Another responsibility was aerodynamics research in the subsonic region for the C-2 project, which was carried out under the management of *Dr.-Ing.* Max Schirmer. After Friedrichshafen was captured by the French 5th Armoured Division on 29 April, evidence indicates that French investigators did not capture the records of the firm's wind tunnel experiments. A dossier of reports pertaining to the C-2 project is recorded as having been recovered by American investigators.[6]

The other company was *Holzbau-Kissing KG* at Sonthofen, the firm that was contracted by *Messerschmitt* to develop the *Enzian*. The French 1st Armoured Division captured Sonthofen on the evening of 30 April. In this case, French investigators immediately missed out on capturing war booty because most, if not all, the hardware and documentation that pertained to the *Enzian* project were removed or destroyed by German personnel prior to the occupation of the town by French troops. A consolation was that both the general manager and technical manager of the firm, *Herr* Kissing and *Herr* Jacobsen respectively, were found and available for interrogation. French investigators may have also acquired information on German flak rockets from a number of the approximately 450 guided missile specialists who had evacuated from the Harz Mountains to Upper Bavaria in early April 1945. Other possible sources were personnel from the AIA, who also evacuated to Upper Bavaria, prior to when Aachen became a battle zone. By the time of the final German surrender on 8 May, French troops had reached as far south as the Tyrol in western Austria, almost to the Italian border.

Just prior to the German capitulation, the French military command again requested closer co-operation and participation with CIOS. What this meant was the exchange of intelligence obtained from investigations. On 8 May, the request was granted by the Anglo-American Combined Chiefs of Staff, subject to two conditions – that the present British and American security policies regarding technical developments and future operations remain in place and any changes in the operation or functioning of the French agencies involved be avoided. In addition, a verbal agreement was made between the 12th Army Group headquarters and the liaison detachment of the French foreign intelligence service, the *Direction Générale des Études et Recherches* (DGER), whereby French counter-intelligence liaison officers attached to the 6th and 12th Army Groups would be recalled to Paris once the control of the *Première Armée Française* passed from American to French command. Subsequently, there was closer association from mid-June 1945 when General Eisenhower agreed to issue the French authorities with the CIOS geographical 'Black List', a location guide to German intelligence targets of the highest priority. However, Eisenhower ordered that an accompanying CIOS technical list not be issued to the French command. A reciprocal agreement was made whereby French investigative teams were permitted to enter the British and American Zones of Occupation to investigate intelligence targets, but were not permitted to remove any enemy documents, hardware or personnel without express permission from the occupation authorities. The same rule applied to American and British investigators who visited targets inside the *Zone Française d'Occupation* (ZFO) in Germany.[7]

The boundaries of the ZFO in Germany were officially announced on 23 June 1945. During late June and July, the control of French territory was transferred from the authority of the *Première Armée Française* to a new military administration. On 7 July, the control of German and Austrian affairs in the ZFO and in the French sector of Berlin was taken over by an inter-ministerial committee based in Paris. The *Première Armée Française* was dissolved on 15 July, two days after the dissolution of SHAEF, and on 31 July, the administration of the *Gouvernement Militaire de la Zone Française d'Occupation* (GMZFO) officially assumed control. By 25 August, the final policy on reciprocal access between the French and British Zones of Occupation was finalised. A similar policy between FIAT (US) and France was already in place by this time. According to the FIAT (US) policy on co-operation with French scientific and technical missions, French investigators were allowed access by FIAT (US) to FIAT non-secret records; reserved documents (those classified Restricted, Confidential and Secret) were to be withheld. Despite these limitations, relations between the French and FIAT (US and British) were described by the head of FIAT (British), Brigadier R. J. Maunsell, as being at this time 'extremely cordial'. A French branch of FIAT was also created, run by Colonel Gaston de Verbigier de Saint-Paul and under the authority of Emile Laffon, the general administrator of the military government. Interrogations of German personnel through these reciprocal agreements produced credible information. For example, the interrogation of one Dr. Braun from the '*Institut de Recherches Aéronautiques*' (probably the LFA near Völkenrode) in June 1945 resulted in the receipt of 'very succinct' information about the *Rheintochter* and the *Feuerlilie* research rocket programme.[8]

The DEFA acquisition of intelligence on the C-2 and *Taifun* from the *WVA Kochelsee*

The French armed forces were very eager to acquire knowledge from German scientists and engineers on the development of the A-4/V-2 at Peenemünde and other guided missiles. For this task, the *Direction des Études et Fabrications d'Armement* (DEFA, Directorate for the Study and Fabrication of Armaments), a new French government department that would eventually be responsible for research, development, manufacturing and procurement of armaments and equipment for the *Armée de Terre*, set up the *Groupe Opérationnel des Projectiles Autopropulsés* (GOPA), led by Professor Dr. Henri Moureu, a chemist and scientific adviser to the French Army, in May 1945. The primary objective of the GOPA was to facilitate, through relations with the relevant Allied agencies, the transfer of knowledge on the A-4/V-2 from occupied Germany to France. In the course of its activities, the GOPA also obtained information about the C-2, the *Taifun* and other German flak rockets. A GOPA mission to the *WVA Kochelsee* in September 1945 was one such case. Two GOPA investigators, *Commandant* (Major) Dr. Demontvignier and *Capitaine* Carrière from the DEFA, were granted access to the facilities by American authorities. By the time of the visit, in addition to intelligence on the *Rheintochter* and *Feuerlilie*, French military intelligence had received information about the *Enzian*, the Hs 117 and also very likely on the C-2. When the two French investigators reached the site, personnel from the US Naval Technical Mission in Europe had already dismantled the wind tunnels for shipment to the Naval Ordnance Laboratory in Maryland. Both men were permitted to access the establishment's archives, which is evidenced in a report produced by them on 20 November 1945 about aerodynamic research with the *Taifun*. In this report, the dimensions of the five different liquid propellant versions of the rocket (as set out in the previous chapter) were studied and recorded, along with the wind tunnel data of the production version at speeds of Mach 0.85, 1.88 and 2.87.[9]

Another purpose of the French mission to the WVA was to recruit German specialists. De Gaulle had officially authorised the recruitment and exploitation of selected German specialists for the benefit of France after receiving a note from the *État-Major Général de la Défense Nationale*, dated 16 May 1945, apprising him of the British and American removal of German specialists to the respective countries and similar French efforts. At this time, there was uncertainty among the German staff at the WVA over whether the US War Department would authorise contracts for the specialists to work in the US. The two French investigators, possibly with the assistance of intelligence officers from the *Deuxième Bureau* of the *Armée de Terre*, or the DGER, exploited this uncertainty by offering the Germans contracts with good conditions, salaries and arrangements for the transport of families. One example was the recruitment of Hans Gessner, who was the head of the technical development department at the WVA. A mechanical engineer by training, Gessner had been with the aerodynamics institute at Peenemünde-East since 1937 and helped construct its first 40cm x 40cm supersonic wind tunnel. The French representatives offered Gessner a salary double what he was paid in Germany and the promise he could take 20 colleagues and their families to France. In another case, Dr. Richard Lenhert, a physicist who was recruited by the US Navy, claimed that a wind tunnel employee was ready to leave for the ZFO on a truck with his entire family and household goods but was stopped at the last minute by an American guard. There were more surreptitious attempts by the French to transfer German specialists. An American report describes the discovery of a French liaison officer attempting to remove 11 personnel and their families in the middle of the night.[10]

Since the Americans could not employ all of the scientists and engineers from the WVA in the US, the situation presented the DEFA personnel with a unique opportunity. The most senior of the WVA staff who was recruited by the DEFA was the administrative director of the facility, Dr. Herbert Graf. In the process of

Graf's recruitment, numerous documents from the facility's archives were removed and ended up in the hands of the DEFA. Some light has been shed on how the DEFA acquired the documents by Dr. Peter Wegener, a physicist who was recruited by the US Navy:

> I am not sure of the exact sequence of events, but the administrative director of the WVA, Herbert Graf, suddenly disappeared with some employees in the direction of France. It was said that certain papers went along with them. Nobody I knew joined him, and none of the scientific staff of the WVA followed him.[11]

Amongst the documents that Graf and his associates removed were 23 original reports from the series of 176 in *Archiv 66* that concerned research that was conducted in the wind tunnels at both Peenemünde-East and the WVA. Five of the 23 reports concerned aerodynamic and thermodynamic research that was undertaken for the C-2 project.[12]

Graf must have been aware that the documents had been microfilmed, and therefore knew he would not be separating the specialists who were recruited by the US Navy and AAF from their wartime research, or be guilty of removing captured assets from the Americans. The removal of the documents was in contravention of paragraph 1b of an order from SHAEF dated 15 June 1945, that forbid the French from removing any personnel, documentation and equipment from the American Zone of Occupation without express authority.[13] In addition, the reports would have fallen into the three categories of classified documents which FIAT had decreed were not to be disclosed to French investigators. The strict orders to exclude French investigators from such classified information were not adhered to at the WVA for apparently amiable reasons. According to a December 1945 report by Moureu to the director of the DEFA, the two French investigators viewed the entire collection of reports at the WVA thanks to the *complaisance* of the Americans.[14] Although there is no evidence to suggest that the two investigators from the DEFA or personnel from the *Deuxième Bureau* of the *Armée de Terre* or the DGER orchestrated the removal of the documents, the case nevertheless shows how determined French agencies were to acquire German knowledge and specialist expertise in the field of guided missiles.

Other sources of intelligence on German SAM development were contacts in the British Air Ministry. Following the mission to the WVA, Moureu, Demontvignier and the engineer-in-chief at the *Laboratoire Municipal de la Ville de Paris*, Paul Chovin, visited London from 19–30 September 1945 to obtain further technical information on German guided missiles. During a meeting between the three Frenchmen and Squadron Leader André Kenny, an Air Ministry liaison officer attached to the MoS, there was an exchange of documents. Perhaps unaware that the British and Americans shared all of the scientific and technical intelligence that was gathered in Germany, the French team handed over reports on the supersonic wind tunnel at the WVA and a LOX plant at Lehesten in Thuringia (which had been investigated before it was turned over to the Soviet Union, as it was located within the boundaries of the Soviet Zone of Occupation). Kenny handed over 10 ADI (K) intelligence reports which contained information about German guided missiles that was obtained from the interrogation of POWs. At least four of the reports concerned flak rockets: one about HFW missiles (the report that was not passed on to the *Ministère de l'Air* in May), one on the development of the C-2, one about the development of the *Enzian* and one concerning the *Rheintochter* and the *Taifun*. The information in these reports was accompanied with a caveat which warned against accepting any of the statements as facts prior to verification. Whether the French were assured of their veracity or not, the reports were nevertheless later translated into French by the *Centre d'Études des Projectiles Autopropulsés* (CEPA, Centre for the Study of Self-Propelled Projectiles), a department within the DEFA that was set up by Moureu initially to reconstitute captured documentation on guided missiles, and later concentrated research into the technology. During the visit, Kenny also invited Moureu to assist in Operation *Backfire*, and the group met with the DGP at the MoS, Sir Alwyn Crow, over dinner.[15]

Meanwhile, the initial contract negotiations between the DEFA and the group of German specialists who went with Graf to the ZFO took place at Emmendingen, about 20km east of the Rhine and 15km north of Freiburg, on 9 November 1945. After this meeting, the first employment contracts, for six months, were signed on 15 November. The DEFA recruited a total of 29 German specialists, mostly from the WVA, to reconstitute the WVA archives and technology at the *Bureau d'Études d'Emmendingen* (BEE, Emmendingen Design Bureau) in the ZFO. Like the American and British recruitment practices, French organisations did not take a strict approach when dealing with specialists who had been former members of Nazi political organisations. Overall, the wind tunnel specialists who were recruited by the DEFA were not of the same calibre as those recruited by the US Navy and the AAF. It is probably fair to say that most brought varying levels of scientific or technical expertise from the experiments that were carried out for the flak rocket projects (one of the personnel recruited from the WVA was a secretary), but it is hard to say how much, except in the cases of three specialists who brought detailed knowledge of the C-2 and *Taifun* projects due to their roles and positions. Two were research staff – one was Dr. Werner Kraus, a physicist, who had been with the institute since 1941. Kraus was the leader of a testing group in the research laboratory, where he was the chief of all thermodynamic research in the supersonic region. Kraus was the co-author of three reports concerning the C-2. He signed a DEFA contract on 15 November 1945. The second was Elsbeth Hermann, a mathematician, who had been with the institute since 1939. By 1945, she was the leader of the test evaluation group in the research laboratory and had co-authored five reports pertaining to research for the C-2 project. Hermann signed a DEFA contract on 1 March 1946. The third specialist was Eckart Finger, a mechanical engineer who had been with the institute since 1943, where he was employed as a test engineer. He was engaged on aerodynamics research concerned with resistance, drag and moments, and co-authored the report on the development history of the *Taifun* with Drs. Rudolf Hermann and Hermann Kurzweg. Finger signed a DEFA contract in Paris on 15 December 1945.[16]

Thereafter, regular meetings took place between the Germans and the representatives from the DEFA. Initially, these meetings were held in the small town of Offenburg in the Black Forest, where the headquarters of the French technical missions in Germany, the *Direction des Missions Techniques* (DMT), was situated, then from February 1946 at the BEE. The fact that the DEFA and other French departments did not have full possession of the archives from where their recruits had worked was a great disadvantage. The situation was particularly acute in regard to German SAM research and development – the archives of Peenemünde-East, the LFA, DFS, HFW and *Rheinmetall-Borsig* were all in the hands of the Anglo-Americans.[17]

Friction with the Anglo-Americans and the DEFA postwar SAM programme

Probably as a result of the aggressive French recruitment efforts, the British government placed further restrictions and prohibitions upon French access to sites in the British Zone of Occupation in early 1946. The policy through which restrictions on foreign access were implemented was initially directed only at the Soviet Union. On 19 November 1945, a list of subjects relating to various advanced technologies and weapons research which Russian investigators were not permitted to view was drawn up by the JIC. It was titled 'Limitations on Allied Access to Information in the British Zone'. The subject 'Control of Guided Missiles and Homing Devices' was included with the locations where former and current research into SAMs was carried out. One location was the badly damaged *Rheinmetall-Borsig* establishment at Unterlüss (the Unterlüss Work Centre), where former members of the *Rheintochter* development group were compiling reports and monographs on their wartime work for the MoS. Whether France was also excluded from the locations associated with each subject and location in the 19 November 1945 document is not mentioned. At the time, French investigators were permitted to visit German industrial firms and research establishments within the British and American zones where advanced aircraft, rocket and guided missile research and development had been carried out. For example, French investigators visited the LFA on at least two occasions. These visits indicate that French investigators were not initially excluded from all sites of high sensitivity. However, a decision was made by the JIC in early February 1946 to exclude French investigators from certain locations.[18]

In the 9 February 1946 document, the categorisation was changed to 'the list of subjects to which Allies, excluding the United States, should not be allowed access'.

The obvious implication in the title was the addition of France to the Soviet Union. In the list, the same 'Control of Guided Missiles and Homing Devices' subject was included, in addition to MoS Experimental Station Trauen and MOSEC at Altenwalde. At each of these locations, German guided missile and rocketry specialists were compiling reports and monographs for the MoS. On 21 February 1946, the JIC made an amendment to the 9 February list, which now included the entire LFA. In a second amendment, dated 7 March 1946, the subject 'Supersonic Aircraft, Missile, and Engine' research by the MoS and MAP was added (although the MAP was in the process of being amalgamated with the MoS). The *Rheinmetall-Borsig* establishment at Unterlüss was re-added, after initially being removed on 9 February. This was because the MoS did not envisage the reporting and research work there to be completed until 30 June 1946. As will be described later in the chapter, the further restrictions that were placed upon French scientific and technical investigations by the JIC did not prevent the flow of information on German flak rockets from the British and American zones to France from drying up.[19]

Returning to the specialists who were recruited by the DEFA, a document declassified by the SHD archives at Châtellerault on 6 September 2016 lists the employment conditions that the DEFA agreed with the German specialists. They were to collaborate in the production of intelligence by the re-establishment of archives and writing memoirs on their previous work; reconstitute their means of research and studies at the end of the war; make progress with pathways of research and studies with the view of obtaining new results; contribute to the education of French technicians who would replace them at a later stage; and furnish a complement of effective and qualified staff, either temporary or permanent.[20]

The 23 wind tunnel reports from the WVA formed the basis of the documented development histories of the missiles that the aerodynamics institute at Peenemünde-East and the WVA had made significant scientific contributions to in the years from 1937–45. The DEFA would have also relied on the available technical knowledge that was captured or acquired from other research establishments and contractor firms in Germany, plus the memories and notes of the specialists. At first, the DEFA worked with the German specialists from the WVA to acquire and compile more data on the aerodynamic research with the flak rockets within the context of guided missile aerodynamics more generally. For example, during meetings at the BEE from 16–18 April 1946, *Capitaine* Joneaux of the DEFA made a request to Gessner, on behalf of *Capitaine* Carrière, to produce a report on the development of fin-stabilised projectiles at Peenemünde-East. At a meeting on 6 May 1946, Gessner presented the DEFA with a '*rapport sur l'état actuel de ces projectiles quand à leur utilisation dans la DCA et des batteries à longue portée*' (Report on the current state of these projectiles when they are utilised in anti-aircraft defence and long-range batteries). Besides research for the *Aggregat* series of rockets, the C-2, and the *Taifun*, the aerodynamics institute carried out research for a number of other projects. These included a 6.7cm fin-stabilised rocket for *Wasag*; a 'Peenemünde finned projectile'; a proximity fuse for fin-stabilised projectiles; and the instability of fin-stabilised projectiles more generally. The experience that was gained in developing these weapons at Peenemünde-East culminated in the *Taifun F*. The information that was supplied by Gessner supplemented the existing data that the GOPA had accumulated on supersonic aerodynamic research done at Peenemünde-East and the WVA. At the same meeting, Carrière requested that the report on the aerodynamic development of the C-2 be reconstructed.[21]

During the same series of meetings, Carrière agreed in principle for Graf to travel to the British zone to recruit German guided missile specialists who were employed at MOSEC. In the ZFO, Graf had recently encountered Dr. Helmut Weiss, a physicist who had worked on the design and application of homing devices (primarily infra-red) for the C-2 project in EW 224 at Peenemünde-East. From November 1944, Weiss was also the chairman of a subcommittee on counter-measures against enemy homing devices (a rather pointless enterprise at that late stage of the war) under the *Sonderkommission für Elektrische Munitionausrüstung* (the so-called 'Gladenbeck Committee', briefly mentioned in Chapter 1) in the Armaments Ministry. As Huwart recounted in 2004,[22] Weiss subsequently played an important role in recruiting specialists from Peenemünde-East for the DEFA. Weiss was one of a group of specialists who signed six-month DEFA contracts at Paris on 6 May 1946, which included *Dr.-Ing.* Rolf Jauernick and *Dipl.-Ing.* Helmut Habermann. In addition to the group at Emmendingen (*Groupe BEE*), another group of German guided missile specialists, whom the DEFA referred to as *Groupe EA* (*Engins-autopropulsés*), were installed at two locations north of Freiburg in the ZFO. At Riegel, there was the *Groupe Engins-autopropulsés-propulsion* (EA/P), headed by Jauernick, and at Denzlingen was the *Groupe Engins-autopropulsés-guidage* (EA/G), headed by *Dr.-Ing.* Otto Müller, a former deputy section chief in EW 222 who had been responsible for the development of steering and remote control technology for the A-4/V-2. These groups helped to form the foundation of a new DEFA facility that would become one of France's leading military research establishments, the *Laboratoire de Recherches Balistiques et Aérodynamiques* (LRBA, Ballistics and Aerodynamics Research Laboratory) at Vernon, west of Paris, which was officially founded on 17 May 1946. One of the first tasks of *Groupe EA* was to reconstitute captured technical reports on the A-4/V-2 and the C-2. The primary objective of the DEFA was to improve each missile design, although the *Armée de l'Air* had the prerogative to develop SAMs in tactical applications.[23]

The DEFA recruited several more specialists from the British zone who had been involved with the C-2 project at Peenemünde-East. One was Heinz Bringer, an engineer who was employed in the planning section of the EMW development department where propulsion systems were designed. In the first 12 months after the war, Bringer was employed by the British, initially by the War Office in Operation

Backfire, and from early 1946 at MoS Experimental Station Trauen. One example of Bringer's work concerned the design of the propellant tanks of the C-2. Since it was envisaged that the missile would be stored fully fuelled for periods up to a year, a suitable propellant tank design had to be found to prevent the nitric acid from corroding through the internal surfaces of the tank, which was of a monocoque constructional design. Bringer designed a system whereby the *Visol* and nitric acid tanks were made of 4 per cent chrome steel. He commenced work with the French on 1 August 1946 and was subsequently employed at the LRBA, where he was instrumental in the development of the propulsion system in the *Veronique* sounding rocket and later the successful Viking engine.[24]

Another specialist with experience from the C-2 project was *Dipl.-Ing.* Wolfgang Pilz, a 34-year-old Austrian. Pilz had worked under *Dr.-Ing.* Martin Schilling (who went to the US) in the testing department at Peenemünde-East, where he helped to design and develop the propulsion system for the C-2. After the war, Pilz was employed at MOSEC before commencing employment with the DEFA on 15 May 1946 but did not sign the first six-month contract until 3 October 1946. He subsequently became the manager of a sub-group in EA/P, which was one of two (the other sub-group worked on launching facilities), and later was also involved in the development of the *Veronique*. Pilz was employed by the LRBA until 17 March 1959, then went back to the Federal Republic of Germany briefly before going to work in Egypt to develop guided missiles for the Nasser regime. Whether the French authorities would have allowed such a move from France to Egypt is not known. Another specialist was 31-year-old engineer *Dr.-Ing.* Robert Schabert, an ethnic German born in Moscow. He was educated at the TH Stuttgart and TH Breslau, where he earned a diploma in engineering and a doctorate in engineering chemistry. During the war, Schabert was first employed at the University of Posen, before being transferred to the *Flakversuchsstelle der Luftwaffe*, where he undertook propellant research for the C-2 project. In early 1944, he was engaged on research with dinitrogen tetroxide, or *U-stoff* as it was referred to at Peenemünde-East, as a possible oxidising agent with hypergolic fuels. Schabert's whereabouts from February 1945, when Peenemünde-East was being evacuated, until the middle of 1946 are unclear. His LRBA personnel dossier states that after the evacuation, he was employed by a *Flakversuchsstelle Arbeitsgruppe* in Berlin, but the next recorded date of his activities is September 1946, when he was working for the DEFA in *Groupe EA/P* at Riegel. On 15 January 1947, Schabert signed a six-month DEFA contract, and on 14 May of that year commenced work at the LRBA Vernon. He remained at the LRBA until retiring in September 1951, and also returned to Germany.[25]

Meanwhile, during 1946 the DEFA continued to accumulate more intelligence on the German flak rockets. In September 1946, some documents that concerned aerodynamic research for the *Rheintochter* project were acquired through the efforts of Dr. Alexander Naumann, formerly of the AIA, who had recently been recruited by the DEFA to join *Groupe BEE*. Naumann made contact with a Dr. Linke, who had contributed to the development of the *Rheintochter* at *Rheinmetall-Borsig*. On 9 September at Sonthofen in the American zone, Naumann met with Linke, who handed over some photocopies of documents, which were probably from his former employer. Linke also provided Naumann with the names of three of the leading men from the projectile development group at *Rheinmetall-Borsig* under *Dr.-Ing.* Werner Fricke (recruited by the AAF), who had similarly been involved with the *Rheintochter* project. None of the three were as yet recruited by the Allies. They were *Dipl.-Ing.* Ernest Mohr of Lübeck, a specialist in fin-stabilised projectiles, who worked on the aerodynamic and constructional design of the *Rheintochter* at Zwickau in Saxony; Dr. Gottfried Ormanns, a mathematician who contributed to the aerodynamic design of the *Rheintochter*; and Dr. Eckel of Aachen, who specialised on a project referred to as the 'Z-Projectile'. The DEFA used the same technique to acquire information as had been done at the WVA – once the German specialists were recruited, they were probably asked to make contact with any former colleagues or associates who had worked on guided missile research and development during the war. By obtaining the documents through German contacts rather than French investigators, the DEFA circumvented the agreement concerning the removal of classified documents from the British and American zones.[26]

The DEFA *Service Technique*, the section responsible for the recruitment of German specialists, essentially concluded its recruitment programme by the end of May 1947. Around 450 German (and Austrian) specialists had been recruited up to that point, of whom about 200 were scientists and engineers.[27] The specialists were to be distributed in six principal groups in France as follows: *Groupe S* (70 specialists), primarily from the Ballistics Institute of the Technical Academy of the *Luftwaffe* at Berlin-Gatow, at the LRBA Saint-Louis annex (shaped charges, detonation phenomena, X-rays and ballistics); *Groupe EA* (73 specialists), mainly from Peenemünde-East, at the LRBA Vernon and Puteaux in Paris (guided missiles); *Groupe BEE* (28 specialists), primarily from the *WVA Kochelsee*, at the LRBA Vernon (wind tunnels); *Groupe M* (73 specialists), primarily from *Maybach Motorenbau GmbH* at Friedrichshafen, also at the LRBA Vernon (tank engines); *Groupe N* (150 specialists), predominately from *Mauser-Werke AG* at Oberndorf am Neckar, at Mulhouse in Alsace-Lorraine (small- and medium-calibre weapons); and *Groupe B* (48 specialists), also from *Mauser-Werke* at Oberndorf am Neckar, at Levallois in Paris (adding machines).

The specialists in *Groupe EA* were recruited to carry out work on liquid propellant guided missiles (*Engins autopropulsés à liquides*). The specialists were not recruited to undertake any research on solid propellant guided missiles (*Engins autopropulsés à poudres*). Fifty-seven of them were divided between sections EA/P and EA/G, with the remaining 16 employed at the *Atelier de Construction de Puteaux* (APX, Puteaux Construction Workshop) in Paris, where prototype missiles were later built. The specialists in *Groupe BEE* remained stationed at Emmendingen, where their knowledge was used to assist the DEFA in the construction of a supersonic wind tunnel at the LRBA Vernon. The range of projects that these two groups of Germans were employed on included the construction of a workshop for tactical missiles; a static test facility for LPREs that were designed for strategic missiles; a guidance laboratory; a rocket propulsion laboratory for the study of chemistry and propellant combinations; a firing range in French North Africa; and in the case of the wind tunnel specialists, a workshop and design bureau. Several of the Germans were seconded to CEPA, which was an auxiliary department of the *Service Technique*. The *Service Technique* also carried out research and development of liquid propellant guided missiles for the DTIA (*Armée de l'Air*) and DCCAN (*Marine Nationale*) of a similar nature to work being done by the DEFA (the DCCAN posted engineers to the DEFA who were attached to the APX). By this time, the DEFA had begun studies of five guided missiles with the assistance of the German specialists in *Groupe EA* and *Groupe S:* an anti-tank missile, a long-range SSM, two ballistic missiles (both strategic) and a long-range SAM.[28]

Table 12. German specialists employed at the LRBA Vernon, 1 April 1949

Service des Engins Autopropulsés/Guidage (EA/G)	
Chef de service:	Dr.-Ing. Otto Müller
Secrétaire:	Hildegard Gleixner
Stabilité:	*Dr.-Ing.* Reinhard Schubert, *Dipl.-Ing.* Helmut Habermann, Friedrich Kauba, *Ing.* Herbert Frey
Servo-Moteurs:	*Dipl.-Ing.* Ferdinand Bodenstein, Siegfried Runge
Gyroscopes:	Karl Graeter, Eduard Heyne
Télécommandes (remote control):	Dr. Horst Lammerhirt, *Dipl.-Ing.* Eugen Just, *Ing.* Felix Boese
Guidage (guidance):	*Dr.-Ing.* Reinhold Strobel, Dr. Hermann Stümke, *Dipl.-Ing.* Alfred Bachmann (plus two others)
Autopoursuite (homing devices):	Dr. Helmut Weiss (*Chef de Groupe*), *Dr.-Ing.* Ernst Deuker, *Dr.-Ing.* Heinz Kleinwächter, *Dr.-Ing.* Wulfo Schmidt, *Dr.-Ing.* Herbert Wojtech, *Ing.* Ernst Haass, Alfred Laebe
Hyperfréquence (microwaves):	–
Service des Engins Autopropulsés/Propulsion (EA/P)	
Chef de Groupe (of German specialists):	*Dr.-Ing.* Rolf Jauernick
Bureau des Études (design bureau):	*Dipl.-Ing.* Wolfgang Pilz
Calcul Trajectoire (trajectory calculation):	*Dipl.-Ing.* Karl Kiefer, Hermann Buhl
Statique (statics):	*Dr.-Ing.* Friedrich Bornscheuer, *Ing.* Willi Walther, *Ing.* Paul Klar, Heinz Keiner
Matières Premières (raw materials):	*Dipl.-Ing.* Johannes Fabian
Chimie (chemistry):	Dr. Ernst Büchner, *Dr.-Ing.* Robert Schabert

(Continued)

Table 12. (Continued)

Groupe Réalisation des Engins	
Section Accessoires (accessories section):	*Ing.* Karl Fölster, *Ing.* Otto Kraehe, Josef Schlotzer
Section Éjecteurs et Turbo-pompe:	*Dipl.-Ing.* Wolfram Goertz, *Ing.* Heinz Bringer,
(turbopump and ejectors section)	Wilhelm Dollhopf, *Ing.* Kurt Menke, *Ing.* Willibald Prasthofer, Franz Seidel, Willi Sohn
Interchargeabilité et réglage:	*Dipl.-Ing.* Walter Lörtsch
Service Essais (EA/E)	
2IC to *Chef de Service*:	*Dipl.-Ing.* Wolfgang Zangl
Point Fixe (fixed point):	*Ing.* Willi Huttenberger, *Ing.* Herbert Lang, Karl Nettersheim, *Ing.* Walter Schuran, *Ing.* Alexander Voigt
Instruments de Mesure:	*Dipl.-Ing.* Rudolf Hackh, Karl Benhke
(measuring instruments)	
Stand d'essai des Accessoires:	Alfred Liesegang, Heinz-Harold Scheidt
(test stand accessories)	
Atelier (workshop):	Georg Gardian (*Chef*), Kurt König, Kurt Billig, Albert Gross
Service Souffleries (S)	
Technicien Allemand (German technician):	André Roda

Source: 'Note sur l'organisation du LRBA Vernon (Personnel)', 23 March 1949 – SHD Châtellerault AA 789 1H1 34

By 1949, the DEFA had either cancelled or scaled down development work on ballistic missiles, and thus from 1949–52 most of the activities of the LRBA and the APX – and the work of the German specialists at both locations – were devoted to the long-range SAM project, which was dubbed the *Projectile Autopropulsé Radioguidé Contre Avion* (PARCA). The staff under Jauernick in EA/P contributed to the design and development of the PARCA propulsion system, which formed the basis of the LPRE in the *Veronique* sounding rocket, while the staff in EA/G under Müller conducted research and development of a command-link guidance system. The PARCA was 4.5 metres in length, with a diameter of 45cm, and weighed 1 ton at launch. The missile was boosted by four solid propellant rockets wrapped around the fuselage, and the sustainer rocket had a nitric acid/gas-oil LPRE, which by 1953 enabled the missile to reach supersonic velocities and travel to an altitude of 32,800ft (10,000 metres). Four fins were fitted towards the rear of the fuselage, and four delta-shaped control surfaces were situated near the nose of the missile in a cruciform configuration that resembled the designs of the *Rheintochter* and the Nike. Fins were also attached to the booster rockets for stability at launching, which resembled British innovations with the designs of the English Electric Thunderbird and Bristol/Ferranti Bloodhound SAMs. The test missiles were built at the APX, and the command-link guidance system was designed in co-operation with *Compagnie*

Française Thomson-Houston (CFTH). Test-launches on ramps attached to captured German 88mm gun platforms commenced in 1950 at the *Camp de Suippes*, and later in the 1950s at the *Centre Interarmées d'Essais des Engins Spéciaux* (CIEES, Interservice Guided Missile Testing Centre) firing range at Colomb-Béchar in Algeria. Consistent with concurrent developments in the US and the UK, the results from research with solid propellants led the designers to replace the LPRE with a SPRE in 1955. The DEFA ultimately did not succeed in developing the PARCA for operational deployment in the air defence of French or NATO territory. The PARCA and several other SAM programmes were eventually terminated for budgetary reasons following a ministerial decree of 4 August 1958. The French government decided to purchase an American missile system, the Raytheon Hawk, manufactured under licence by firms in NATO countries including – besides France – Belgium, the Netherlands, Italy and the Federal Republic of Germany.[29]

Experimental guided missiles for the *Armée de l'Air*

Scientific and technical intelligence from the German SAM development programme was also exploited in the first experimental guided missile projects that were developed for the *Armée de l'Air*. The appearance of German guided missile technology in World War II prompted debates within the higher echelons of the *Armée de l'Air* over the future role of aircraft in the service's warfighting doctrine, as Claude d'Abzac-Epezy discussed in 1990.[30] The roles of fighter aircraft and guided missiles in the French air defence architecture was one such debate. Within this context, an extensive effort to investigate tactical guided missiles on behalf of the *Armée de l'Air* was undertaken during the late 1940s and 1950s. The development of armaments and equipment for the *Armée de l'Air* in French industry was co-ordinated by the *Direction Technique et Industrielle de l'Aéronautique* (DTIA, Aeronautical Technical and Industrial Directorate), a department within a succession of service ministries that was responsible for research, development and procurement for the *Armée de l'Air*. To develop guided missiles, the technical service of the DTIA, the *Service Technique de l'Aéronautique* (STAé), created a special section for the purpose, the *Section des Engins Spéciaux* (STAé/ES).

In many respects, French scientists and engineers continued the development of guided missiles from where the Germans left off in 1945, and this was largely done on the basis of German advances from the war. As Carpentier has stated, it was through the knowledge of German work and the transfer to France of German specialists that French studies of tactical missiles started in 1946.[31] Soon after the war, the STAé received some samples of captured German guided missile and ancillary technology – in a number of instances from the generosity of the British – which included a complete *Enzian* prototype, HWK 109-501 and 109-502 hydrogen peroxide ATO units, one Fi-103/V-1, one or two *Ruhrstahl* X-4 AAMs, components from the A-4/V-2, *Telefunken Würzburg* radars and *Askania* cinetheodolites. In the autumn of 1946, the STAé entrusted the hardware from Germany and some documents to a government research establishment controlled by the DTIA, *Arsenal de l'Aéronautique*, in Paris. One of the first efforts by French industry to exploit German LPRE technology was in a contract awarded by the STAé to the firm *Société Civile d'Étude de la Propulsion par Réaction* (SCEPR, founded in 1944, later renamed *Société d'Étude de la Propulsion par Réaction*, SEPR) on 20 June 1946 to improve the design, aerodynamic qualities and facilities of assembly of the 109-502 LPRE, fitted in the *Enzian*. French pilots had a particular interest in the engine in its original purpose as an ATO unit, but there appears to be no evidence that the STAé was interested in fitting the engine in a SAM.[32]

On 1 July 1946, the general staff of the *Armée de l'Air* established a tactical guided missile development programme and issued specifications for missiles in the surface-to-air, air-to-air, surface-to-surface and air-to-surface categories. In the SAM category, there were three requirements. The SA-10 programme called for the development of guided missiles for purely experimental purposes. The specifications were for a missile that could travel to an altitude of between 5,000 and 10,000 metres, using the beam-rider guidance technique with automatic control in the latter stages of flight. The two subsequent programmes were the SA-20 and SA-30 to develop operational missiles. The latter was planned to be an anti-missile system.[33]

One of the contracts that the STAé signed with an industrial firm to develop a guided missile for the SA-10 programme led to a missile design that was based on the aerodynamic shape of the C-2 and fitted with a LPRE that was the product of knowledge acquired from BMW. The contractor was *Société Générale de Mécanique, Aviation, Traction* (MATRA), a privately owned firm that was founded during the Occupation in 1941. The contract was planned to begin on 1 October 1946 but was not finalised until 4 February 1947. MATRA designated the missile R.04. The shape of the fuselage and the design of the fins in the mid-section were similar to the C-2, as was the vertical launching method. The propulsion system was designed by SEPR.[34]

SEPR had acquired several samples of BMW LPREs with accompanying parts and information. The transfer of BMW LPRE technology to France took place within the context of the recruitment of engineers from the firm by the DTIA, most notably the turbojet engine specialists in *Groupe O* led by *Dr.-Ing.* Hermann Oestrich, the former chief of the jet engine development department at BMW. All the propulsion systems that SEPR designed for the missiles in the SA-10 programme were LPREs that used a propellant combination of nitric acid and a fuel referred to as TX, which is a hypergolic mixture of triethylamine and xylidine that was created at BMW under the direction of chief chemist Dr. Hermann Hemesath. At BMW, the mixture was referred to as *Tonka* 250 or *R-stoff* and was used in the 109-548 LPRE designed for the *Ruhrstahl* X-4 by Hans Ziegler. According to a CIOS report,

Tonka 250 was also used in the more powerful 109-558 LPRE for the Hs 117, also designed by Ziegler. The information that French chemists obtained from German sources indicated that German chemists utilised a mixture ratio that corresponded to approximately five parts nitric acid to one part *Tonka* 250, the fuel consisting of 50 per cent triethylamine and 50 per cent xylidine. The design of the 109-548 was studied at SEPR in a contract that was awarded to the firm by the STAé *Section des Moteurs* in 1946, which also involved the fabrication of 25 of the engines. SEPR further developed the 109-548 under contract with the STAé, commencing on 4 August 1947, to manufacture 200 engines derived from the German design, which was designated the SEPR 16.[35]

The work at SEPR was assisted by three German rocket engineers from BMW. They were *Dipl.-Ing.* Helmut von Zborowski, who was employed as a research engineer at SEPR until 1950 (possibly, if so with irony for von Zborowski, at the firm's facilities at Villejuif, 'Jewtown', in Paris), and two assistants. The first was *Dr.-Ing.* Hans Schneider, formerly employed in the experimental and design sections of BMW at Munich-Allach and Bruckmuhl, and designer of the 109-718 auxiliary propulsion unit, which accompanied the 109-003R jet engine to provide additional acceleration for the Me-262. The other assistant was Elmar Mucha, also from the experimental section at Munich-Allach. In the two-stage propulsion system of the R.04, the boosters were SEPR 43s, each of which could produce 1,250kg of thrust for 14 seconds, and the sustainer rocket was the SEPR 44, which could produce 1,500kg of thrust for 40 seconds. These rockets enabled the missile to attain supersonic velocities of around Mach 1.5.[36]

At first glance, the shape of another missile that was developed for the SA-10 programme, the NC-3500 by the state-owned aircraft manufacturer *Société Nationale de Constructions Aéronautiques du Centre* (SNCAC), appears to be based on the Hs 117. But upon closer inspection, one can actually recognise the originality of the French aerodynamic design. The development of the NC-3500 officially began on 19 May 1947, when the *Président Directeur Général* of SNCAC, Marcel Bloch, signed a contract with the director of STAé, chief engineer Guy du Merle (the former chief of the SCIT mission to Germany), to make a preliminary resolution of the general problems and definition of a missile responding in principle to the SA 10 programme, and to undertake experimentation in-flight of various methods of guidance. The NC-3500 was developed by a team of 15 personnel led by Louis Besson at the firm's Billancourt factory in Paris. German influences can also be seen in the SEPR nitric acid/*Tonka* LPREs for the missile. The booster rockets (two) were the SEPR 4, each of which produced 1,250kg of thrust for 10 seconds, while the sustainer rocket was the SEPR 2 (specifically designed for SAMs) that produced 1,250kg of thrust for 36 seconds. The SEPR 2 used a simple gas pressure feed system to force the propellants into the combustion chamber, a characteristic of all World War II German LPRE designs. In a contract signed with the STAé on 14 December 1948,

SEPR was to supply 100 SEPR 4s and 20 SEPR 2s for the NC-3500 and another prototype missile, the NC-3501, in addition to 20 SEPR 12s for the MATRA M.04, an air-to-air missile. Although it appears the design of the BMW 109-558 LPRE for the Hs 117 was not exploited at SEPR, German knowledge from BMW nevertheless had a profound influence on the engineers at the firm who designed LPREs for the missiles in the SA-10 programme.[37]

On 23 September 1948, SNCAC was awarded a second contract by the STAé for Besson and his team to develop a missile for the *Armée de l'Air* guided missile programme, in this instance a surface-to-surface bombardment weapon for the SS-40 programme, which called for a SSM with a range of 30km to destroy tactical objectives of small dimensions, and to be armed with a 100kg warhead. Aspects of the design of the first of two prototypes, the NC-3510, powered, in a uniquely French innovation, by a ramjet, bore the hallmarks of German influence from the *Enzian*. The idea to base the NC-3510 on the *Enzian* may have originated from the ADI (K) report on the development of the *Enzian* that was acquired by the GOPA mission. The German POW who was interrogated by Anglo-American investigators referred to plans by *Messerschmitt* to build a simplified version of the *Enzian* for use as a SSM, but by the end of the war no detailed research on the proposal had been undertaken. If Besson did not read the ADI (K) report, he must have had access to the captured technical documents from the *Messerschmitt AG Oberbayerische Forschungsanstalt* at Oberammergau that were obtained by French scientific and technical investigators either towards the end of 1945 or early 1946.[38]

The recovery of the *Messerschmitt* archives from Upper Bavaria was one of the most important finds for the French scientific and technical missions in Germany. The archives contained the technical documentation on the development of the firm's jet aircraft projects and the *Enzian*. By early July 1945, the French had only recovered a quarter of the archives, which contained all the technical information on the Me 262 jet fighter but nothing on the *Enzian*. US Army G-2 intelligence knew the French had found the documents but could not gain access; Brigadier-General Eugene Harrison, a G-2 officer with Sixth Army Group, reported in early August 1945 that attempts to reach 40 cases of *Messerschmitt* documents hidden in the ZFO were frustrated. The other three-quarters were eventually found by American investigators, which resulted in an agreement between the two countries to share the archives. The negotiations also involved a huge unfinished transonic wind tunnel at Ötztal in the Austrian Tyrol region that was being built for the *Luftfahrtforschungsanstalt München*. The tunnel was situated in the ZFO in Austria about 40km south by south-east of Garmisch-Partenkirchen, but the two firms that made the sections of the tunnel were located in the American Zone of Occupation in Germany. American and French negotiators eventually agreed to exchange their captured *Messerschmitt* documents; the Americans acceded to having the remaining parts for the tunnel manufactured by the two firms in the American zone, and the AAF was allowed access to the tunnel when it was completed at the new *Centre de Modane-Avrieux*, operated by the *Office National d'Études et de Recherches Aéronautiques* (ONERA) from 1952.[39]

In 1949, the contracts for the NC-3500 and NC-3510 were taken over by another state-owned aircraft manufacturer, *Société Nationale de Constructions Aéronautiques du Sud-Est* (SNCASE) after SNCAC was forced into liquidation. Besson's group were transferred to the SNCASE plant at Cannes to continue the development of both missiles. The NC-3500 became the SE-4100 series, while the NC-3510 became the SE-4200 series. The first test-launch of the SE-4100 is presumed to have taken place on 29 September 1949. The missile was launched vertically and intended to reach an altitude of 32,800ft (10,000 metres). German technology was utilised in the programme – the missiles were guided using the beam-rider technique with a *Würzburg* radar that operated on a wavelength of 50cm, and flight measurements were recorded by *Askania* cinetheodolites. The guidance, control and recording equipment that was used in the SE-4100 programme – gyroscopes, servomechanisms, radio transmitters and receivers, telemetry instrumentation and various other electronic components – were supplied by the firm *Société Française d'Équipements pour la Navigation Aérienne* (SFENA), which from 1948 (one year after the firm was founded) employed seven German specialists formerly of *Askania* and *Siemens*. The designers of the SE-4200 retained the aerodynamic shape based on the *Enzian* for the prototypes throughout the development period.[40]

None of the experimental SAMs under development by MATRA and SNCASE for the DTIA led to the realisation of a first-generation SAM system by 1960. In the mid-1950s, the French government also decided to purchase an American SAM system for the *Armée de l'Air*, the Nike-Ajax. In 1959, also due to budgetary reasons, the government cancelled the contracts of all the SAM projects by SNCASE and MATRA. With the termination in the same year of the SSM project, SNCASE abandoned all tactical missile development.[41]

Considering how comprehensively the German SAM development programme was documented by the Anglo-Americans, it is surprising to find evidence of the possible use of one of the weapons against Allied aircraft in French documents. A German engineer searching for employment in France provided a story which, if true, would dramatically alter the orthodox history of German attainments during the war. He was 31-year-old *Dipl.-Ing.* Ewald Müller, an electrical engineer and former *Luftwaffe* anti-aircraft artillery officer. In a statement to the French occupation authorities in the state of Rhineland-Palatinate in 1949, Müller wrote that during the war he had conducted research into electro-magnetic proximity fuses for flak rockets whilst serving in the *Luftwaffe*. In March 1945, he was evacuated to Berlin from Latvia, where he had served with the 6th Flak Division in the battle for Kurland and was again evacuated in April 1945 to an anti-aircraft instruction unit in Munich. Müller stated that a successful test of a flak rocket against Anglo-American heavy bombers had taken place near the end of the war. This information he purportedly received in late April 1945 from *Ministerialrat* (assistant secretary) Dr. Lautner, the head of the experimental air armaments department in the RLM, at Schongau in Upper Bavaria, whom Müller claimed to have first met in May 1944 at the RLM in Berlin. According to Müller, during a conversation about flak rocket development and production, Lautner told him about an extraordinary experiment:

> In March 1945, in the Alps, a test of great scope took place when 14 four-engine bomber aircraft in an approaching formation were shot down [by guided flak rockets]. In April 1945, the first anti-aircraft batteries were equipped with these weapons, but since German industry was overwhelmed, it was impossible to realise production. Therefore, the scope of the test was not surpassed. We manufactured these rockets in three different dimensions, of which the smallest was used against individual aircraft, and the largest launched against formations. If I have a good memory, the largest rockets weighed 700 kilograms and measured 3.5 metres in length. The rockets were projected vertically, and having reached the height of the objective, were steered horizontally towards the target. Within the interior of a periphery of 300 metres, an electromagnetic conduit began to function, bringing the rocket to the point of magnetic attraction, where a primer was triggered subject to reaction of either a noise or an electric magnet.[42]

This account was at best highly embellished, and at worst entirely fabricated. The scientific and technical information provided by Müller was mostly inaccurate. Firstly, the only flak rockets with a similar length and mass were early prototypes of the Hs 117, but the planned operational version had a length of around 4.3 metres and a weight at launch of 430kg. Secondly, the homing technology did not exist. However, the developmental status and geographical location of the *Taifun* project during March and April 1945 were as described by Müller. A British intelligence report states that after the evacuation of Peenemünde-East to central Germany in February 1945, test-launchings of the *Taifun* were continued at Woffleben in the Harz Mountains, and by the end of March 1945, the rocket had reached the production stage. Also, soon before American troops occupied the area around Nordhausen and Bleicherode during the second week of April 1945, the project was again evacuated, this time to Gmunden, south-west of Linz in present-day Austria. The veracity of this information was substantiated by the discovery, in 1961 and 1962, of parts of *Taifun* rockets that were unearthed at a burial site underneath a shed at Pinsdorf near Gmunden. Whether any rockets were actually fired against a bomber formation from the Bavarian Alps is much harder to determine. By April 1945, neither SHAEF nor MI15 had received any reports about the successful use of such weapons against Allied aircraft, and there are no postwar reports about their operational use. The US Fifteenth Air Force did fly sorties of B-17s and B-24s from Italy over the Bavarian Alps to attack targets in Germany. But as Kevin Mahoney stated in *Fifteenth Air Force Against the Axis: Combat Missions over Europe during World War II*, the only major losses in March 1945 were on the 22nd and 24th, when 20 B-17s were shot down on both days by Me-262 jet fighters near Berlin. There was also a mission on 31 March in which six B-24s and a Lockheed P-38 Lightning were shot down as a result of flak that was encountered at Linz in Austria and two other locations.

It is possible these aircraft were downed with rockets of the *Taifun* type. Despite the circumstantial evidence, Müller probably invented the story to obtain more meaningful employment in France. The prohibition of military-related industries in Germany was a cogent factor that led engineers such as Müller to seek work in the French defence industry.[43]

The *Marine Nationale* makes a late start in exploiting German flak rocket technology

After the war, the pace of development of experimental weapons for the *Marine Nationale* was influenced by at least three primary factors. Firstly, there was the extent to which German knowledge could be exploited – it was contingent on the success of the French scientific and technical missions in Germany and Austria and diplomatic means through connections with British and American agencies. Secondly, the advent of guided missiles during the war did not make naval guns for anti-aircraft defence obsolete until well into the 1950s. French engineers therefore had time to study rockets and guided missiles in order to develop the weapons for future naval applications. Thirdly, during this period there were tough economic realities that the *Marine Nationale* had to accept. In 1945, the supreme decision-making body of the *Marine Nationale*, the *Conseil Supérieur de la Marine*, embarked on an ambitious postwar rebuilding programme, but by early 1946 the costs of reconstructing the French economy and the escalating expenses of the colonial war in Indochina forced the government to impose budgetary restrictions on the armed forces.[44]

The regulations for exploiting foreign intelligence in documentary form pertinent to the *Marine Nationale* were contained in a ministerial decree of 31 October 1945. The decree charged a division within the *Direction Centrale des Constructions et Armes Navales* (DCCAN) – the government directorate that had overall responsibility for research, development, production and procurement for the *Marine Nationale* – called the *Centre de Documentation, d'Information et d'Instruction des Constructions et Armes Navales* (CDIICAN), with the collection and methodical conservation of all documents, from both domestic and foreign sources, and the exploitation and diffusion of documents of interest to naval weapons and construction. Exploitation involved the selection of documents, translation, analysis and copying. Another responsibility of the CDIICAN was the instruction of management and technical personnel in naval weapons and construction. The CDIICAN also produced publications on naval weapons and construction, in the forms of weekly bibliographic bulletins (without analysis), monthly bulletins (with analysis) and periodical publications on specialised subjects.[45]

The CDIICAN comprised three sections. The first was the *Section de Documentation*, which had an autonomous function, that included a central library on subjects that pertained to naval weapons and construction, a printing workshop and microfilm reproduction. It centralised and co-ordinated at the level of the DCCAN the activities of *Services de Documentation des Constructions et Armes Navales* in the *Service Technique des Constructions et Armes Navales* (STCAN), the technical service of the DCCAN, in ports and at naval establishments. The second section was the *Section d'Exploitation et d'Instruction*, which was divided into eight groups according to broad subjects: *Architecture Navale, Machines et Chaudières* (machines and boilers), *Armes* (weapons), *Équipements, Transmissions et de la Détection Radioélectriques* (radar and radio technology), *Ateliers et Services à Terre* (shore workshops and services), *Aéronautique Navale* (*Aéronavale*), and also *Administration et d'Organisation Industrielle et Sociale*. Each of these groups was under the orders of a marine engineer (*Ingénieur du Génie Maritime*, IGA (M)), who was responsible for informing and instruction on all the materials that corresponded to the various subjects. He co-ordinated, from the point of view of the exploitation of documents, the activities of all the specialist engineers within his domain and the teaching of instructors from different fields. The third section was the *Section Diffusion et Publications*, which was responsible, in particular, for the production of the periodic and aperiodic bulletins of the DCCAN.[46]

The exploitation of German flak rocket technology for the benefit of the *Marine Nationale* took comparatively more time than for the *Armée de Terre* and the *Armée de l'Air*. The immediate interest of the *Marine Nationale* was in the documentation and exploitation of air-to-surface guided weapons of the Hs 293 and Fritz X types for the *Aéronavale*. To develop guided weapons, the DCCAN had the STCAN create an *Engins Spéciaux* section. It was headed by IGA (M) Jean-Jacques Peyrat, whose work included the transfer of a factory (which was accompanied by several German specialists) from Germany to the Brest arsenal in 1946 as war reparations. A precision-guided bomb based on the Fritz X was subsequently developed by the STCAN for the *Aéronavale*. The STCAN collected scientific and technical information on German research and development of guided weapons and related electronics technology at a number of locations in the ZFO. The most important was an operation centred at Hechingen, about 20km south of Tübingen on the edge of the Black Forest. Following a ministerial decree in mid-1945, a special unit was established at the town called the *Groupe d'Aéronautique Navale d'Engins Spéciaux* (GANES), commanded by *Lieutenant de Vaisseau* Patrice Desaix, the former commander of the *Sixième Flottille d'Exploration* (6FE) that was based at Agadir in Morocco. Desaix took command of GANES on 20 July 1945. The *Marine Nationale* already had a presence in Hechingen – officials from the service were the main advocates for exploiting the remnants of the Kaiser Wilhelm Institute for Physics near the town and removing the apparatus and equipment from the laboratories to France. Although the activities of GANES were concerned with the study and production of air-to-surface guided weapons, information on German flak rockets would almost certainly have also been gathered during the process. In November 1946, GANES was relocated to the naval aviation base at Saint-Raphaël on the Mediterranean coast just west of Cannes. The group's activities were dissolved on 1 March 1947, with its functions replaced by the *Section des Engins Spéciaux et Torpilles* in the *Commission d'Études Pratiques d'Aviation* (CEPA) at Saint-Raphaël. The following year, CEPA set up the *Groupe Technique des Engins Spéciaux* (GTES), led by IGA (M) Maurice Natta, to develop and test guided weapons for the *Marine Nationale*.[47]

The STCAN acquired scientific and technical information on ancillary technology for German guided weapons at a number of other establishments under its control. The application of television to guidance systems was studied at the *Post Deinstelle F* (Post Office F) of the *Forschungsanstalt der Deutschen Reichspost* at Aach. Radio technology for the *Kriegsmarine* was researched and developed here, in addition to radio-controls for the Hs 293 and seekers for flak rockets in a laboratory headed by Dr. Georg Weiss, who was a member of the *Arbeitsstab Dornberger* in 1945. At Constance, the development of control systems for guided weapons was carried out at the premises of *Askania-Werke*, the construction of aerofoils was done at *Schwartzwald Flugzeugbau*, and studies and manufacturing of German radar technology for naval shipping and the *Aéronavale* were undertaken at *Funkstrahl Gesellschaft für Nachrichtentechnik mbH*. During the war, *Funkstrahl* manufactured electronic and radio equipment for the *Wehrmacht*, including VHF (30–300 MHz) radio devices and radar components. The DCCAN operated the firm until 1947, then transferred a number of German specialists to the naval research laboratory at Marcoussis near Paris, where their expertise in electronics, guidance and servomechanisms was further exploited, along with several *Telefunken Würzburg* radars that had been acquired from the UK. A German specialist whom the STCAN recruited from *Funkstrahl* for employment at Marcoussis was *Dipl.-Ing.* Franz Sernatinger, who was subsequently employed for about 15 years at Marcoussis before being sent to work on SAMs at the *Fonderie Nationale de Ruelle*, a DCCAN facility near Angoulême in the department of Charente, about 100km north of Bordeaux. Information was also acquired through liaisons and contacts with the Americans and the British. For example, an Anglo-American intelligence report about an infra-red seeker code named '*Paplitz*', being designed for the C-2 by *Electroacoustic AG* of Kiel, was obtained by the *Marine Nationale Deuxième Bureau Liaison Recherches* through a *Mission d'Information Technique* liaison attached to FIAT.[48]

The development of the first experimental SAM for the *Marine Nationale* did not commence until 1949. In that year, the DCCAN contracted the *Fonderie Nationale de Ruelle* to undertake the task, with Jean-Jacques Peyrat as head of the project. As the engineers at MATRA and SNCAC had done for the *Armée de l'Air* SA-10 programme, Peyrat chose a German SAM design as a starting point. In 2006, Jean Robert *et al.* mentioned in their paper about the missile, dubbed the MARUCA (*Marine Ruelle Contre Avions*) in 1951, that the engineers at Ruelle completed drawings of a first prototype in October 1949, which was identical to the Hs 117 A2 type VI production version of January 1945. There were good reasons to utilise the Hs 117 design as a test vehicle for a naval SAM – it was the lightest of the five guided flak rockets, was fired from a zero-length launcher, was the simplest to manufacture and was at the most advanced stage of development in 1945. In 1949, DCCAN engineers were in a similar position to the US Navy in May 1945, when counter-intelligence officers with the NavTechMisEu in Germany saw the weapon's potential for use against Japanese aircraft in the Pacific theatre.[49] Two questions therefore must be asked – when and how did the CDIICAN acquire the schematic plans of the Hs 117? There were several sources from where the plans may have originated. In 1946, an intelligence report was produced on the weapon, most probably by the DEFA,[50] and there was also the British MoS, the GANES mission centred at Hechingen or ex-employees from HFW, most of whom were not recruited by either the US or the UK. Another possible source was the group of BMW rocket engineers recruited by the DTIA, who would have known the dimensions of the Hs 117.

Subsequently, the original Hs 117 design was considerably altered during development. However, as Robert *et al.* diagrammatically elucidated in their 2006 paper, a number of structural design characteristics were retained until 1957. The first five prototypes (A1 to A5) had a fuselage diameter of 356mm, an increase of only 6mm, and both the swept wing (which had a wingspan of 2 metres) and tailplane (1 metre in width) designs were also retained with only slight modifications. The retention of these design features – which had undergone wind tunnel testing at the DVL in Berlin as late as March 1945 – would seem to indicate that the French engineers decided to save time and resources by not engaging in a time-consuming and possibly expensive series of wind tunnel experiments. The development experience from the MARUCA programme was built upon in the subsequent MASURCA (*Marine Supersonique Ruelle Contre Avions*) project which commenced in 1955. Like the DEFA with the PARCA, the DCCAN could not successfully develop an operational first-generation missile system for the *Marine Nationale* in time with the US Navy and the Royal Navy. That outcome can be attributed to several factors – there was the lack of scientific and technical intelligence from Germany, the late entry into the field compared with the British and Americans, and limited resources. As a result, the first SAM system for the *Marine Nationale* was also American – the French government decided to purchase the Terrier, which was developed from the Bumblebee programme and had entered service with the US Navy in 1956.[51]

Conclusion

The DEFA, the DTIA and the DCCAN all recruited German scientists, engineers and technicians who were subsequently engaged on research and development work in French postwar guided missile programmes, not only with SAMs. SAM research and development absorbed much of the specialists' time during the late 1940s. A major setback to these efforts was the lack of scientific and technical documentation recovered from Germany. As a result, there were significant gaps in the knowledge that the French scientific and technical missions brought back to France, embodied in the archives of Peenemünde-East and West, the LFA and the DFS, which contained thousands of reports, books and engineering drawings that pertained to guided missile research and development that was amassed during the war years.

The sheer volume and importance of the scientific and technical intelligence that the US and the UK obtained from Germany gave these countries a very great advantage over the French. The difficulties that the *Armée de Terre* and the *Armée de l'Air* experienced in trying to master the technology was one consequence. As was shown with the case of the MARUCA programme, by the early 1950s, the STCAN had still not surpassed some aspects of the extent of German advances in May 1945. The French defence industry could not design and develop tactical weapons systems in time to provide suitable air defence of French territory and the borders of NATO. When the French government made the decisions to purchase the Nike-Ajax, Hawk and Terrier missile systems from the US, the Americans and the Soviets had already introduced their first-generation SAM systems into service, followed by the British Army and the RAF before the end of the decade.

CHAPTER SEVEN

Conclusion

What meaning can be derived from this historical account? The historical events are salient examples of where war (not necessarily armed conflict, economic warfare for example) can stimulate scientific and engineering progress. The rapid evolution of SAM technology between 1939 and 1945 was directly attributable to armed conflict, and during the period of the Cold War from 1945 to 1960 the possibility of armed conflict between the American-led West and the communist bloc brought further rapid developments. The SAM technology which appeared during these periods was not designed to have associated applications that could benefit both science and mankind, unlike ballistic missile technology that can be adapted for scientific experiments and space exploration. Nevertheless, the science that was applied to engineer the technology in these SAM systems, and the technology itself, was later used for non-military purposes. For instance, data and experience that was acquired from firings of the British RTV 1 at Woomera in South Australia during the 1950s was of importance to the subsequent British Black Knight sounding rocket programme, the purpose of which was to acquire scientific data from the upper atmosphere. Also, in the US Wernher von Braun proposed the use of several solid propellant Loki unguided anti-aircraft rockets as the upper stage of a rocket in Project Orbiter, the American programme to place a scientific satellite into orbit around the earth during the International Geophysical Year (1 July 1957 – 31 December 1958).[1]

Another meaning that is derived is the application of science to warfare, with the associated axiom that scientific knowledge can translate into military power. The former has been understood since at least World War I, and the latter became widely accepted during World War II when a whole range of new, sophisticated, and revolutionary technologies were developed on the basis of scientific research, for example, atomic fission, radar, jet aircraft, rocket propulsion, and guided missiles. Generally speaking, scientific knowledge led to the development and production of better and more advanced weapons and weapons systems, provided there was effective co-ordination of research and development by governments, the armed forces and industries, and the financial resources and technology to engineer the components for the armaments were available.

During World War II, Germany was ahead of the Allies in SAM development for numerous reasons. Early German progress with liquid propellant rocket technology was a significant factor. The great investment by the National Socialist regime in the German aeronautical research complex and the direction of scientific and engineering resources towards the development of highly advanced experimental aeronautical technologies were also important factors. Foresight on the part of the *Luftwaffe* in recognising that surface-to-air guided missiles would eventually replace anti-aircraft artillery can be considered as another reason. Most importantly, there was the conduct of the aerial war by the Western Allies, particularly from 1942 onwards, which engendered and stimulated efforts across the German military-industrial complex to develop more effective anti-aircraft armaments. Yet to attribute the astonishing German lead over the Allies with the technology by 1945 to German scientific and engineering 'superiority' would be inaccurate and does not represent the facts. The Allies (the US and the UK in particular) successfully applied their scientific knowledge to the development and production of certain categories of armaments where Germany was demonstrably far behind, and in some cases where such German technology was non-existent. For instance, the *Luftwaffe* did not build up a strategic bombing capability; the *Kriegsmarine* did not possess any aircraft carriers; microwave radio research in Germany was neglected until the last two years of the war, when it was too late; and then there was, of course, the atomic bomb.

The Allies' exploitation of scientific and technical intelligence on German SAM research and development, and the benefits derived from the employment of German specialists after World War II, went some way to compensate for the political, military, economic and, most of all, the human costs that were incurred as a result of fighting Nazi Germany. By the early 1950s, the Allies had completed the exploitation of wartime German technology and were in the process of developing far more sophisticated weapons systems. The result of the transfer of German SAM technology to the US and the UK after 1945 was that German knowledge made significant, although not vital, contributions to the development of SAM systems in the two countries during the early period of the Cold War up to around 1960. In the cases of France and the Soviet Union, the exploitation of German scientific and engineering knowledge of SAM technology was comparatively much more consequential.

Another 18 years would pass before Germany resumed the development of SAMs. In 1954, there was an unsuccessful attempt by an association of prominent German scientists and engineers – a number of whom had been involved with guided missile research and development during the war, and some who had recently been repatriated from the Soviet Union – to revive German work in the field of SAMs through a proposed collaborative project with the British.[2] A resumption did not take place until 1963, when a Franco-West German programme to develop a SAM system for use against low-altitude targets was initiated, called 'Roland'. However, the missile system did not enter service with the French and West German armies until 1977. The Roland was one of the few foreign armaments that were purchased by the US, previous examples including the French anti-tank guided missiles SS 10 and SS 11 by *Arsenal de l'Aéronautique/Nord-Aviation* and the DEFA ENTAC (*Engin Téléguidé Antichar*).[3]

Enduring reminders of the transfer of German SAM technology to the Allies after 1945 can be found today in partial or complete prototypes of flak rockets that are held by museums in the US, the UK, Germany, Australia and possibly elsewhere. In the US, the USAF Museum transferred a *Rheintochter* R-I to the Smithsonian National Air and Space Museum (NASM) in Washington DC in 1966. In 1983, the artefact was loaned to the Deutches Museum in Munich, where it was refurbished and is still on display. In 1969, the US Navy Supply Centre at Williamsburg in Virginia transferred another captured R-I to the Smithsonian Institution. It was on display in the NASM from 1976 until the early 1980s. In 2002, the specimen was restored to its original condition and paint scheme and placed on display in the James S. McDonnell Space Hangar at the NASM Stephen F. Udzar-Hazy Center in Chantilly, Virginia. Also on display at the same museum is an Hs 117 that was transferred to the Smithsonian from the US Army Ordnance Museum at Aberdeen in Maryland in 1988. In the UK, the captured flak rockets that were kept in a hangar at the RAE/RPD at Westcott near Aylesbury in Buckinghamshire can today be found on display at the RAF Museum at Cosford in Shropshire. Finally, an *Enzian* prototype that was sent to Australia in 1947, destined for the Long Range Weapons Establishment in South Australia, is today held by the Australian War Memorial at Canberra.[4]

APPENDICES

General Aerodynamic, Structural and Performance Specifications of World War II German *Flak-Raketen*[1]

Luftfahrtforschungsanstalt Hermann Göring Feuerlilie F-25 research rocket

Length:	2 metres
Span:	0.9 metres
Fuselage diameter:	0.25 metres
Weight at launch:	115kg
Rocket engine:	1 x *Rheinmetall-Borsig* RI-502 solid propellant ATO unit, with an impulse of 3,000kg/sec for six seconds. A number of variations of the RI-502 design were also built that had similar power and burning durations of 2.8–9 seconds.
Maximum velocity:	Mach 0.65
Number of test-launches:	19
Guidance system:	gyroscope stabilised for roll, with fixed control surfaces

Luftfahrtforschungsanstalt Hermann Göring Feuerlilie F-55 research rocket

Length:	4.8 metres	
Span:	2.5 metres	
Fuselage diameter:	0.55 metres	
	Pulverantrieb (SPRE)	*flüssigantrieb* (LPRE)
Weight at launch:	473kg	655kg
Rocket engine:	4 x *Rheinmetall-Borsig* RI-503 solid propellant ATO units, each with an impulse of 3,000kg/sec for 6 seconds.	
Rocket engine, booster:	–	*Rheinmetall-Borsig Pirat*

		Wilhelm Schmidding SG 20
Rocket engine, sustainer:	–	LOX/petrol-alcohol engine
Total impulse:	5,400kg/sec	25,000kg/sec for 25 seconds
Maximum velocity:	Mach 1.25	–
Number of missiles built:	1	2
Number of test-launches:	1	1 (launch failed)
Guidance system:	–	gyro stabilised for roll
Maximum altitude:	4,800 metres (15,750ft)	–
Maximum range:	7,500 metres (24,600ft)	–

Rheinmetall-Borsig AG Rheintochter	*Rheintochter* R-I	*Rheintochter* R-IIIf
Length:	6.30 metres	5 metres
Span:	2.75 metres	3.18 metres
Fuselage diameter:	0.54 metres	0.54 metres
Weight at launch:	1,750kg	1,570kg
Weight at target:	750kg	685kg
Booster rockets:	by *Rheinmetall-Borsig*	2 x R-B RI-503 ATO units
Propellants, booster:	*Diglykol* powder	*Diglykol* powder
Propellants, sustainer:	*Diglykol* powder	*Visol* and mixed acid
Launching impulse:	80,000kg/sec	105,000kg/sec
Launching thrust:	65,000kg	26,700kg
Booster impulse:	40,000kg/sec	25,000kg/sec
Booster thrust:	65,000kg	–
Booster duration:	0.6 seconds	1 second
Rocket engine, sustainer:	by *Rheinmetall-Borsig*	LPRE by *VfK*, Berlin
Sustainer thrust:	–	1,700–2,300kg
Sustainer duration:	60 seconds	45 seconds
Velocity at burnout:	Mach 1.06	Mach 1.2
Velocity at target:	Mach 0.88	Mach 0.59–1.18
Number of test-launches:	at least 88	six or seven
Guidance system (R-I):	*Elsass* cartesian co-ordinate command-link with target and missile tracking by optical means, or by radar with either the *Telefunken Würzburg Riese* (*Funk-Meßgerät* FuSE 65) or *Mannheim* (*Funk-Meßgerät* FuSE 64) radars. *Telefunken/*	

Stassfurter Rundfunk Kehl-Strassburg radio-control system (S 203/E 230-V), upgraded in 1945 to the *Telefunken Kogge-Brigg* system (S 510/S 512) at a frequency of 1.25 GHz.

Proximity fuse (R-I and R-III):   –

Seeker (R-I):   infra-red device by *Rheinmetall-Borsig*

Peenemünde/EMW C-2 *Wasserfall* W-3

Length:	7.83 metres
Span (at gas vanes):	2.51 metres
Fuselage diameter:	0.88 metres
Weight at launch:	3,540kg
Mass at burnout:	1,615kg
Rocket engine:	designed at Peenemünde-East
Propellants:	*Visol* or *Optolin* with mixed acid
Thrust:	8,000kg
Duration:	45 seconds
Velocity at burnout:	2,500–2,600ft/sec (Mach 2.3)
Maximum operational ceiling:	58,000ft
Maximum operational range:	26,500ft
Number of test-launches:	between 40 and 45
Guidance system:	the *Einlenkrechner* to guide the missile onto a line-of-sight trajectory, and thereafter the same cartesian co-ordinate command-link system as the *Rheintochter*.
Proximity fuse:	the FuG 570 (*Kakadu*), designed and developed by *Donauländische Apparatenbau GmbH* (*Donag*) of Vienna; the *Marabu* by *Siemens und Halske* of Berlin, a smaller version of an electronic aircraft distance-meter designed by the same company; the *Fox*, designed and developed by Dr. Hilpert and *Dipl.-Ing.* Herzog of AEG in Berlin; the *Trichter*, developed by a team under Dr. Georg Güllner at *Blaupunkt GmbH* of Berlin, to be used in conjunction with the *Max A* seeker by the same firm; the *Kugelblitz* by *Patent-Werkungs Gesellschaft* of Salzburg; and a photoelectric device under development at the *WVA Kochelsee*. All electrical systems.

Seeker:	infra-red device by Dr. Helmut Weiss at Peenemünde-East and *Electroacoustic AG* of Kiel; *Netzhaut* infra-red device by Dr. Richard Orthuber at AEG.

Peenemünde/EMW *Taifun F* (*flüssigantrieb*, *Visol* and mixed acid)

	Taifun F	Variation 1	Variation 2	7cm	5cm
Length (mm):	1,920	1,920	1,920	2,150	1,500
Fuselage diameter (mm):	100	100	100	70	50
Total mass at launch (kg):	21	21	21	14	8
Empty weight (kg):	10	10	10	6.5	4.5
Thrust (kg):	840	1,300–1,500	3,000	1,500	750
Burning time (seconds):	2.5	1.5	0.8–1	1	0.5–0.7
Maximum velocity (m/sec):	1,000 (M. 3)	–	1,000	–	–
Velocity at operational height:	300	300	300	300	–
Operational height (ft):	49,200	49,200	49,200	49,200	49,200

Proximity fuse: a 'graze fuse' supplied by *Rheinmetall-Borsig* with a warhead that contained 0.5 or 1.3kg of high explosive as the standard device; also a 20-second pyrotechnic self-destruction time fuse which would detonate in the event of a miss.

Henschel Flugzeug-Werke AG Hs 117 *Schmetterling* A2 type VI (production version)

Length:	4.29 metres
Span:	2 metres
Fuselage diameter:	0.35 metres
Weight at launch:	430kg
Weight at target:	175kg
Rocket engine, booster:	2 x *Wilhelm Schmidding* 109-553 rockets
Propellants, booster:	*Diglykol* powder (DGDN/nitrocellulose)
Rocket engine, sustainer:	BMW 109-558 or HWK 109-729 LPREs

Propellants, sustainer:	109-558: *Tonka* 250 or 500 and mixed acid 109-729: petrol and mixed acid
Launching impulse:	26,000kg/sec
Launching thrust:	3,400kg
Booster impulse:	13,600kg/sec
Booster thrust:	3,400kg
Booster duration:	4 seconds
Sustainer impulse:	12,500kg/sec
Sustainer thrust:	380kg, declining to 60kg
Sustainer duration:	57 seconds
Velocity at burnout and target:	Mach 0.71
Number of test-launches:	at least 59
Guidance system:	*Burgund* polar co-ordinate command-link with target and missile tracking by optical means. The first radio receiver that was used was the *Telefunken* S-203 *Kehl*, superseded by two subsequent versions, the S-203-1 and S-203-3, which were to be replaced by the S-510 *Kran*. At first, the radio receiver was the E-232 *Colmar* by *Friesecke und Hopfner GmbH* that operated at a frequency of around 50 MHz, which was superseded by an updated version, E-232-1, of approximately 38 MHz, and again by the *Stassfurter Rundfunk* E-230-3 *Strassburg*, of approximately 70 MHz. The E-230-3 was to be replaced by a far more powerful receiver, the *Telefunken* S-512 *Brigg*, that functioned at a microwave frequency of 1.25 GHz for use with the *Kran* transmitter (the *Kogge-Brigg* system).
Proximity fuse:	FuG 570 *Kakadu*, *Marabu* or *Fox*. Two other devices were an acoustical device code named *Meise* (titmouse), designed at *Neumann und Borm Gesellschaft* of Berlin, and the Fuse S1, which was designed by HFW in conjunction with *Elektromechanik GmbH* of Reichenberg in Bohemia-Moravia and *Preh* of Neustadt (two versions were designed, one for the Hs 117 and another for the Hs 298 air-to-air missile).
Seeker:	an infra-red device by *Electroacoustic AG* of Kiel.

Messerschmitt AG Enzian E-1 to E-3

Length:	3.65 metres
Span:	4 metres
Fuselage diameter:	0.90 metres
Weight at launch:	1,600kg
Weight at target:	690kg
Rocket engine, booster:	4 x *Wilhelm Schmidding* 109-553 rockets
Propellants, booster:	*Diglykol* powder (DGDN/nitrocellulose)
Rocket engine, sustainer	HWK RI-209 (109-502) and RI-210b ATO units
Propellants, sustainer:	RI-209: aqueous solution of calcium permanganate and hydrogen peroxide RI-210b: petrol and mixed acid
Launching impulse:	124,000kg/sec
Launching thrust:	6,000kg
Booster impulse:	24,000kg/sec
Booster thrust:	4,000kg
Booster duration:	6 seconds
Sustainer impulse:	93,000kg/sec
Sustainer thrust:	1,000–2,000kg
Sustainer duration:	62 seconds
Velocity at burnout:	Mach 0.79
Velocity at target:	Mach 0.73
Number of test-launches:	38
Guidance system:	polar co-ordinate command-link with target and missile tracking by optical means using the same radio-control systems as for the *Rheintochter* and C-2.
Proximity fuse:	–
Seeker:	*Madrid* infra-red device by *Kepka* of Vienna in conjunction with the Institute for Aeronautical Equipment at the DFS *Ernst Udet* at Ainring; a television device by Dr. Werner Rambauske of *Askania-Werke/Institut für Physikalische Forschung* at Stargard in Pomerania.

Key Personalities in the German SAM Development Programme

Dr.-Ing. Gerhard Braun

Technical director of the *Feuerlilie* research rocket programme at the *Luftfahrtforschungsanstalt Hermann Göring* at Völkenrode. Emigrated to the US in September 1945 through Project Overcast.

Flieger-Oberstabsingenieur Rudolf Brée

Luftwaffe engineering officer in the RLM who was responsible for directing air-to-air guided missile development. His duties overlapped with Halder's when a variant of the Hs 117 was developed as an AAM, and after the *Arbeitsstab Dornberger* was created in January 1945 when he apparently shared responsibility for SAM development with Halder. Remained in Germany.

Generalmajor Dr.-Ing. Walter Dornberger

German Army officer in charge of liquid propellant rocket development in the HWA until September 1943, and thereafter was *BzbV Heer* (Army Commissioner for Special Tasks) under *Generaloberst* Friedrich Fromm, the chief of army armaments. Emigrated to the US in 1947 through Project Paperclip.

Dr.-Ing. Werner Fricke

Technical director of the *Rheintochter* project at *Rheinmetall-Borsig AG* in Berlin-Marienfelde. Emigrated to the US through Project Overcast or Project Paperclip.

Dr. Friedrich Gladenbeck

Head of a committee, under the control of the development group for guided missiles, that was responsible for the industrial development of proximity fuse and seeker technology for German guided missile projects. He was the director of the *Forschungsanstalt der Deutschen Reichspost* until 1942, and thereafter was employed by the electrical engineering firm *Allgemeine Elektrische Gesellschaft* (AEG). Remained in Germany.

Reichsmarschall Hermann Göring

Reichsminister for Aviation and commander in chief of the *Luftwaffe*. Göring authorised the advanced flak armament development programme on 1 September 1942. Committed suicide at Nuremberg after being sentenced to death by the International Military Tribunal in 1946.

Oberstleutnant Dr. Friedrich Halder

Luftwaffe officer in the RLM and the military director of the German SAM development programme. The programme was directed from the flak armament development section in the RLM (GL/Flak-E 5) until August 1944, from OKL-TLR-Flak-E 5 between August 1944 and January 1945, and thereafter by the *Arbeitsstab Dornberger*, January–April 1945. Remained in Germany.

SS-Obergruppenführer Dr.-Ing. Hans Kammler

Head of all German guided missile development from late January to April 1945. He constructed the *Mittelwerk* underground factory in the Harz Mountains where the A-4/V-2 rocket was manufactured. Was thought to have been killed in Prague in May 1945, but recent research has found that Kammler was in American custody after the war and subsequently disappeared, whereabouts apparently unknown.[1]

Major Dr. Hermann König

Commanding officer of the *Flakversuchsstelle der Luftwaffe* at Peenemünde. Remained in Germany.

Dr. Wilhelm Runge

Head of the development group for guided missiles in the Reich Ministry of Armaments and Munitions/War Production that was responsible for the co-ordination of radio, electrical and radar developments for guided missiles in German industry. Employed by *Telefunken* until 1944 and thereafter at the DVL at Adlershof in Berlin. Remained in Germany.

Dipl.-Ing. Klaus Scheufelen

Luftwaffe officer and designer of the *Taifun* barrage rocket at Peenemünde-East. Emigrated to the US through Project Overcast.

Dr.-Ing. Johannes Schmidt

Director of rocket engine development at *H. Walter KG*. Emigrated to the UK through the DCOS scheme in November 1946. Schmidt was killed on 14 November 1947 when a rocket engine exploded during a test at the RAE in Westcott.

Dr. Theodor Sturm

Physicist and chief engineer at the electronics firm *Stassfurter Rundfunk GmbH* of Stassfurt, and director of a work commission for remote control techniques linked to the development group for guided missiles in the Reich Ministry of Armaments and Munitions/War Production. Emigrated to the US through Project Overcast or Paperclip.

Dr. Wernher von Braun

The civilian technical director of the HWA installation at Peenemünde until August 1944, afterwards head of the development works at EMW until the evacuation to central Germany in February 1945, and chairman of the *Arbeitsstab Dornberger*, January–March 1945. Technical director of the A-4/V-2 and C-2 projects. Emigrated to the US in September 1945 through Project Overcast.

Dipl.-Ing. Helmut von Zborowski

Director of the rocket engine development department at BMW, Munich. Emigrated to France in 1947.

Prof. *Dr.-Ing.* Herbert Wagner

Chief of the Development Department for Guided Missiles at *Henschel Flugzeug-Werke AG* at Schönefeld in Berlin, and technical director of the Hs 117 *Schmetterling* project. Emigrated to the US in May 1945, and later returned to Germany.

Dr.-Ing. Hermann Wurster

Technical director of the *Enzian* project at *Messerschmitt AG*, 1943–45. Remained in Germany.

The Transfer and Exploitation of German SAM Technology by the Soviet Union After World War II

Introduction

After World War II, the armed forces of the communist Soviet Union – like the British, Americans, and French – sought to acquire German scientific and technical intelligence from the development of SAMs. The Soviet activities were undertaken as one element of broader operations to exploit the advances that were made by Germany in aeronautical, guided missile and rocket developments during the war. From the outset, the Soviet Union was at a disadvantage because many of the German research and experimental establishments and industrial concerns, along with their staffs, were either beyond the reach of the forward units of the Red Army or had evacuated westwards towards the end of the war to prevent their knowledge falling into the hands of the Soviet Union. In the territories that the Red Army occupied, both before and after the transition to the four zones of occupation had been completed, Soviet investigators had to piece together what remnants could be found of the missile projects and their ancillary equipment with the assistance of German specialists.

There was a vast contrast in the amount of documentation that was captured, and the calibre of the specialists recruited, by the Western Allies and the Soviets. Soviet investigators recovered only a small fraction of the records from the SAM development programme – the archives of the DVL in Berlin and a documents cache in Bohemia and Moravia probably represented the most significant acquisitions. The Soviet Union also largely missed out on opportunities to recruit the best scientists, engineers and technicians who were involved with the programme. This was not only due to the more attractive career prospects and the working and living conditions in the three western countries, but also because the British and American governments, through their intelligence organisations, implemented operations to deny German 'war potential' specialists to the Soviet Union. As a result, the best specialists were subsequently recruited by the Western Allies during the period from 1945–48.

What is remarkable about the Soviet Union's exploitation of German SAM technology was that despite the American and British lead over the Soviet Union in SAM research and development at the end of the war, and the success of the Anglo-American investigations in Germany and the technology transfers to both countries, Soviet scientists and engineers were able to develop a guided SAM system based on German technology and bring it to operational status just over a year after the Western Electric Nike SAM system entered service with the US Army in December 1953, and over three years before the first British system, the Bristol/Ferranti Bloodhound, became operational with the RAF in 1958.

The following account was produced from Russian sources translated into English, English-language secondary sources, and British and American intelligence reports. The memoirs of Boris Chertok, an engineer at the centre of Soviet guided missile and space rocket development during the Cold War, which have been translated into English, and works in English by Asif Siddiqi, significantly account for the technology transfer. However, technical details have been added thanks to British and American intelligence reports on the activities of the organisations created by the Soviets to reconstruct and exploit German SAM technology: the *Institut Berlin*, in existence until October 1946, and its successor, the *Technisches Büro 11*, in existence from October 1946 until the second half of 1948. These reports were produced on the basis of information that was provided by Germans who were employed in both organisations and reveal the methods that were used by the Soviets to assemble and modify the German technology with the limited resources at their disposal. No Russian or German archival sources were consulted, and therefore the depth of the inquiry is limited. Nevertheless, the facts that have been drawn from the various historical sources provide enough material to write this appendix, which overall paints an accurate picture of the Soviet transfer and exploitation of German SAM technology during the second half of the 1940s.

Soviet scientific and technical missions and the capture of German intelligence targets

During World War II, Soviet ground-based anti-aircraft weaponry, in common with the other belligerents, comprised anti-aircraft artillery and fire-control radars. The Soviet Union had a number of heavy anti-aircraft guns in service. These included a 76.5mm gun; the 85mm Model 1939 gun, which was capable of firing 15–20 9kg shells per minute at a velocity of 2,625ft per second to an effective ceiling of 27,000ft; and the successor to the Model 1939, the 85mm Model 1944 gun, which was capable of firing the same number of shells per minute at a velocity of 2,950ft per second to an effective ceiling of 31,000ft. In the way of radar and fire-control technology, the Soviets were in possession of the most modern developments in the BTL SCR-584 microwave gun-laying radar that the Americans provided to the Soviet Union through the Lend-Lease programme. Rocket development in the Soviet

Union during the war – with origins stretching back to the 1920s – consisted of small powder projectiles fired from *Ilyushin* Il-2 *Shturmovik* ground-attack aircraft, numerous versions of the *Katyusha* rocket system that was used with such devastating effect by the Guards Mortar Units of the Red Army on the Eastern Front, and an experimental rocket-powered aircraft with a nitric acid/kerosene LPRE. The *Katyusha* rockets, which were unguided and fin-stabilised, were fuelled with a double-base solid propellant that enabled them to attain ranges between 4 and 12km. Due to the exigencies of the conflict with Nazi Germany, no development or experimentation with liquid propellant missiles was carried out in the Soviet Union during World War II.[1]

In the search for German military secrets, due to reasons of geography and the decisions made by German specialists to evacuate westwards and southwards away from the Red Army, the Soviet Union was largely denied the human and material assets from the German SAM development programme. An additional factor that contributed to this outcome was the absence of a formal intelligence-sharing agreement with the US and the UK. When it came to exploitation, Soviet experts from the aviation and artillery branches of the armed forces had a great interest in rocket propulsion and guided missiles. The communist regime established a 'Special Technical Commission' to co-ordinate and supervise all investigations and operations related to German rocket technology. It was chaired by Artillery Major-General Lev Gaidukov, who during the war was a member of the Guards Mortar Units Military Council, one of two bodies that co-ordinated *Katyusha* development (the Military Council determined strategy and policy, whereas the Main Directorate of Armaments handled issues related to development and procurement). Gaidukov was also a prominent political figure, serving as a department chief within the Central Committee of the Communist Party of the Soviet Union. Later, Gaidukov was assisted by another prominent personage in rocket engineer Sergey P. Korolev. Soviet investigators probably made the first discoveries of German flak rocket technology in eastern Germany, where there were a number of sites linked to the *Feuerlilie* and *Rheintochter* programmes. The *Luftwaffe* and *Rheinmetall-Borsig* had facilities at Leba in eastern Pomerania, where prototypes of both missiles were test-launched; *Ardelt-Werke GmbH* at Breslau in Lower Silesia manufactured the fuselage of both missiles; and *Seyffarth* of Eberswelde, 30km north-east of Berlin in Mark Brandenburg, manufactured the LPRE for the *Rheintochter* R-IIIf prototype.[2]

Czechoslovakia was another possible location where the Soviets first discovered German anti-aircraft rocket technology. Near Prague, Gaidukov found a collection of documents on rocket developments in a 30-car train. The documents probably belonged to the firm *Versuchsanstalt Grossendorf*, which originally had its premises at Grossendorf, 80km east of Leba near Danzig on the Baltic coast, under the direction of SS officer Rolf Engel. The company had a number of solid propellant rocket and multiple rocket projector (*Nebelwerfer*) contracts with the *Luftwaffe* (air-to-air and surface-to-air applications), *Kriegsmarine* (anti-ship), *Heer* (anti-tank) and the *SS Waffenamt*. These included a new powerful booster rocket for the *Rheintochter* R-I and the surface-to-air variant of the R4M air-to-air rocket, code named *Orkan* (hurricane), that was designed for use against low-flying aircraft. During the winter of 1944/45, the firm was evacuated to the state-owned company *Waffen-Union Škoda-Brünn* at Příbram, 50km south-west of Prague. In April 1945, the centre was dissolved and the staff were permitted to leave.[3]

One of the most important German intelligence targets that was captured by the Soviets was Germany's oldest aeronautical research establishment, the *Deutsche Versuchsanstalt für Luftfahrt* (DVL) at Berlin-Adlershof in the south-east of the city, which was founded in 1912. The target was significant not only for the Soviet aircraft industry, but also for Soviet investigators searching for records on German research and development of guided weapons. Amazingly, the establishment had not suffered any bombing damage and was captured virtually intact, along with its archives and a number of the leading staff. Furthermore, according to a CIOS report, a full-scale model of the Hs 117 was buried in the grounds of the DVL before the establishment was captured but had not been tested. Among the archives were records of the wind tunnel experiments that were done for the *Ruhrstahl* and HFW guided weapon projects. Below is a table that sets out the wind tunnel experiments for the Hs 117 project at the DVL. It is highly probable that all of the materials at the DVL were removed and sent to Russia during 1945 and 1946.[4]

Another important intelligence target that was captured by the Soviets was the inter-service installation at Peenemünde. On 5 May 1945, two groups of investigators reached the base – one was a team of aviation investigators led by Dr. Genrikh Abramovich, a recognised expert in jet propulsion and gas dynamics and deputy director of the Scientific-Research Institute 1 (NII-1) in Moscow; the other a team of artillery investigators led by Major-General Andrei Sokolov, who was involved in *Katyusha* development.[5] Little remained as a result of bomb damage and the evacuations earlier that year, although some rocket parts, such as A-4/V-2 combustion chambers, were left behind.

The Red Army remained in control of the areas that were occupied on 8 May 1945 until the Council of People's Commissars authorised, on 6 June 1945, the creation of the Soviet Military Administration in Germany to administer the future Soviet Zone

Table 13. Aerodynamics research for the Hs 117 project in the wind tunnels of the DVL

Series	Prototype	Tunnel speed	Date	Scale/material of model
1	III	Low-speed	May 1944	1:2 scale wood
2	IV	Low-speed	August 1944	1:2 scale wood
3	V	High-speed (Mach 0.85)	October 1944	1:2 scale aluminium
4	VI	High-speed (Mach 0.85)	March 1945	1:2 scale aluminium

Sources: TNA, DSIR 23/15145, 80; DSIR 23/15888

of Occupation, which had its headquarters at Karlshorst in East Berlin.[6] On 1 July 1945, the Western Allies began to move into the areas within the demarcation lines of the four zones of occupation and sectors of Berlin. At the Potsdam Conference from 17 July to 2 August 1945, the leaders of the US, the UK and the Soviet Union agreed to the total disarmament and demilitarisation of Germany. The Soviet Union, like the Western Allies, did not strictly adhere to this agreement.

German flak rockets and the establishment of postwar Soviet SAM development

The techniques that Soviet departments used to exploit the scientific and engineering knowledge on SAMs that had accumulated in Germany, and on German guided weapons and rocketry more generally, were quite different to the methods employed by the Western Allies. The Soviet military authorities centralised the SAM intelligence-gathering operations in the Soviet sector of Berlin, with a number of satellite stations in the Soviet zone. The Soviets induced German specialists with offers of decent pay and food supplies in order to acquire their services. In the Harz Mountains, the NKGB rounded up hundreds, and possibly thousands, of former employees of Peenemünde-East and the *Mittelwerk* who decided to remain in the Nordhausen and Bleicherode areas after the transition to the zones of occupation. Some of these German personnel would certainly have been among the 1,300–1,400 who worked on the C-2 and *Taifun* projects at Peenemünde-East, although they were probably technicians and skilled workers rather than scientists and engineers. The central organisation was called the *Institut Berlin*, one of at least two institutes that were tasked with rebuilding the German guided missile industry with the assistance of German specialists, another being the well-documented *Institut Nordhausen* (of which Gaidukov was chief) that was concerned with reconstructing the testing and manufacturing facilities from the A-4/V-2 programme and assembling the missiles for trials in the Soviet Union.

The organisation which became the *Institut Berlin* was set up by Gaidukov in September 1945, and by 1946 it was located at Hohenschönhausen in the Soviet sector of Berlin. Its main purpose was to study the *Wasserfall, Rheintochter* and *Schmetterling*, and this involved scavenging for specimens of components in the missiles and ground equipment so drawings could be made and the missiles reconstructed. There was little or no interest in the *Enzian*. The *Institut Berlin*, like the *Institut Nordhausen*, comprised five 'objects', or centres of operation. The main site in Berlin was under the command of Dmitrii Diatlov, who was assisted by a chief engineer, Vladimir Barmin, and was divided into several bureaus and departments. Two secondary sources are relied upon for the organisational structure of the institute, one by Siddiqi and the other by Lardier and Barensky, both published in 2010. They identified three design bureaus and six construction sectors made up of Soviet and German personnel. There was a *Wasserfall* design bureau, headed by Yevgeniy Sinilshchikov, assisted by guidance chief I. S. Aralov (Design Bureau 2, KB-2); a *Schmetterling* and *Rheintochter* design bureau, headed by Semyon E. Rashkov (KB-3); a third design bureau (KB-4, concerned with a 283mm ramjet-powered shell, headed by N. A. Soudakov); Sector No. 5, headed by N. I Krupnov, which dealt with the solid propellant booster rockets for the *Schmetterling* and *Rheintochter*, the solid propellant version of the *Taifun* and the *Ruhrstahl* X-7 anti-tank guided missile code named *Rotkäppchen*; a department for SAM LPREs, headed by Naum L. Umanskiy (Sector No. 6); and a number of other departments concerned with radio control (Sector No. 7, headed by Vladimir Goviadinov), stabilisation/computation (Sector No. 8, headed by K. P. Kliaritsky), ground equipment (Sector No. 9, headed by V. A. Timofeiev), a chemical laboratory (Sector No. 10, headed by A. K. Polevik, concerned with the synthesis of liquid propellants, such as *Tonka 250*) and a ballistics section. The four satellite objects were at Peenemünde under V. K. Shitov, Zwickau in Saxony under A. N. Vlasov, Leipzig (also in Saxony) and Leuna in Halle-Merseburg. A central directorate in Berlin monitored the work of the two institutes in Germany, headed by chief engineer Iurii Pobedonostsev. This body was subordinate to an interdepartmental commission under Gaidukov.[7]

The discoveries in Germany prompted the Bolshevik regime in Moscow to initiate a large-scale guided missile development programme, signalled in a special decree number 1017-419, dated 13 May 1946, of the Central Committee and Council of Ministers. The decree created a 'Special Committee for Reactive Technology', also known as 'State Committee No. 2', to discuss the organisation and distribution of responsibilities among the Soviet ministries, scientific research establishments and the industrial sector for the experimental development, production and operation of guided missiles for purely military purposes, and for the utilisation and exploitation of German specialists. The Special Committee had to submit its plan of scientific research and experimental operations for 1946–48 to the Chairman of the Council of Ministers for approval. The decree stated that the top-priority tasks were the reproductions of the A-4/V-2 and *Wasserfall* using domestic materials. In Germany, this involved as top priorities:

a) the complete restoration of the technical documentation and models of the A-4/V-2 and *Wasserfall*, *Rheintochter* and *Schmetterling*;

b) the restoration of the laboratories and test rigs with all the equipment and instrumentation required to perform research and experimentation on the A-4/V-2, *Wasserfall*, *Rheintochter*, *Schmetterling* and other rockets;

c) the training of Soviet specialists who would master the design of the A-4/V-2, surface-to-air guided missiles and other rockets, testing methods and production processes for rocket parts, components and their final assembly.

A provision was included to transfer the facilities in Germany with their equipment, along with the German specialists, to the Soviet Union at an undetermined date before the end of 1946.[8]

Soviet efforts to reconstruct German SAM technology at the *Institut Berlin*

In 1946, the *Institut Berlin* was transferred from Hohenschönhausen to a plant that had formerly belonged to the electronics firm *Gesellschaft für Elektroakustische und Mechanische Apparate GmbH* (GEMA) at Wendelschloßstraße 3 in Köpenick near Oberschöneweide, in the south-east of Berlin. During the war, GEMA designed and manufactured a number of early warning and surveillance radars for the *Luftwaffe*, including the *Freya*, *Wassermann*, *Mammut* and *Jagdschloß*. At this facility, Aralov established a new department to reconstruct the control system of the *Wasserfall* from sample parts found in Germany, with the objective of furnishing a complete set of production drawings and a sample missile. This new department was headed by a Russian named Nikolai Kapirin. German engineers were employed to set up the new department and its subsections, although by this time the *Institut Berlin* already had high-frequency and gyroscope departments. These departments worked on the reconstruction of a sample missile of each of the *Wasserfall*, *Rheintochter* and *Schmetterling*, and plans for assembly line production of the control systems of each missile.[9]

According to a US Army intelligence report dated 19 October 1948, parts from a number of different *Wasserfall* production series were found, which sometimes made the differences in the individual types and modifications difficult for the Soviet engineers to understand. Some of the fundamental physical connections of the technology were often quite unclear to the Soviet and German engineers and technicians and had to be clarified by painstaking laboratory work. In addition, the Soviet engineers also demanded modifications to the missiles. In September 1946, the Soviet authorities urgently asked for the complete sets of technical drawings and samples of each missile to be finished by the middle of October, in anticipation of the impending Operation *Osoaviakhim*, the mass transfer of German specialists and their families to the Soviet Union on 22 October 1946. The reproduction room at the institute subsequently worked 24 hours a day to complete the task.[10]

The same intelligence report states that by the time of Operation *Osoaviakhim*, the progress with reconstructing the three guided flak rockets was as follows. One old *Rheintochter* R-I prototype had been found, and it was reconditioned, tested, found useable and completely reconstructed. Production drawings of the fuselage and the control system were completed. One *Schmetterling* prototype was also found and completely reconstructed. The fuselage and five control systems were assembled from old parts. All of these parts were tested and found useable, except the housing for the gyroscope, which was considered too weak. Production drawings of the fuselage and the control system were also completed. Some parts of the *Burgund* command-link guidance system were found, and drawings of the system were made on paper. At an anti-aircraft ordnance shop in Velten, 25km north-west of Berlin, a '*Schmetterling A*' calculator developed by *Kreiselgeräte GmbH* in Berlin was found, reconditioned, and readjusted. This piece of technology was designed to calculate the parallax angle between the direction of a radio guide beam and the direction of the missile launcher, so the missile, after being fired, would cut across the guide beam to pick up radio control and travel towards the airborne target. Meanwhile, a definite prototype of the *Wasserfall* could not be found. A fuselage of the missile was constructed from recovered parts, but these parts were from various production types. Parts of the control system were found and tested, including the gyroscopes, a mixing unit (*Mischgerät*), relay plates and rudder servomotors. Drawings of the control system were produced from artefacts sent to the *Institut Berlin* from Moscow. A calculator for the *Wasserfall* ground equipment was 'being worked on' but was not completed. Other finds included several *Strassburg* receivers at a *Stassfurter Rundfunk* plant (which is interesting, because the main production plant at Stassfurt was under American control from April–July 1945) and two 4m Em R-40 range finders at a *Carl Zeiss AG* plant.[11]

Prior to Operation *Osoaviakhim*, Soviet engineers at the *Institut Berlin* decided to modify the propulsion system in the *Wasserfall*. An insight into these activities was provided to British Air Intelligence by a German rocket engineer employed at the Institute, *Dipl.-Ing.* Hermann Zumpe, who was subsequently recruited to work in the UK through the DCOS scheme. From May–August 1946, the Soviets employed Zumpe in a section at Oberschöneweide where he was charged with completing a set of incomplete technical drawings of the *Wasserfall*. On 15 August 1946, Zumpe began work at the *Institut Berlin* in Berlin-Köpenick, in a section within the LPRE department under Umanskiy, to make improvements to the *Wasserfall* using a nitric acid/petrol engine that incorporated principles of his own design. According to Zumpe, the construction of 10 engines for the *Wasserfall* had commenced at the *Institut Berlin*, but none had been completed by 22 October 1946. The work had progressed to the stage of completing the manufacturing of various parts, but none of the engines had been tested owing to a lack of nitric acid, which the Soviet authorities had promised to make available but did not deliver. By 22 October, a number of improvements had been made to the *Wasserfall* propulsion system. One of the main improvements was a reduction in the dead weight of the LPRE by eliminating the compressed nitrogen gas bottles that were used to force the propellants into the combustion chamber and substituting them with a new propellant feed technique (during the war, the Germans on the project also designed lighter expulsion systems, one of which they claimed could have reduced the weight from around 295kg to as low as around 62kg). The other main improvements were a lightening of the mass of the combustion chamber and a reduction in propellant consumption during the missile's turn onto the trajectory by metering the propellant supply, thus reducing the amount of thrust during the turning period. The intended effect of the latter modification was a smaller radius during the turn, completed by the time the missile would reach an altitude of 2km, instead of the 5km previously. Also, the flight time could be increased to 71 seconds, from the previous time of between 40 and 45 seconds. The thrust produced by the engine was originally almost 8 tons, but in the new design would be divided into four stages, namely 4 tons for three seconds to get the gyroscopes stabilised, 8 tons for six seconds at launch, 4 tons for 20 seconds during the turning phase and 6 tons for 42 seconds during the period of controllable flight.[12]

Operation *Osoaviakhim* saw all the Soviet department chiefs at the *Institut Berlin* return to the Soviet Union except Nikolai Kapirin, subsequently the leader of the low-frequency laboratory then technical head of *Technisches Büro 11*, and a man named Solovjev of the high-frequency group. A *Wasserfall* experimental laboratory was dismantled and transferred to Moscow, either to the Ministry of Armaments State Union Scientific-Research Institute No. 88 (NII-88) in the outer Moscow suburb of Podlipki (now called Korolev), about 30km north-east from the centre of Moscow, or the NII-885, situated in inner Moscow. These two establishments were responsible for research and development of ballistic missiles based on the A-4/V-2 and SAM systems based on the German advances, in accordance with the decree of 13 May 1946. Some of the departments at the *Institut Berlin* were transferred to the Special Design Bureau (SKB) at the NII-88, under chief designers. They were Department No. 4, headed by Sinilshchikov, which was responsible for modifying the *Wasserfall* and designing long-range surface-to-air guided missiles with a seeker (R-101); Department No. 5, headed by Rashkov, which was responsible for reconstructing the *Rheintochter* and *Schmetterling*, and designing a medium-range surface-to-air guided missile (R-102); Department No. 6, headed by P. I. Kostin (not a department head at the *Institut Berlin)*, which was responsible for designing solid and liquid propellant unguided surface-to-air rockets with an altitude of up to 15km, using as a basis the solid propellant *Taifun* (R-103 and R-110); and Department No. 8, headed by Umanskiy, which was a special department responsible for SAM LPREs using storable propellants as oxidizers (such as nitric acid). At the NII-88, there was a test station and an experiment shop. At the NII-885, departments were set up to develop guidance systems, which were in the charge of Vladimir Govyadinov, the former head of the department at the *Institut Berlin* which was concerned with radio-control systems.[13]

The continued Soviet exploitation of German SAM technology at *Technisches Büro 11*

After Operation *Osoaviakhim* was completed, the *Institut Berlin* was renamed *Technisches Büro 11* (TB-11, Technical Bureau 11). The German personnel remaining at TB-11 were engaged in packing up the records, equipment and instruments of the deportees and their departments for dispatch to Russia. Naimark mentioned in 1995 that the Soviet authorities subsequently had difficulty recruiting German specialists due to fear they may be later deported to the Soviet Union. This fear existed at the *Institut Berlin*. According to an American intelligence report, specialists by the names of Gudakovski and Eitzenberger avoided deportation by going into hiding for a long time. In Zumpe's case, he fled to the British sector of Berlin on the day the operation was executed, seeking asylum for himself and his family in the British zone. Fear of deportation was probably also why the group of six former *Telefunken* employees at the institute decided to seek employment in the UK in late 1946 (see Chapter 5). As late as 1948, German employees at the institute still feared being deported – an electronics specialist by the name of Springstein, also a former *Telefunken* employee, who headed the low-frequency laboratory at TB-11 from December 1947, reportedly made preparations that year to escape to western Germany.[14]

The next phase of the Soviet exploitation of German SAM technology was the manufacture and assembly of test missiles in Germany for trials in the Soviet Union. This enterprise mirrored the operation to manufacture and assemble A-4/V-2 missiles at the *Institut Nordhausen*. TB-11 did not manufacture the components on a large scale; instead these tasks were undertaken by German firms located within the Soviet Zone of Occupation in Germany, which were supplied with raw materials by TB-11. This seems to indicate that Soviet industry did not yet have the capability or enough manufacturing capacity to produce the components. A design bureau established at TB-11, headed by a German specialist, received the original technical drawings of the parts from the *Institut Berlin*. Two copies of the drawings were made, one copy going to the TB-11 archives and another to the manufacturing firms. After the completion of the copying work, the engineers were ordered by the laboratories to design testing instruments and to develop modifications. For security reasons, the firms were not told what parts they were manufacturing or what their purpose was. The first requests from the Soviet Union to TB-11 were for:

a) procurement of the missing parts of the *Burgund* command-link guidance system;

b) production of 110 radio-control sets (presumably the control equipment in the missile) for both the *Schmetterling* and *Wasserfall*;

c) production of three *Kehlheim* radio transmitters;

d) development and production of testing instruments for the control equipment and the transmitters; and

e) modification of the control systems for simpler production and better operational safety.

It appears the *Rheintochter* was cancelled in preference of the other two missiles. Once the components were manufactured, they were sent to TB 11 for testing. The test missiles were then assembled in a workshop staffed with mechanics and electricians before being dispatched to the Soviet Union.[15]

At TB-11, cover numbers were used to designate the parts of the control systems. The control system in the *Wasserfall*, which was redesigned at TB-11, was numbered 1.00, with the subassemblies numbered 1.01–1.20. The *Schmetterling* control system was designated 2.00. In the following table, the cover numbers of the subassemblies, with descriptions, in both control systems are listed. The data was sourced from two US Army intelligence reports held by the UK National Archives. The data concerning the *Wasserfall* control system, which was more complex than the *Schmetterling*, is complete; however, the data on the *Schmetterling* is almost certainly incomplete. The veracity of the information appertaining to the redesign of the *Wasserfall* propulsion system provided to British Air Intelligence by Zumpe in November 1946 is shown in subassembly 1.19, the four-stage thrust control.

Evidently, Soviet investigators did not recover any samples of the latest microwave radio-control technology for the *Schmetterling* and the *Wasserfall*, namely the *Kogge* system in development by *Telefunken*, consisting of the transmitter (*Kran* FuG 512) and receiver (*Brigg* E 531); it appears the Western Allies did not find any samples either.[16] The components in the command-link guidance systems were similarly assigned cover numbers, and are listed in the table below. In this case, the data in one of the intelligence reports includes the components of the *Rheintochter* guidance system. The components roughly correspond to those in the *Burgund* system.

By mid-1948, the TB-11 high-frequency laboratory had completed the order for the production of the three *Kehlheim* transmitters for shipment to the Soviet Union, and was still engaged in developing the testing instruments for the transmitters.[17] The low-frequency laboratory had received enough parts to fulfill the order for 110 sets of the *Wasserfall* control system, and almost enough parts for the 110 sets of that for the *Schmetterling*.[18] TB-11 was dissolved in the second half of 1948, and its equipment, comprising parts and test instruments for the *Wasserfall* and *Schmetterling*, was sent to the Soviet Union.[19] The experimental development in the Soviet Union of the modified *Wasserfall* apparently proceeded at a faster pace than that of the *Schmetterling*. According to Lardier and Barensky, 30 flights of the *Wasserfall* took place in the Soviet Union during 1948 and 1949, while 17 flights of the *Schmetterling* were made from October 1949 to December 1949.[20] In 2010,

Table 14. The cover numbers used by TB-11 for the subassemblies in the redesigned control system in the *Wasserfall*, and also in the *Schmetterling*

Wasserfall control system (1.00)	
1.01	*Stassfurter Rundfunk Strassburg* superheterodyne receiver, two channels (I for height, II for course), one commutating contact each.
1.02	Elevator unit, power supply for 1.01, two relay regulators.
1.03	Two course-giving servo motors for two of the control surfaces.
1.04	Three gyroscopes, I, II and III (III was for turn) and unlocking device.
1.05	Main distributing box and test circuit relays.
1.06	Commutator switch, thrust stage, 13 contacts.
1.07	Mixing unit (*Mischgerät*) containing differentiating, modulating, mixing, amplifying and demodulating units and power supply.
1.08	Relay plate, four relay regulators (two for channel I, two for channel II).
1.09	Two low-inertia Ferrari servo motors for the two control surfaces.
1.10	Rudder position potentiometers, returning reconduction potentiometers, and end-marking contacts.
1.11	Edison type battery, 27 cells (33V to 22V within 90 seconds, about 15 amp).
1.12	Dynamo: 180 watts, three times 36V, three times two amps, 500 c/s (the power supply for the *Mischgerät*).
1.13	Stabiliser, Buchhold type, 500 c/s + 0.5%, for 1.12.
1.14	Ground test set cable termination panel and automatic switch.
1.15	Complete cable harness.
1.16	Missile detonator relay.
1.17	Jettisoning detonators for the four gas vanes. During the war, the German designers fitted explosive bolts to the gas vanes that were detonated 15 seconds after launch using a time switch. This method was subsequently dispensed with because the jet would burn the gas vanes away.
1.18	Special coldproof dry battery, 50-60V, 250 amps.
1.19	Jet control, four thrust stages (for the redesigned propulsion system).
1.20	Two radio receiver aerials in outer edges of two control surfaces, and low-loss HF co-axial cables.
Schmetterling (2.00)	
2.01	*Friesecke und Höpfer Colmar* superheterodyne receiver (either the E 232 or E 232-1).
2.04	Gyroscope.
2.06	Commutator switch.
2.07	Relay plates.
2.09	Rudder magnet for elevation (for elevator flaps, horizontal tail surfaces).
2.09a	Rudder magnet for deviation.
2.12	Battery in pressurised box.

Sources: TNA, AIR 20/8638, 'Soviet Research on Radio Control System of Guided Missile "Wasserfall"', 28.7.1949; AIR 40/2543; DEFE 15/216

Table 15. The cover numbers used by TB-11 for the parts in the ground apparatuses of the guidance and control systems

	Wasserfall	*Schmetterling*	*Rheintochter*
Kehlheim transmitter	4.19	4.19	4.19
Ground control set	6.00	7.00	8.00
Ground radar set	6.01	7.01	8.01
Range finder	6.02	7.02	8.02
Vision control unit	6.03	7.03	8.03
Calculator unit	6.04	7.04	8.04
Launching control desk	6.05	7.05	8.05
Command desk	6.06	7.06	8.06
Dynamo	6.07	7.07	8.07
Remote control set	6.08	7.08	8.08
Launching device/table	6.10	7.10	8.10

Source: TNA, AIR 40/2543

The liquid propellant V-300 missile in the Soviet S-25 *Berkut* SAM system. (Source: GlobalSecurity. org, https://www.globalsecurity.org/military/world/russia/s-25-pics.htm)

Krag, citing German sources, discussed the technical development of these two missiles and a number of subsequent derivations in the Soviet Union. It is therefore not necessary to duplicate Krag's account other than to mention that tests with the Soviet versions of the *Wasserfall*, designated R-101, and the *Schmetterling*, designated R-102, continued in the Soviet Union until 1951 and 1950 respectively, and formed the basis for further experimental designs of surface-to-air guided missile systems which eventually led to the first and second operational Soviet systems, the S-25 and S-75.[21]

The Soviet realisation of the German SAM development programme

On 9 August 1950, Stalin signed a decree which authorised the construction of the KB-1, a special facility to develop SAM air defence systems. It was attached to a facility called SB-1 that was tasked with developing air-to-sea precision-guided munitions. KB-1 inherited the work on the exploitation of the *Wasserfall* from NII-88 and the NII-885, but neither of the latter two institutes were assigned the task of developing the first SAM air defence system. Instead, this task was assigned to Semyon Lavochkin of Factory 301 at Khimki, under the Ministry of Aviation Industry. NII-88 was thereafter only responsible for developing the liquid propellant propulsion system, until that institute was shut down. The exclusion of the German specialists from any major Soviet guided missile developments, along with a resulting lack of enthusiasm, prompted a decision by the Council of Ministers on 13 August 1950 to regulate the process whereby the German guided missile specialists in the Soviet Union were to be returned to East Germany (*Deutsche Demokratische Republik*, DDR). By November 1953, the last contingent of Germans had departed the Soviet Union for the DDR.[22]

The design, development and operational service of the first Soviet surface-to-air guided missile system, the S-25, which armed the Moscow air defence network from 1955, has been well documented. In summary, the missile, the V-300, resembled the *Wasserfall* in a number of ways. It was a single-stage missile, with an LPRE fitted with a pressurised gas propellant feed system (not the system in Zumpe's design), and was vertically launched without the need for a launching apparatus. According to one source, the design of the guidance and control system of the S-25 was also assisted by German expertise.[23] Meanwhile, the two versions of the R-110 were developed, based on the *Taifun*. The liquid propellant version, called *Chirok*, was entrusted to D. D. Sevruk of Experimental Design Bureau 3 (OKB-3) at NII-88 in March 1952, which would later build the *Korchoun* system, derived from the *Taifun*; the solid propellant version, called *Strij*, was assigned to A. D. Nadiradze of Design Bureau 2 (KB-2) in 1950, and became part of the RZS-115 system which was abandoned in 1953.[24] The S-25 was followed by the S-75 in 1957, a design more the product of Soviet ideas. The latter system, arguably inferior to the American and British SAM systems, was credited with shooting down a number of American-supplied Martin B-57 reconnaissance aircraft (a derivative of the English Electric Canberra) of the Taiwanese air force by the communist Chinese air force in the late 1950s, the U-2 reconnaissance aircraft piloted by Gary Powers over the Soviet Union in the spring of 1960 and another U-2 that was shot down over Cuba in October 1962.

Chronology of Events

Spring 1939	British 3-inch rockets are test-fired for the first time, in Jamaica.
1940	In Germany, the use of radio technology to remotely control aircraft-launched guided weapons is proven feasible.
13 May 1941	Flak rocket research programme established by the RLM.
May–June 1942	RAF Bomber Command carries out three 'thousand bomber' raids against German cities.
1 September 1942	Hermann Göring, commander in chief of the *Luftwaffe*, formally establishes an advanced flak armament development programme which includes the development of surface-to-air guided missile systems.
12 October 1943	The first test-launch of a solid propellant *Rheintochter* R-I prototype, at Leba in Pomerania.
29 February 1944	The first test-launch of a liquid propellant C-2 prototype, on the Greifswalder Oie, off the coast of Peenemünde.
February 1945	The last test-launch of a C-2 prototype at Peenemünde.
March–April 1945	The last test-launch of a *Taifun* and possible use against Allied aircraft.
Late July 1945	The first test-launch of a British LOP/GAP dummy round at the Ynyslas Experimental Gunnery Establishment range in Wales.
4 October 1945	Project Nike development programme approved by the US Army Ordnance Department.
24 September 1946	First test-launch of a Nike prototype at the White Sands Proving Ground in New Mexico.
1947	In France, the DEFA initiates the *Projectile Autopropulsé Radioguidé Contre-avions* (PARCA) SAM project for the French Army.
28 November 1947	With German assistance, the American company Bendix Aviation Corporation initiates development of a SAM system based on the *Taifun* for the US Army.

1949	Commencement of the *Marine Ruelle Contre Avions* (MARUCA) experimental SAM programme for the French Navy.
August 1949	Captured *Taifun* rounds are test-fired at the Long Range Weapons Establishment in South Australia.
27 November 1951	The first reported successful destruction of an aerial target by a guided SAM, when a Nike prototype detonates within lethal range of a QB-17G drone at 29,000ft above White Sands Proving Ground.
December 1953	The Nike-Ajax becomes the first guided SAM system to enter service, with the US Army.
1955	The first Soviet guided SAM system to enter service, the S-25 *Berkut*, becomes operational in the Moscow air defence network.
1958	The ramjet-powered Bristol/Ferranti Bloodhound, Britain's first operational guided SAM system, enters service with the RAF.
1959	In the US the BOMARC, also powered by ramjets, is the first nuclear-capable SAM system to enter service, with the US Air Force.

Glossary of Terminology[1]

aerodynamics:
: the field in dynamics (the branch of mechanics in physics concerned with those forces which cause or affect the motion of bodies) which deals with the motion of air and other gaseous fluids and of the forces acting on solids in motion relative to such fluids.

aileron:
: a hinged or movable surface on an airframe, the primary function of which is to induce a rolling moment on the airframe. It usually is part of the trailing edge of a wing.

altimeter:
: an instrument that measures elevation above a given datum plane.

antenna:
: a device for transmitting or receiving radio waves, exclusive of the means of connecting its main portion with the transmitting or receiving apparatus.

automatic pilot:
: an automatic control mechanism for keeping an aircraft in level flight, on a set course, or for executing desired manoeuvres.

azimuth:
: an angle measured clockwise from the south or north.

ballistic missile:
: a vehicle whose flight path from termination of thrust to impact has essentially zero lift. It is subject to gravitation and drag and may or may not perform manoeuvres to modify or correct the flight path.

booster:
: an auxiliary propulsion system which travels with the missile and may or may not separate from the missile when its impulse has been delivered.

canard:
: a type of airframe which has the stabilising and control surfaces forward of the main supporting surfaces.

cartesian co-ordinates:
: the co-ordinates of a point in a plane (or in a space) defined by the perpendicular (vertical) distances of the point from two (or three) intersecting axes which are at right angles to each other.

centre of gravity:
: the point at which all the mass of a body may be regarded as being concentrated, so far as motion of translation is concerned.

cinetheodolite:
: an optical tracking instrument designed for measuring horizontal and vertical angles that combines the functions of a cine-camera and a theodolite.

diffuser:
: a duct of varying cross-section designed to convert a high-speed gas flow into low-speed flow at an increased temperature.

Doppler effect:
: the apparent change in frequency of a sound or radio wave reaching an observer or a radio receiver, caused by a change in distance or range between the source and the observer or the receiver during the interval of reception.

drag:
: that component of the total air forces on a body, in excess of the forces owing to static pressure in the atmosphere, and parallel to the relative gas stream but opposing the direction of the motion.

dynamo:
: any rotating machine in which mechanical energy is converted into electrical energy.

frequency band: in communications and electronics, a continuous range of frequencies extending between the two limiting frequencies.

g-force: the gravitational acceleration of terrestrial bodies towards the centre of the earth, which is about 32.16ft per second per second.

guidance: the entire process of determining the path of a missile and maintaining the missile on the path.

guided missile: an unmanned vehicle moving above the earth's surface, whose trajectory or flight path is capable of being altered by a mechanism within the vehicle.

gyroscope: a wheel or disc, mounted to spin rapidly about an axis and also free to rotate about one or both of two axes perpendicular to each other and to the axis of the spin. A gyroscope exhibits the property of rigidity in space.

homing: a system in which a missile steers toward a target by means of radiation which the missile receives from the target, either by reflection (radar or visible light) or by emission from the target (infra-red or acoustic energy).

homing, active: a form of guidance wherein both the source for illuminating the target and the receiver are carried within the missile.

homing, passive: a system of homing guidance wherein the receiver in the missile utilises natural radiation from the target.

homing, semi-active: a system of homing guidance wherein the receiver in the missile utilises radiation from the target which has been illuminated from a source other than the missile.

hypergolic: capable of igniting spontaneously upon contact.

infra-red radiation: radiation emitted from all bodies between the temperatures of absolute zero and 3,000 degrees Celsius.

Mach number: the ratio of the velocity of a body to that of sound in the medium being considered. At sea level in air at the Standard US Atmosphere, a body moving at a Mach number of one (Ma = 1) would have a velocity of approximately 1,116.2ft per second, the speed of sound in air under those conditions (named after Ernst Mach, 1838–1916, Austrian physicist).

missile: a self-propelled unmanned vehicle which travels above the earth's surface.

monocoque: a form of fuselage or motor vehicle body construction in which all or most of the stresses are carried by the skin. In guided missiles, the design is characterised by the missile fuselage serving as the part of the interior surfaces of the propellant tanks.

nozzle: a duct of changing cross section in which the fluid or gas velocity is increased.

physics: the science dealing with natural laws and processes, and the states and properties of matter and energy, other than those restricted to living matter and chemical changes.

pitch: an angular displacement about an axis parallel to the lateral axis of an airframe.

polar co-ordinates: the co-ordinates of a point in a plane (or in a space) defined by the distance of the point from the origin and the degrees from the vertical axis.

propellant: material consisting of fuel and oxidiser, either separate or together in a mixture or compound, which if suitably ignited changes into a larger volume of hot gases, capable of propelling a rocket or projectile.

proximity fuse: a mechanical or electronic device designed to detonate an explosive charge near an object.

radar beacon: generally, a non-directional radiating device, containing an automatic radar receiver and transmitter, that receives pulses from a radar and returns a similar pulse or set of pulses. The beacon response may be on the same frequency as the radar or on a different frequency.

radar clutter:
the visual evidence on the radar indicator screen of sea- or ground-return which tends to obscure the target indication.

ramjet:
a compressorless jet propulsion device which depends on the air compression accomplished by the forward motion of the unit for its operation.

rocket:
a thrust-producing system or a complete missile which derives its thrust from ejection of hot gases generated from material carried in the system, not requiring the intake of air or water.

roll:
an angular displacement about an axis parallel to the longitudinal axis of an airframe.

seeker:
a receiving device on a missile that receives signals emitted from or reflected off the target that is used in guiding the missile towards the target.

servomechanism:
a mechanism which is used to convert a low-powered mechanical motion into one which requires considerably greater power. The output power is usually proportional to the input power and the device is often electronically controlled.

servomotor:
any motor which provides the power for a servomechanism.

sonic:
velocity that is equal to the local speed of sound.

specific impulse:
the ratio of thrust to the fuel mass flow.

speed of sound:
the velocity at which sound waves are transmitted through a medium.

subsonic:
a velocity less than the local speed of sound, or a Mach number of one.

supersonic:
a velocity greater than the local speed of sound, or a Mach number of one.

sustainer:
a propulsion system which travels with and does not separate from the missile, usually distinguished from an auxiliary motor, or booster.

telemetry:
the use of equipment and instruments to measure, transmit, receive, indicate, and record remote information.

thermodynamics:
the science concerned with the relations between heat and mechanical energy or work, and the conversion of one into another.

thrust:
the resultant force in the direction of motion, owing to the components of the pressure forces in excess of ambient atmospheric pressure, acting on all inner surfaces of the vehicle parallel to the direction of motion. Thrust less drag equals accelerating force.

transonic:
the intermediate speed in which the flow patterns change from the subsonic flow to supersonic, i.e., from Mach numbers 0.8–1.2, or vice-versa

wind tunnel:
a tunnel-like device through which a controlled airstream can be drawn at various speeds, in order to subject scale models of parts of aircraft or missiles, or complete aircraft or missiles, to aerodynamic tests.

yaw:
an angular displacement about an axis parallel to the 'normal' axis of an aircraft.

Notes

Frontmatter

1. Wernher von Braun, 'Survey of Development of Liquid Rockets in Germany and Their Future Prospects', in *The Story of Peenemünde, or What Might Have Been* (also known as *Peenemünde-East, through the Eyes of 500 Detained at Garmisch*) (US Army Ordnance Department, 1945), 247–254.
2. Theodore von Kármán, 'Where We Stand: A Report of the AAF Scientific Advisory Group', Headquarters AAF Air Materiel Command, Dayton (1946), 16, Defense Technical Information Center (DTIC), https://www.discover.dtic.mil.

Introduction

1. René Carpentier, *Un demi-siècle d'aéronautique en France: Les missiles tactiques de 1945 à 1995*, 9, Academie de l'air et de l'espace, https://www.academie-air-espace.com.
2. CIOS report XXVI-30, 'Gas turbine development by BMW' (1945).
3. The works include the multi-volume memoirs of Boris Chertok, *Rockets and People*, first published in the 1990s in Russian; an article by Zaloga in 1997, 'Defending the capitals: The first generation of Soviet strategic air defense systems 1950–1960'; a three-article series by Fiszer published from 2004–06, entitled 'Moscow's Air-Defense Network'; a book chapter by Krag in 2010, 'Special features of anti-aircraft rockets with swept or low aspect ratio wings' (mostly based on German sources); and the book *The Red Rockets' Glare: Spaceflight and the Soviet Imagination, 1857–1957* by Siddiqi, also from 2010.

Chapter 1

1. Michael J. Neufeld, *The Rocket and the Reich* (New York: The Free Press, 1995), 151.
2. 'Technical Report on Strassburg-Kehl Radio Controlling System for Bombs and Rockets', Enemy Equipment Intelligence Section (EEIS), Signal Corps, U.S. Ninth Army, 22.5.1945, Records of the Office of the Chief of Naval Operations (CNO), Record Group 38, National Archives at College Park, College Park, Maryland.
3. Walter Wernitz, 'Research and Development of the Guided Missile "Feuerlilie"', in *History of German Guided Missiles Development*, ed. W. Quick and Theodore Benecke (Brunswick: Verlag E. Appelhans and Co., 1957), 419–420; Bernd Krag, 'Special Features of Anti-Aircraft Rockets with Swept or Low-Ratio-Aspect Wings', in *German Development of the Swept Wing: 1939–1945*, ed. Hans-Ulrich Meier (Reston: AIAA, 2010), 545–614.
4. 'General Report on Guided Missiles', Intelligence Report GDM-2, Signal Corps, USFET, 11.7.1945, Records of the Office of the CNO, RG 38, National Archives College Park; TNA, AIR 40/2458, 'Chef der Technischen Luftrüstung', Air P/W IU (BAFO), 1.8.1945; Hermann

Vüllers, 'Design and Development of the Solid Fuel Rocket', in *History of German Guided Missiles Development*, ed. W. Quick and Theodore Benecke (Brunswick: Verlag E. Appelhans and Co., 1957), 253–262; Neufeld, *The Rocket and the Reich*, 150–151.

5. See Edward B. Westermann, *Flak: German Anti-aircraft Defenses, 1914–1945* (Lawrence, Kansas: University Press of Kansas, 2001).

6. Vüllers, 'Design and Development of the Solid Fuel Rocket', 253–262; Charles Webster and Nobel Frankland, *The Strategic Air Offensive Against Germany 1939–1945, Volume II* (London: HM Stationery Office, 1961), 355, 417, 473 and 490–491; Neufeld, *The Rocket and the Reich*, 153. For a more detailed account of the beginnings of anti-aircraft missile studies in 1941 and the events leading up to the 1 September 1942 decision, see Horst Boog *et al.*, *Germany and the Second World War, Volume VI: The Global War* (New York: Oxford University Press, 2001).

7. Neufeld, *The Rocket and the Reich*, 231 and 284.

8. TNA, DSIR 23/15145, CIOS report XXXII-125, 'German guided missile research' (1945), 15, 102 and 114–117; Webster and Frankland, *The Strategic Air Offensive Against Germany 1939–1945, Volume II*, 206.

9. US Naval Technical Mission in Europe, 'German Mechanical Time Fuzes', Technical Report No. 491-45, September 1945, DTIC, https://www.discover.dtic.mil; Werner Fricke, 'Report on a meeting held in Leba on 12 May 1944 (a letter setting requirements of the German Navy in regard to use of guided missiles of the "R" series)', in 'Bibliography on German Guided Missiles', Headquarters AAF Air Materiel Command, Dayton (July 1946), 66, Hathitrust, https://www.catalog.hathitrust.org.

10. Horst Boog, Gerhard Krebs and Detlef Vogel, *Germany and the Second World War, Volume VII: The Strategic Air War in Europe and the War in West and East Asia 1943–1944/5* (Oxford: Clarendon Press, 2006), 319.

11. CIOS report XXXII-66, 'Deutsche Forschungsanstalt für Segelflug Ainring' (1945), 6–9; Ernst Heinrich Hirschel, 'The High Rating of Aeronautical Research During the Third Reich', in E. H. Hirschel, H. Prem and G. Madelung, *Aeronautical Research in Germany: From Lilienthal Until Today* (Berlin: Springer Verlag, 2004), 71–98.

12. CIOS report XXXII-66, 6–9; Hirschel, 'The High Rating of Aeronautical Research During the Third Reich', 71–98.

13. BIOS Final Report 160, 'Luftfahrtforschungsanstalt Hermann Göring Völkenrode, Brunswick' (1945), 1–3.

14. CIOS report XXXII-66, 77–79, 93, 152 and 180–181; Hirschel, 'The High Rating of Aeronautical Research During the Third Reich', 71–98.

15. Michael J. Neufeld, 'Hitler, the V-2, and the Battle for Priority, 1939–1943', *Journal of Military History* 57, no. 3 (1993): 511–538; Neufeld, *The Rocket and the Reich*, 203 and 287.

16. *The Story of Peenemünde*, 79–80; H. Stokes and W. Hausz, 'Interrogation of Dr. Oskar [Oswald] Lange (EW 2222)', 5.6.1945, in *The Story of Peenemünde*, 376–377; CIOS report XXXI-71, 'Interrogation of Helmut Gröttrup, Dipl. Ing., Elektromechanische Werke' (1945), 3–4.

17. Stokes, 'Interrogation of Dr. Netzer, EW 224, re *Wasserfall*', 24.5.1945, in *The Story of Peenemünde*, 387.

18. Krag, 'Special Features of Anti-Aircraft Rockets with Swept or Low-Ratio-Aspect Wings', 545–614.

19. CIOS report XXXI-2, 'Research work undertaken by the German universities and technical high schools for the Bevollmachtigter für Hochfrequenztechnik; independent research on associated subjects' (1945), 15.

20. The company that manufactured the *Zeppelin* airships that attacked England during World War I.

21. Krag, 'Special Features of Anti-Aircraft Rockets with Swept or Low-Ratio-Aspect Wings', 545–614.

22. TNA, DSIR 23/15145, CIOS report XXXII-125, 80; DSIR 23/15888, J. J. Henrici and D. Mandel, 'Aerodynamics of the Butterfly', 7.3.1946; BIOS Final Report 160, 15.

23. Leslie E. Simon, *German Scientific Establishments* (New York: Mapleton House, 1947), 196, Hathitrust, https://www.catalog.hathitrust.org; Hirschel, 'The High Rating of Aeronautical Research During the Third Reich', 71–98.

24. Eric Burgess, 'German Guided and Rocket Missiles', *The Engineer*, no. 184 (31 October 1947): 407–409.

25. CIOS report XXXII-38, 'Explosives summary of capacity and production in Germany' (1945), 4; Dryden *et al.*, 'Technical Intelligence Supplement: A Report of the AAF Scientific Advisory Group', Headquarters AAF Air Materiel Command (1946), 4, DTIC, https://www.discover.dtic. mil; Vüllers, 'Design and Development of the Solid Fuel Rocket', 253–262; Boris Kit and Douglas S. Evered, *Rocket Propellant Handbook* (New York: Macmillan, 1960), 21.

26. Hellmuth Walter, 'Development of Hydrogen Peroxide Rockets in Germany', in *History of German Guided Missiles Development*, ed. W. Quick and Theodore Benecke (Brunswick: Verlag E. Appelhans and Co., 1957), 263–280; Kit and Evered, *Rocket Propellant Handbook*, 16; Rowland Pocock, *German Guided Missiles* (London: Ian Allan, 1967), 70. The VfK was evidently one of the institutes that was created during the Nazi four-year economic plan for autarky, launched by Hitler in 1936, and headed by Hermann Göring.

27. TNA, DSIR 23/15145, 106; CIOS report XXVI-30, 'Gas turbine development at BMW', 3–12; Otto Lutz, 'Some Special Problems of Power Plants', and Helmut von Zborowski, 'BMW Developments', in *History of German Guided Missiles Development*, ed. W. Quick and Theodore Benecke (Brunswick: Verlag E. Appelhans and Co., 1957), 238–252 and 297–324; Pocock, *German Guided Missiles*, 60; John D. Clark, *Ignition! An Informal History of Liquid Rocket Propellants* (New Jersey: Rutgers University Press, 1972), 14.

28. Rheinmetall-Borsig, Bericht LS 333, 'Auswertung der Schüsse 1 bis 7 von R-I vom 12.10.43 bis 13.1.44 in Leba' (Report LS 333: Report and evaluation of firings 1 to 7 of R-I from 12.10.1943 to 13.1.1944 in Leba), in 'Bibliography on German Guided Missiles', 141.

29. TNA, FO 1031/236, 'Nach Absetzung der Bv 246 als Gleitkörper sollte sie als Flakzielgerät verwendet warden' (The use of the Bv 246 glide bomb as a *Flak* target device after its cancellation), 3.10.1944, in 'List of Documents found at Wesermünde'; TNA, AIR 40/1151, ADI (K) report 321/1945, 'German Flak', 26.5.1945; Olivier Huwart, *Du V2 à Véronique. La naissance des fusées françaises* (Rennes: Marines Editions, 2004), 86–89.

30. Werner Fricke, 'Schuss von Gerät R-1', July 1944 (English translation of a German report), DTIC, https://www.discover.dtic.mil; R. V. Jones, *Most Secret War* (London: Hamish Hamilton, 1976), 332–375 and 414.

31. R. W. Porter, 'Control of the Wasserfall', in *The Story of Peenemünde*, 147–160; TNA, DSIR 23/15145, 103; DSIR 23/15888; Karl-Heinz Schirrmacher, 'Guidance of Surface-to-Air Missiles by Means of Radar', in *History of German Guided Missiles Development*, ed. W. Quick and Theodore Benecke (Brunswick: Verlag E. Appelhans and Co., 1957), 187–200; Neufeld, *The Rocket and the Reich*, 233.

32. 'Burgund Control Equipment for the Rocket Schmetterling', Intelligence Report GDM-1, Intelligence Branch, Technical Liaison Division, US Army, 28.6.1945, and 'General Report on Guided Missiles', Intelligence Report GDM-2, Signal Corps, USFET, 11.7.1945, RG 38, National Archives College Park; Bernhard R. Kroener *et al.*, *Germany and the Second World War, Volume V, Part II: Organisation and Mobilization in the German Sphere of Power: War Administration, Economy, and Manpower Resources 1942–1944/5* (Oxford: Clarendon Press, 2003), 723.

33. Horst Boog *et al.*, *Germany and the Second World War, Volume VI: The Global War* (New York: Oxford University Press, 2001).

34. Ibid, 196–199.
35. 'Burgund Control Equipment for the Rocket Schmetterling', 28.6.1945, RG 38, National Archives College Park.
36. 'Wasserfall, Control System – Dipl.-Ing. Klein, Dr. Geissler', in *The Story of Peenemünde*, 323–328.
37. R. W. Porter, 'Control of the Wasserfall', in *The Story of Peenemünde*, 147–160; TNA, DSIR 23/15145, 47, 55 and 104–108; Neufeld, *The Rocket and the Reich*, 66–98, 100–107 and 252.
38. CIOS Team 183, 'The Aerodynamic-Ballistics Research Station, Kochelsee', in *The Story of Peenemünde*, 661; 'General Report on Guided Missiles', 11.7.1945, RG 38, National Archives College Park; *The Story of Peenemünde*, 327–328.
39. TNA, DEFE 15/216, 'Examination of a German "Wasserfall" (C2) Guided Anti-aircraft Rocket', 18.
40. CIOS report XXVIII-41, 'Institut für Physikalische Forschung Neu Drossenfeld' (1945), 3–6; TNA, DSIR 23/15145, 122–128; Stokes, R. W. Porter and Sharpe, 'Interrogation of Dr. Helmut Weiss', 21.5.1945 and 25.5.1945, in *The Story of Peenemünde*, 468; Simon, *German Scientific Establishments*, 158–163; Georg Güllner, 'Summary of the Development of High-Frequency Homing Devices', and Edgar W. Kutzscher, 'The Physical and Technical Development of Homing Devices', in *History of German Guided Missiles Development*, ed. W. Quick and Theodore Benecke (Brunswick: Verlag E. Appelhans and Co., 1957), 162–172 and 208–216; Pocock, *German Guided Missiles*, 93.
41. TNA, AIR 40/2458, 'Chef der Technischen Luftrüstung', Air P/W Interrogation Unit (BAFO), 1.8.1945; Harold Faber (ed.), *Luftwaffe: An Analysis by Former Luftwaffe Generals* (London: Sidgwick and Jackson, 1979), 57–60.
42. CIOS report XXII-19, 'I.G. Farbenindustrie AG, Leuna' (1945), 7; CIOS report XXIV-12, 'I.G. Farbenindustrie-Oppau Works Ludwigshafen' (1945), 3–7; TNA, DSIR 23/14848, CIOS report XXVII-67, 'Aerodynamics of rockets and ramjets research and development work at Luftfahrtforschungsanstalt Hermann Göring' (1945), 21; DSIR 23/15145, 15 and 114–117; DSIR 23/15888; Neufeld, *The Rocket and the Reich*, 240 and 253.
43. 'Notes on Conference on 14.10.44 at TLR/Flak-E about the Organisation and the Presumed Supply Points for Flak-Rakete "Schmetterling"', TLR/Flak-E. 5/II, Az. 144 Nr. 703/44, 16.10.1944, and TLR/Flak-E. 5/II B, Az. 103 Nr. 745/44, 22.10.1944, *Deutsche Luftwaffe*, https://www.deutscheluftwaffe.com; TNA, AIR 20/8677, ADI (K) Report No. 331/1945, 18.6.1945; Pocock, *German Guided Missiles*, 72; Boog et al., *Germany and the Second World War, Volume VII: The Strategic Air War in Europe and the War in West and East Asia 1943–1944/5*, 233.
44. TNA, DEFE 15/217, 'German Non-Guided Flak Rocket Taifun'; Neufeld, *The Rocket and the Reich*, 240 and 254–255.
45. TNA, AIR 20/8677, ADI (K) Report No. 331/1945, 18.6.1945; AIR 20/8773, BIOS Final Report 1110, 'Some Aspects of German Rocket Developments' (1946); Michael J. Neufeld, 'Rolf Engel vs the German Army: A Nazi Career in Rocketry and Repression', *History and Technology* 13 (1996): 53–72.
46. Webster and Frankland, *The Strategic Air Offensive Against Germany 1939–1945, Volume IV*, 369.
47. R. W. Porter, 'Control of the Wasserfall', in *The Story of Peenemünde*, 147–160; BIOS Final Report 867, 'Television Development and Application in Germany' (1946), 10; TNA, FO 1031/12, 'Brief Interrogation Report on Prof. Dr. Wernher von Braun', 8.3.1947, 10, 13 and 16; Walter Dornberger, *V-2* (London: Hurst and Blackett, 1954), 244.
48. 'Notes on German Weapons Developments', Seventh Army Interrogation Centre, 3.6.1945, Cornell University Law Library, https://www.lawcollections.library.cornell.edu; Dornberger, *V-2*, 264; Heinz Höhne, *The Order of the Death's Head: The Story of Hitler's SS* (London: Penguin, 2000), 407.

49. Dryden *et al.*, 'Technical Intelligence Supplement', 70; 'Bibliography on German Guided Missiles', 21.

50. TNA, DSIR 23/14848, CIOS report XXVII-67, 7; DSIR 23/15145, 108–109; CIOS report XXXII-66, 152. See Dryden *et al.*, 'Technical Intelligence Supplement', 70, and 'Bibliography on German Guided Missiles', 21.

51. Stokes and Porter, 'Interrogation of Dr. Guntram Haft', 21.5.1945, in *The Story of Peenemünde*, 370; Stokes, 'Interrogation of Dr. Netzer, EW, re Wasserfall', 24.5.1945, in *The Story of Peenemünde*, 387; Stokes and Porter, 'Interrogation of Dr. Theodor Netzer', 18.5.1945, in *The Story of Peenemünde*, 391; CIOS report XXVI-30, 12; CIOS report XXXIII-38, 19; TNA, FO 1031/12, 'Brief Interrogation Report on Prof. Dr. Wernher von Braun', 16; Dieter Huzel, *Peenemünde to Canaveral* (Englewood Cliffs, N. J.: Prentice Hall, 1962), 129–136; Neufeld, *The Rocket and the Reich*, 255.

52. TNA, AIR 20/8677, ADI (K) Report No. 331/1945, 'Remotely Controlled A.A. Projectiles "Rheintochter" and "Taifun"', 18.6.1945; ibid, ADI (K) Report No. 359/1945, 'Radio Control of German A.A. Projectiles', 26.7.1945.

53. CIOS report XXX-80, 'Bavarian Motor Works (BMW) – A Production Survey' (1945), 18 and 51; TNA, DSIR 23/15145, 22, 60, 69–70 and 80; Herbert Wagner, 'Guidance and Control of the Henschel Missiles', in *History of German Guided Missiles Development*, ed. W. Quick and Theodore Benecke (Brunswick: Verlag E Appelhans and Co., 1957), 8–23; Daniel Uziel, *Arming the Luftwaffe: The German Aviation Industry in World War II* (Jefferson: MacFarland and Company, 2012), 127.

54. TNA, DSIR 23/15145, 52.

55. Wagner, 'Guidance and Control of the Henschel Missiles', 8–23.

56. TNA, AIR 20/8677, ADI (K) Report No. 331/1945; DSIR 23/15145, 52, 60 and 70; DEFE 15/217, 'German Non-Guided Flak Rocket Taifun'.

57. TNA, AIR 20/8677, ADI (K) Report No. 331/1945; *The Story of Peenemünde*, 523.

Chapter 2

1. Basil Collier, *The Defence of the United Kingdom* (London: H.M. Stationery Office, 1957), 278 and 322; M. M. Postan, D. Hay and J. D. Scott, *Design and Development of Weapons: Studies in Government and Industrial Organisation* (London: H.M. Stationery Office and Longmans Group Limited, 1964), 290–293; James McGovern, *Crossbow and Overcast* (London: Hutchison & Co., 1965), 9–10.

2. David Zimmerman, *Top Secret Exchange: The Tizard Mission and the Scientific War* (Montreal: McGill-Queen's University Press, 1996), 70.

3. TNA, AVIA 6/15499, J. Clemow, 'History of the Development of the RTV 1', RAE Technical Memo GW 102 (1950), 4 and 10; Collier, *The Defence of the United Kingdom*, 349; Neufeld, *The Rocket and the Reich*, 149.

4. TNA, AVIA 6/15499, 'History of the Development of the RTV 1', 4.

5. Ibid, 4–5.

6. Ibid, 7, 45–46; R. G. Lee *et al.*, *Guided Weapons* (London and Washington: Brassey's, 1998), 15.

7. TNA, AVIA 6/15499, 32, 44–46 and 96.

8. Ibid, 88; AVIA 48/3, 'A Comparison Between the Brakemine and LOP/GAP Guidance Systems', W. J. Challens, GPE Trials Wing, 18.3.1947; Collier, *The Defence of the United Kingdom*, 38; Bill Gunston, *The Illustrated Encyclopedia of the World's Rockets and Missiles* (London: Salamander Books, 1979), 164–165; Stephen Twigge, *The Early Development of Guided Weapons in the United Kingdom, 1940–1960* (Chur, Switzerland: Harwood Academic Publishers, 1993), 103–104.

9. TNA, AVIA 6/15499, 10.

10. Ibid.

11. D. W. Holder, 'The High-Speed Laboratory of the Aerodynamics Division, NPL', ARC Reports and Memoranda No. 2560, December 1946, AERADE, https://www.reports.aerade.cranfield.ac.uk.

12. TNA, AVIA 6/15499, 47.

13. W. A. Mair (ed.), 'Research on High-Speed Aerodynamics at the Royal Aircraft Establishment from 1942 to 1945', ARC Reports and Memoranda No. 2222, September 1946, AERADE, https://www.reports.aerade.cranfield.ac.uk.

14. TNA, AVIA 6/15499, 10–11 and Figures.

15. Ibid, 10–11, 14, 54–55 and 71.

16. Ibid, 12, 33 and 51–56.

17. Ibid, 51.

18. Ibid, 83; Twigge, *The Early Development of Guided Weapons in the United Kingdom, 1940–1960*, 105.

19. CIOS report I-1, 'Radar and Controlled Missiles Paris Area' (1944); Rexmond C. Cochrane, *The National Academy of Sciences: The First Hundred Years 1863–1963* (Washington DC: National Academy of Sciences, 1978), 403; Twigge, *The Early Development of Guided Weapons in the United Kingdom, 1940–1960*, 102–105; Guy Stever, *In War and Peace: My Life in Science and Technology* (Washington DC: Joseph Henry Press, 2002), 42–54.

20. Mary T. Cagle, *Development, Production and Deployment of the Nike-Ajax Guided Missile System 1945–1959* (Huntsville: US Army Rocket and Guided Missile Agency, 1959), 1; J. D. Gerrard-Gough and Albert B. Christman, *History of the Naval Weapons Center, China Lake, California, Volume 2: The Grand Experiment at Inyokern* (Washington DC: Naval History Division, 1978), 1–3 and 15.

21. James W. Bragg, *Development of the Corporal: The Embryo of the Army Missile Program, Volume 1* (Huntsville: Army Ballistic Missile Agency, 1961), 7 and 18–35; Elliott V. Converse III, *Rearming for the Cold War 1945–1960* (Washington DC: Historical Office, Office of the Secretary of Defense, 2012), 205.

22. Bragg, *Development of the Corporal: The Embryo of the Army Missile Program, Volume 1*, 36–41.

23. Ibid, 42–43; John W. Bullard, *History of the Redstone Missile System* (Huntsville: Army Missile Command, 1965), 3.

24. Bullard, *History of the Redstone Missile System*, 7.

25. McGovern, *Crossbow and Overcast*, 101; Bullard, *History of the Redstone Missile System*, 7–8.

26. H. L. Dryden, G. A. Morton and I. A. Getting, 'Guidance and Homing of Missiles and Pilotless Aircraft: A Report of the AAF Scientific Advisory Group', Headquarters AAF Air Materiel Command (1946), 22, DTIC, https://www.discover.dtic.mil.

27. 'Nike: the US Army's Guided Missile System', Western Electric book rack service for employees (date of publication unknown); Cagle, *Development, Production and Deployment of the Nike-Ajax Guided Missile System 1945–1959*, 1–3 and 224; Bullard, *History of the Redstone Missile System*, 9; Converse III, *Rearming for the Cold War 1945–1960*, 627.

28. 'Ground to Air Pilotless Aircraft', Air Materiel Command, 1.10.1945, DTIC, https://www.discover.dtic.mil; Cagle, *Development, Production and Deployment of the Nike-Ajax Guided Missile System 1945–1959*, 4.

29. DTIC, 'Ground to Air Pilotless Aircraft'; Cagle, *Development, Production and Deployment of the Nike-Ajax Guided Missile System 1945–1959*, 4–7; Max Rosenberg, *The Air Force and the National Guided Missile Program 1944–1950* (USAF Historical Division Liaison Office, 1964), 3–4 and 9–13.

30. TNA, AVIA 6/15499, 10–11; Cagle, *Development, Production and Deployment of the Nike-Ajax Guided Missile System 1945–1959*, 4–5 and 17.

31. TNA, AVIA 6/15499, 10; Cagle, *Development, Production and Deployment of the Nike-Ajax Guided Missile System 1945–1959*, 4–5 and 26.

32. Cagle, *Development, Production and Deployment of the Nike-Ajax Guided Missile System 1945–1959*, 8–10.

33. 'Tentative Military Characteristics of the Anti-Aircraft Guided Missile System', Ordnance Committee Meeting Item 29012, 13.9.1945, in Cagle, *Development, Production and Deployment of the Nike-Ajax Guided Missile System 1945–1959*, 218; M. M. Postan *et al.*, *Design and Development of Weapons: Studies in Government and Industrial Organisation*, 290–293; Jones, *Most Secret War*, 427-428; Kenneth P. Werrell, *Archie, Flak, AAA and SAM: A Short Operational History of Ground-Based Air Defense* (Alabama: Air University Press, 1988), 52.

34. 'Tentative Military Characteristics of the Anti-Aircraft Guided Missile System', 13.9.1945, in Cagle, *Development, Production and Deployment of the Nike-Ajax Guided Missile System 1945–1959*, 218; DTIC, 'Ground to Air Pilotless Aircraft'.

35. Von Kármán, 'Where We Stand', 31; Frank H. Winter and George S. James, 'Highlights of 50 Years of Aerojet, A Pioneering American Rocket Company, 1942–1992', in *History of Rocketry and Astronautics, AAS History Series 22*, ed. Philippe Jung (San Diego: American Astronautical Society, 1998), 53–104; Converse III, *Rearming for the Cold War 1945–1960*, 336.

36. Winter and James, 'Highlights of 50 Years of Aerojet, A Pioneering American Rocket Company, 1942–1992', 53–104.

37. L. L. Cronvich, 'Aerodynamic Development of Fleet Guided Missiles in the Navy's Bumblebee Program', presented at the Seventeenth Aerospace Sciences Meeting, 15–17 January 1979, American Institute of Aeronautics and Astronautics (AIAA), https://www.aiaa.org; Gerrard-Gough and Christman, *History of the Naval Weapons Center, China Lake, California, Vol. 2: The Grand Experiment at Inyokern*, 281–287.

38. Von Kármán, 'Where We Stand', 31; Johns Hopkins University Applied Physics Laboratory, 'Semi Annual Report of Bumblebee Project July–December 1949', March 1950, 2, DTIC, https://www.discover.dtic.mil; Cronvich, 'Aerodynamic Development of Fleet Guided Missiles in the Navy's Bumblebee Program'; Werrell, *Archie, Flak, AAA and SAM: A Short Operational History of Ground-Based Air Defense*, 87; Winter and James, 'Highlights of 50 Years of Aerojet, A Pioneering American Rocket Company, 1942–1992', 53–104.

39. See Daniel Baucom, 'Eisenhower and Ballistic Missile Defence: The Formative Years, 1944–1961', *Air Power History* 51, no. 4 (2004): 4–17.

40. Von Kármán, 'Where We Stand', 13.

Chapter 3

1. TNA, AIR 34/702, 'Interpretation of N.R.8., Trial Photographs of Flak', 3.5.1944; AIR 34/702, 'Rocket Flak', 1.4.1945.

2. TNA, AIR 34/702, memorandum from ADI (Ph), Air Ministry, to CO, RAF Medmenham, 'Investigation of Flak Rockets', 14.5.1944.

3. Jones, *Most Secret War*, 431.

4. TNA, CAB 176/3, JIC/968/44, 'Combined Intelligence Priorities Committee', 8.7.1944; CAB 176/4, JIC/1065/44, R. P. Linstead, 'Report on the work of the Combined Intelligence Priorities Committee and of the Intelligence Priorities Committee for May, June, and July 1944', 2.8.1944; FO 1031/51, 'Basic Directive, as Amended 1 November 1944, CIOS'.

5. *The Conference at Yalta and Malta, 1945, Volume 1* (Washington DC: United States Department of State (Historical Division), 1955), 113.

6. 'Patents, Designs, Copyright and Trade Marks (Emergency) Act, 1939', https://www.legislation. gov.uk/ukpga/Geo6/2-3/107/enacted.

7. Stephen Pericles Ladas, *Patents, Trademarks, and Related Rights: National and International Protection, Volume 1* (Cambridge, Massachusetts: Harvard University Press, 1975), 1,844.

8. *The Conference at Yalta and Malta, 1945, Volume 1*, 113–117.

9. TNA, FO 1050/1417, CIPC, Black List, Description of Items, 10.9.1944.

10. Ibid.

11. James Mills, 'Pandora's box closed: the Royal Air Force Institute of Aviation Medicine and Nazi medical experiments on human beings during World War II', *Studies in History and Philosophy of Science Part C: Studies in History and Philosophy of Biological and Biomedical Sciences*, 79 (2020).

12. Memorandum from Commander in Chief, United States Fleet and Chief of Naval Operations, to various parties, subject: US Naval Technical Mission in Europe, 26.12.1944, National Archives and Records Administration (NARA), https://www.nara.org.

13. 'Historical Data on US Naval Technical Mission in Europe, First Narrative', 1.11.1945, Fischer-Tropsch Archive, https://www.fischer-tropsch.org.

14. TNA, DSIR 23/15145, 1–3; *The Story of Peenemünde*, 547.

15. TNA, FO 1050/1417, CIPC, Black List, Description of Items, 4.8.1944.

16. TNA, AIR 40/1151, memorandum from MI15(a), War Office, to AI3(e), Air Ministry, 5.10.1943. MI15 (Military Intelligence Section 15, Directorate of Military Intelligence, War Office) was created in July 1943 to collate and distribute all intelligence on German air defences as an inter-service and inter-Allied organisation. Peter Gudgin, *Military Intelligence: A History* (Stroud: Sutton, 1999), 70.

17. TNA, AIR 34/702, 'Flak Prospects in 1945', 15.1.1945.

18. TNA, AIR 34/702, memorandum from P. I. Team No. 8 to Team No. 48, HQ US Army XV Corps, 25.2.1945.

19. TNA, AIR 40/1151, 'MI15 Periodical AA Intelligence Summary No. 18', 12.3.1945.

20. Ibid.

21. TNA, AIR 40/2162, G. E. F. Proctor, 'German Airborne Controlled Missiles for Anti-aircraft Employment', AI2(g) Report No. 1765, 29.3.1945.

22. TNA, AIR 34/702, 'Rocket Flak', 1.4.1945.

23. See James Mills, 'The transfer and exploitation of German air-to-air rocket and guided missile technology by the Western Allies after World War II', *The International Journal for the History of Engineering & Technology* 90, no. 1 (2020): 75–108.

24. TNA, AIR 40/2542, Wolfgang B. Klemperer, 'Survey of Facilities in Germany for Development of Guided Missiles Part I', ALSOS, 2.7.1945; BIOS Final Report 160, 1; Simon, *German Scientific Establishments*, 28A and 38.

25. TNA, DSIR 23/14848, 4–5.

26. 'Technical Report on Strassburg-Kehl Radio Controlling System for Bombs and Rockets', Enemy Equipment Intelligence Section, Signal Corps, US Ninth Army, 22.5.1945, and 'Burgund Control Equipment for the Rocket Schmetterling', Intelligence Report GDM-1, Intelligence Branch, Technical Liaison Division, US Army, 28.6.1945, RG 38, National Archives College Park; CIOS report XXXII-88, 'Stassfurter Rundfunk, Stassfurt' (1945).

27. Jones, *Most Secret War*, 223–230.

28. CIOS report XXI-1, 'Organization of Telefunken' (1945), 3–19.

29. CIOS report XXVI-30, 5–8, 12, 29 and 34; TNA, AVIA 6/15882, 'A Review of Rocket Developments at BMW Until November 1944' (Über die R-Entwicklung bei BMW November 1944) (HEC 10610), trans. R. C. Murray (January 1947).

30. CIOS report XXVI-30, 12; CIOS report XXVI-83, 'Bayerische Motor Werke', 3–5; US Naval Technical Mission in Europe Report No. 236-45, 'General Survey of Rocket Motor Development

in Germany', 5.9.1945, The Foundation for German Communications and Related Technologies, https://www.cdvandt.org.

31. CIOS report XXVI-3, 'Seefliegerhorst Wesermünde (Evacuation from Erprobungsstelle der Luftwaffe, Karlshagen)' (1945), 3–4 and 16.

32. CIOS report XXVIII-53, 'Walterwerke Kiel' (1945), 2–3; SHAEF G-2, 'Interrogation of Albert Speer, former Reich Minister of Armaments and War Production', 29.5.1945, Cornell University Law Library, https://www.lawschool.cornell.edu; Nicholas Rankin, *Ian Fleming's Commandos: The Story of 30 Assault Unit in WWII* (London: Faber and Faber, 2012), 304–318.

33. TNA, AVIA 40/4466–4516, *Schmetterling* engineering drawings; Fischer-Tropsch Archive, 'Historical Data on US Naval Technical Mission in Europe, First Narrative'; McGovern, *Crossbow and Overcast*, 164–165; Clarence G. Lasby, *Project Paperclip: German Scientists and the Cold War* (New York: Atheneum, 1971), 4 and 40; Frederick I. Ordway and Mitchell R. Sharpe, *The Rocket Team* (London: Heinemann, 1979), 280–281.

34. TNA, DSIR 23/15145, 18 and 80; Lasby, *Project Paperclip: German Scientists and the Cold War*, 28.

35. TNA, DSIR 23/15145, 80; Simon, *German Scientific Establishments*, 56.

36. McGovern, *Crossbow and Overcast*, 164–167.

37. Ibid, 173–174; Lasby, *Project Paperclip: German Scientists and the Cold War*, 70.

38. E. H. Hull, 'Interrogation of Wernher von Braun', 5.6.1945, in *The Story of Peenemünde*, 232; TNA, DSIR 23/15145, 87–91; Ordway and Sharpe, *The Rocket Team*, 280–281.

39. 'Interrogation of Dr. Phys. N. J. H. David, 18 May 1945, by Dr. R. W. Porter and F/Lt. Stokes, at Partenkirchen', in *The Story of Peenemünde*, 277–280; *The Story of Peenemünde*, 504–522.

40. *The Story of Peenemünde*, 597–598; Peter Wegener, *The Peenemünde Wind Tunnels: A Memoir* (New Haven and London: Yale University Press, 1996), 105.

41. TNA, DSIR 15/15140, 'Interrogation Report No. 2, Meeting at Messerschmitt Plant, Oberammergau', 20.5.1945; ibid, 'Interrogation No. 1, subject: Dr. Wurster on Rocket Motors and the Enzian Rocket', CIOS, 19.6.1945.

42. TNA, DSIR 15/15140, Hugh L. Dryden, 'Interrogation Report No. 11. Interrogation of Dr. Wurster', AAFSAG, 15.6.1945; ibid, H. T. Sparrow, 'Interrogation Report No. 20, subject: "Enzian Rocket"', USSTAF, 15–21.6.1945.

43. CIOS report XXXII-66, 153.

44. Ibid, 82–83, 93–94, 152 and 181; TNA, DSIR 23/15145, 92–100.

45. TNA, ADM 256/140, 'Calculator "Schmetterling A"', British Naval Gunnery Mission, Special Technical Report Number 7 (1945).

46. TNA, AIR 20/8677, ADI (K) Report 359/1945, 'Radio Control of German A.A. Projectiles'; DSIR 23/15145, 3; AIR 20/8773, BIOS Final Report 1110, 5; AVIA 20/8640, Heinrich Klein *et al.*, '"Rheintochter" Rocket Motor System of Drive with Liquid Propellants', 15.12.1946.

47. For the evolution of US government policy concerning the recruitment of German specialists, see John Gimbel, 'Project Paperclip: German Scientists, American Policy, and the Cold War', *Diplomatic History* 14, no. 3 (1990): 343–365.

48. US Naval Technical Mission in Europe Technical Report No. 229-45, 'Trainers for Operators of Guided Missiles' (September 1945) and Technical Report No. 257-45, 'Nitroglycerin, Diethylene Glycol Dinitrate and Similar Explosive Oils – Manufacture and Development in Germany' (September 1945), DTIC, https://www.discover.dtic.mil; Fischer-Tropsch Archive, 'Historical Data on US Naval Technical Mission in Europe, First Narrative'.

49. TNA, FO 1031/51, 'BIOS, General', note by L. King-Slater, Secretary, Offices of the Cabinet and Minister of Defence, 30.7.1945; ibid, BIOS Directive, 30.7.1945; ibid, 'List of Officially Appointed Members of BIOS', 7.9.1945; Michael S. Goodman, *The Official History of the Joint*

Intelligence Committee, Volume 1: From the Approach of the Second World War to the Suez Crisis (New York and London: Routledge, 2014), 160.

50. TNA, FO 1031/51, BIOS Directive, 30.7.1945; ibid, appendix B to BIOS (45) 11, 'Reallocation of Responsibility for Exploitation of Intelligence Objectives and Dissemination of Information', 12.9.1945; AVIA 48/1, Report by Sir Alwyn Crow, DGP, MoS (November 1945).

51. 'Report on Operation Backfire', Volume 1, United Kingdom War Office (1946), 12, NARA, https://www.nara.org; Michael J. Neufeld, *Von Braun: Dreamer of Space, Engineer of War* (New York: Vintage Books, 2008), 208.

52. TNA, AVIA 12/82, 'Operation Surgeon', memorandum by HQ Air Division (REAR), Detmold, 23.11.1946. See Matthew Uttley, 'Operation Surgeon and Britain's Post-War Exploitation of Nazi German Aeronautics', *Intelligence and National Security* 2, no. 17 (2002): 1–26 for details about Operation *Surgeon*.

53. BIOS Final Report 163, 'Some German aircraft armament projects with particular reference to fire control developments' (1945), 9; TNA, DSIR 23/15888; AVIA 6/13225, R. C. Murray, 'Developments in liquid bi-fuel rocket technique at the MOS Establishment at Trauen, Germany. Jan–June 1946' (August 1946); Uttley, 'Operation Surgeon and Britain's Post-War Exploitation of Nazi German Aeronautics', 1–26.

54. BIOS Final Report 163, 8; TNA, AIR 20/8640, Heinrich Klein *et al.*, 'Der Fluessigkeitsantrieb für die Ferngestenerte Flakrakete "Rheintochter"', 15.12.1946; AVIA 54/1407, memorandum from Brigadier G. Hinds, DWR (D), to MoS, 8.10.1946; ibid, A. Donovan, 'Report on the Writing of Scientific and Technical Papers at the Unterlüss Work Centre, on the Subject of Rockets, Automatic Rocket Launchers, Guided Projectiles and their Accessories', 7.2.1947; ibid, memorandum from P. Dawson, Operations Group 2 BIOS, MoS, to numerous agencies, 'Return visit to Unterlüss Work Centre to progress reports upon Rockets and Guided Projectiles', 2.5.1947; DEFE 43, FO and MoD, Record Cards of Alien Specialists, 1946–1964, 'KLEIN, Dr.-Ing. Heinrich'; Neufeld, *Von Braun: Dreamer of Space, Engineer of War*, 212.

55. James Mills and Graeme Johanson, 'Project Abstract: an Anglo-American intelligence operation in 1947 to recover guided weapon technical documentation buried in Germany', *Intelligence and National Security* 34, no. 1 (2019): 129–148.

56. Ibid.

Chapter 4

1. Intelligence Branch Technical Liaison Division, US Army Signals Corps, 'Guided Missiles for German Naval Use', Intelligence Report GDM-9, 24.7.1945, RG 38, National Archives College Park; DTIC, 'Semi-annual Report of Bumblebee Project July–December 1949'.

2. TNA, AIR 40/2875, ADI (K) Report No. 312/1945, 'Remotely controlled missiles (Henschel Flugzeugwerke A.G.)', 18.5.1945; Intelligence Branch Technical Liaison Division, US Army Signals Corps, 'Burgund Control Equipment for the Rocket Schmetterling', Intelligence Report GDM-1, 28.6.1945, RG 38, National Archives College Park; Lasby, *Project Paperclip: German Scientists and the Cold War*, 4, 66–67 and 77.

3. Fischer-Tropsch Archive, 'Historical Data on US Naval Technical Mission in Europe, First Narrative'; Lasby, *Project Paperclip: German Scientists and the Cold War*, 99–100.

4. 'The Aerodynamic Ballistics Research Station, Kochelsee', in *The Story of Peenemünde*, 554; F1 Kochel (Personal), Inventaires – List de personnels, 12.9.1945 – SHD, *Châtellerault, AA 910 6F1 11*; Tom Bower, *The Paperclip Conspiracy: The battle for the spoils and secrets of Nazi Germany* (London: Michael Joseph, 1987), 254.

5. 'The Aerodynamic Ballistics Research Station, Kochelsee', in *The Story of Peenemünde*, 554–555; F1 Kochel (Personal), Inventaires – List de personnels, 12.9.1945 – *SHD, Châtellerault, AA 910 6F1 11*; Wegener, *The Peenemünde Wind Tunnels: A Memoir*, 123 and 147; Neufeld, *Von Braun: Dreamer of Space, Engineer of War*, 216.

6. DTIC, 'Semi-annual Report of Bumblebee Project July–December 1949'.

7. Elizabeth Babcock, *History of the Navy at China Lake, California, Volume 3, Magnificent Mavericks: Transition of the Naval Ordnance Test Station From Rocket Station to Research, Development, Test, and Evaluation Center, 1948–1958* (Washington DC: Naval Historical Center and the Naval Air Systems Command, 2008), 178 and 220.

8. Ibid, 216–224 and 537.

9. CIOS report XXVI-30, 29–30; William C. House, 'Project Squid: Liquid Propellant Rockets Field Survey Report', Volume II Part 2, Princeton University, 30.6.1947, DTIC, https://www.discover.dtic.mil.

10. Fischer-Tropsch Archive, 'Historical Data on US Naval Technical Mission in Europe, First Narrative'; Bower, *The Paperclip Conspiracy*, 99–101.

11. 'Bibliography on German Guided Missiles'.

12. Ibid.

13. TNA, AVIA 6/15499, 52; Eduard Fischel, 'Contribution to the Guidance of Missiles', in *History of German Guided Missiles Development*, ed. W. Quick and Theodore Benecke (Brunswick: Verlag E. Appelhans and Co., 1957), 24–38; Rosenberg, *The Air Force and the National Guided Missile Program 1944–1950*, 79; Kenneth M. Keisel, *Wright Field* (Charleston: Arcadia Publishing, 2016), 114–116.

14. Michael J. Neufeld, 'Overcast, Paperclip and Osoaviakhim: Looting and the Transfer of German Military Technology', in *The United States and Germany in the Era of the Cold War, 1945–1990, A Handbook, Volume I: 1945–1968*, ed. Detlef Junker (New York: German Historical Institute, Washington DC, and Cambridge University Press, 2004), 197–203.

15. TNA, FO 1031/236, 'List of Documents found at Wesermünde'; Lasby, *Project Paperclip: German Scientists and the Cold War*, 259.

16. *The Story of Peenemünde*, 320–321; D. Singelmann and H. Müller, 'BMW Rocket-engine Development', HQ AMC, Wright-Patterson AFB (May 1948), DTIC, https://www.discover.dtic.mil; McGovern, *Crossbow and Overcast*, 252–253; *American Men and Women of Science: The Physical and Biological Sciences, Volume 2* (Jacques Cattell Press, 1982), 1,153.

17. American Power Jet Company, 'Analysis and Evaluation of German Attainments and Research in the Liquid Rocket Engine Field', Central Air Documents Office (Army-Navy-Air Force), Dayton (1952), DTIC, https://www.discover.dtic.mil.

18. TNA, DEFE 15/216, 'Examination of German *Wasserfall* (C2) Guided Anti-Aircraft Rocket' (1946); Rosenberg, *The Air Force and the National Guided Missile Program 1944–1950*, 75–76.

19. Ordway and Sharpe, *The Rocket Team*, 313–314.

20. Simon, *German Scientific Establishments*, 125.

21. L. D. White, 'Final Report, Project Hermes V-2 Missile Program', General Electric Guided Missiles Department (September 1952), 51, Smithsonian Libraries, https://www.library.si.edu.

22. Wernher von Braun, 'The Redstone, Jupiter, and Juno', *Technology and Culture* 4, no. 4 (1963): 452–465.

23. *The Story of Peenemünde*, 79–83; Neufeld, *Von Braun: Dreamer of Space, Engineer of War*, 216.

24. 'Targets of Opportunity Connected with Darmstadt Technische Hochschule Vorhaben Peenemünde', in *The Story of Peenemünde*, 130.

25. Cagle, *Development, Production and Deployment of the Nike-Ajax Guided Missile System 1945–1959*, 82.

26. McGovern, *Crossbow and Overcast*, 210–211; Bullard, *History of the Redstone Missile System*, 27–28; Neufeld, *Von Braun: Dreamer of Space, Engineer of War*, 217, 238–239 and 248–250.

27. Quick and Benecke (ed.), *History of German Guided Missiles Development*, 6.

28. Dryden *et al.*, 'Technical Intelligence Supplement', 174.

29. Porter, 'Control of the Wasserfall', in *The Story of Peenemünde*, 150; Cagle, *Development, Production and Deployment of the Nike-Ajax Guided Missile System 1945–1959*, 19–43.

30. Dryden *et al.*, 'Technical Intelligence Supplement', 173–174.

31. TNA, DSIR 23/15145, 51; Cagle, *Development, Production and Deployment of the Nike-Ajax Guided Missile System 1945–1959*, 95–98.

32. Von Kármán, 'Where We Stand', 34.

33. Cagle, *Development, Production and Deployment of the Nike-Ajax Guided Missile System 1945–1959*, 24 and 32; Bragg, *Development of the Corporal: The Embryo of the Army Missile Program, Volume 1*, 44, 50–51 and 59–61.

34. Cagle, *Development, Production and Deployment of the Nike-Ajax Guided Missile System 1945–1959*, 37–39; Winter and James, 'Highlights of 50 Years of Aerojet, A Pioneering American Rocket Company, 1942–1992', 53–104.

35. Cagle, *Development, Production and Deployment of the Nike-Ajax Guided Missile System 1945–1959*, 30–31 and 35.

36. Ibid, 30–31, 35, 45–46 and 53–55.

37. TNA, DSIR 23/14848, 3; Cagle, *Development, Production and Deployment of the Nike-Ajax Guided Missile System 1945–1959*, 17, 22–23, 30 and 33.

38. TNA, DSIR 23/14848, 3; Cagle, *Development, Production and Deployment of the Nike-Ajax Guided Missile System 1945–1959*, 32; Hirschel *et al.*, *Aeronautical Research in Germany: From Lilienthal Until Today*, 271.

39. Mary T. Cagle, *Loki Antiaircraft Free Flight Rocket, December 1947–November 1955* (Huntsville: Redstone Arsenal, 1957), 2–3.

40. Cagle, *Loki Antiaircraft Free Flight Rocket, December 1947–November 1955*, 3–4 and 67; Ordway and Sharpe, *The Rocket Team*, 117 and 350–351; Neufeld, *Von Braun: Dreamer of Space, Engineer of War*, 231.

41. Cagle, *Loki Antiaircraft Free Flight Rocket, December 1947–November 1955*, 3–12 and 43–44; Kit and Evered, *Rocket Propellant Handbook*, 21; Ordway and Sharpe, *The Rocket Team*, 117 and 375; US Department of the Interior and Building Technology Incorporated, 'Historic Properties Report. White Sands Missile Range, New Mexico, and Subinstallation Utah Launch Complex, Green River, Utah', Final Report (July 1984), 78, DTIC, https://www.discover.dtic.mil.

42. Memorandum from Colonel Holger Toftoy to Secretary, Ordnance Technical Committee (OTC), 29.4.1948, in Cagle, *Development, Production and Deployment of the Nike-Ajax Guided Missile System 1945–1959*, 246–249; General Electric, 'Project Hermes Interim Report Number 1', film (date unknown), Internet Archive, https://www.archive.org.

43. William C. House, 'Project Squid: Liquid Propellant Rockets Field Survey Report', Volume II Part 2, Princeton University, 30.6.1947, DTIC, https://www.discover.dtic.mil.

44. General Electric, 'Project Hermes Interim Report Number 1'.

45. Memorandum from Toftoy to Secretary, OTC, 25.9.1951, in Cagle, *Development, Production and Deployment of the Nike-Ajax Guided Missile System 1945–1959*, 250–254; Bullard, *History of the Redstone Missile System*, 9–10.

46. Memorandum from Toftoy to Secretary, OTC, 29.4.1948, in Cagle, *Development, Production and Deployment of the Nike-Ajax Guided Missile System 1945–1959*, 246–249; Frederick I. Ordway and Ronald C. Wakeford, *International Missile and Spacecraft Guide* (New York: McGraw-Hill

Book Company, 1960),182–183; Bragg, *Development of the Corporal: The Embryo of the Army Missile Program, Volume 1*, 123.

47. Cagle, *Development, Production and Deployment of the Nike-Ajax Guided Missile System 1945–1959*, 98–100.

48. 'Missiles 1959', *Flight* (6 November 1959).

Chapter 5

1. TNA, DSIR 23/15140, H. Dryden, 'Interrogation of Dr. Wurster', AAF SAG, 15.6.1945; DEFE 15/194, 'Examination of Enzian Rocket Motor', ADD Technical Report 18/45 (July 1945); AVIA 41/388, 'Tentative results to date of chemical analysis of German rocket propellants', PDE Note No. 1946/2 (February 1946); AVIA 6/15499, 57, 67, 89 and 97; 'Rocketry at Westcott. A Visit to the Ministry of Supply Rocket-propulsion Establishment', *Flight* (21 August 1947); Estelle Campbell, 'The V2 rocket – how it worked and how we acquired it', AWM, https://www.awm. gov.au/articles/blog/v2rocket.

2. TNA, AVIA 49/10, G. Sherman, 'Review of activities of G5c in connection with BIOS' (August 1948); AVIA 54/1404, 'Notes on procedure for the exploitation of foreign documents', HEC (August 1947); ibid, D. W. Bartington, 'Halstead Exploitation Centre', 18.12.1947.

3. TNA, DEFE 15/216, 'Examination of German *Wasserfall* (C2) Guided Anti-Aircraft Rocket' (1946).

4. TNA, AVIA 48/2, Guided Projectiles Progressing Committee, minutes of the third meeting held at GPE Westcott, 21.10.1946; L. Wiseman, 'The Suitability of the Oxidants Liquid Oxygen, Hydrogen Peroxide and Nitric Acid in Liquid Propellant Systems for Operational Use', ERDE Technical Memorandum No. 13/M/48 (October 1948), and A. Baxter, 'Further Considerations on Selection of an Oxidant for Rocket Motors', RAE Technical Note RPD 11 (February 1949), DTIC, https://www.discover.dtic.mil; Peter Morton, *Fire Across the Desert: Woomera and the Anglo-Australian Joint Project 1946–1980* (Canberra: Australian Government Publishing Service, 1989), 189.

5. 'Our Latest on Show. Impressive Display at Farnborough: Newest British Aircraft and Engines Alongside Germany's Best: An Outline Description of an Outstanding Exhibition', *Flight* (1 November 1945).

6. Pocock, *German Guided Missiles of the Second World War*, 72–73.

7. Twigge, *The Early Development of Guided Weapons in the United Kingdom, 1940–1960*, 107–121.

8. Ibid, 107.

9. Ibid, 84.

10. TNA, AVIA 6/15499, 9.

11. TNA, AVIA 48/3, 'Minutes of the Tenth Brakemine Progress Meeting', 22.10.1946; ibid, W. Challens, 'A Comparison Between the Brakemine and LOP/GAP Guidance Systems', 18.3.1947; AVIA 6/15499, 75.

12. TNA, AVIA 6/15499, 15.

13. Kenneth Gatland, *Development of the Guided Missile* (London: Iliffe, 1952), 16.

14. TNA, AVIA 6/15499, 73 and 89.

15. Ibid, 15, 85 and 92.

16. Ibid, 20 and 23; L. Broughton and H. Frauenberger, 'A Comparison of Liquid and Solid Propellant Boost Rocket Motors', RPD Technical Note 25, DTIC, https://www.discover.dtic.mil.

17. TNA, AVIA 6/15499, 17 and 19–20; Wiseman, 'The Suitability of the Oxidants Liquid Oxygen, Hydrogen Peroxide and Nitric Acid in Liquid Propellant Systems for Operational Use'; J. Harlow, 'Alpha, Beta, and RTV 1: The Development of Early British Liquid Propellant Rocket Engines', in

History of Rocketry and Astronautics, AAS History Series, Volume 22, ed. Philippe Jung (San Diego: American Astronautical Society, 1998), 173–201.

18. TNA, AVIA 6/15499, 23–25.
19. Ibid, 17.
20. TNA, AVIA 54/1295, extract from paper DCOS (45) 61 (Final), 'Policy for the Exploitation of German Science and Technology', Annex.
21. TNA, AVIA 54/1295, 'Russian Enticement of German Scientists and Technicians', 3.9.1946.
22. Kerrie Dougherty, 'A German rocket team at Woomera? A lost opportunity for Australia', *Acta Astronautica* 5 (2004): 741–751.
23. 'The Royal Naval Scientific Service', *Nature* 154, no. 3908 (1944): 398–399.
24. TNA, ADM 1/20374, note by the Director of Research Programmes and Planning, 9.8.1947.
25. John Becklake, 'German Engineers: Their Contribution to British Rocket Technology after World War II', in *History of Rocketry and Astronautics, AAS History Series, Volume 22*, ed. Philippe Jung (San Diego: American Astronautical Society, 1998), 157–172.
26. TNA, AVIA 54/1295, 'Scientists Employed Under DCOS Scheme by MoS. 30.8.1947'; DEFE 43/15, record cards of German and other scientists and technicians, 1946–1964, SCHLO–SEI, 'SCHONHEIT, Werner'; AVIA 54/1828, 'G. Fiedler (Revision of Salary)' (November 1952); Bower, *The Paperclip Conspiracy*, 168–170; John Becklake, 'German Rocket Engineers in Britain – Their Influence Revisited', *Acta Astronautica* 59 (2006): 510–515.
27. 'Rocketry at Westcott. A Visit to the Ministry of Supply Rocket Propulsion Establishment', *Flight* (21 August 1947).
28. Ibid; W. Long, 'The Design of Emplacements for Rocket Motor Testing', RPD Technical Note 16 (June 1949), DTIC, https://www.discover.dtic.mil; Becklake, 'German Engineers: Their Contribution to British Rocket Technology after World War II', 157–172; Charlie Hall, *British Exploitation of German Science and Technology, 1943–1949* (Abingdon and New York: Routledge, 2019), 130.
29. TNA, DEFE 43/3, record cards of German and other scientists and technicians, 1946–1964, BEY–BY, 'BUSEMANN, Adolf'.
30. TNA, CAB 122/350, memorandum from J. Wilson, British Joint Staff Mission, to Director, JIOA, 1.2.1947.
31. TNA, DEFE 43/17, record cards of German and other scientists and technicians, 1946–1964, THI–WEN, 'VOEPEL, Heinrich'.
32. Krag, 'Special Features of Anti-Aircraft Rockets with Swept or Low-Ratio-Aspect Wings', 545–614.
33. TNA, AVIA 6/13225, R. C. Murray, 'Developments in liquid bi-fuel rocket technique at the MOS Establishment at Trauen, Germany. Jan–June 1946' (August 1946); AVIA 54/1295, 'Scientists Employed Under DCOS Scheme by MoS. 30.8.1947'.
34. TNA, AVIA 54/1403, memorandum from J. Thwaites, British Consulate, Munich, to H. Moneypenny, British Consulate General, Frankfurt, 23.1.1947; ibid, telegram from BAFO, Berlin, to Control Office, London, 15.3.1947; ibid, memorandum from E. Boyce, MoS, to A. Hitchcock, Control Office for Germany and Austria (23 April 1947); ibid, memorandum from J. Graham, CCG(BE), to Dr. F. Hollingworth, MoS, 22.1.1948; Stephen Dorril, *MI6: Inside the Covert World of Her Majesty's Secret Intelligence Service* (New York: Simon & Schuster, 2002), 144.
35. CIOS report XXI-1, 'Organisation of Telefunken'; TNA, AVIA 54/1295, 'Scientists Employed Under DCOS Scheme by MoS. 30.8.1947'.
36. TNA, AVIA 54/1295, 'Scientists Employed Under DCOS Scheme by MoS. 30.8.1947'.
37. TNA, DEFE 43, record cards of German and other scientists and technicians, 1946–1964: 43/3, BEY–BY, 'BÜCHS, Karl'; 43/10, LA–MAN, 'LINKE, Josef'; 43/13, PLA–RIX, 'PROST, Hans'; 43/14, ROB–SCHLI, 'ROCKSTUHL, Fritz' and 'ROTHE, Paul'; 43/18, WER–ZUS,

'WILHELM, Karl'; DEFE 41/62, signal from D. Evans, D/STIB, to CCG(BE), 13.3.1947; ibid, letter from Evans to Secretary, JIC (Germany), 18.3.1947; ibid, extract from minutes of 40th JIC (Germany) meeting, 24.3.1947; AVIA 54/1295, 'Scientists Employed Under DCOS Scheme by MoS. 30.8.1947'; AVIA 54/1827, TPA3 to Est.4(S), 22.4.1950.

38. TNA, AIR 20/8677, ADI (K) Report 359/1945, 'Radio Control of Anti-Aircraft Projectiles'; DEFE 43/14, record cards of German and other scientists and technicians, 1946–1964, ROB–SCHLI, 'SCHIRRMACHER, Karl-Heinz'; AVIA 54/1295, 'Scientists Employed Under DCOS Scheme by MoS. 30.8.1947'; HO 213/2255, 'German scientists whom the MoS would wish to retain provided they acquire British nationality', 13.1.1953.

39. TNA, AVIA 54/1295, 'Scientists Employed Under DCOS Scheme by MoS. 30.8.1947'.

40. Stokes and Porter, 'Interrogation of Dr. Wilhelm Elfers', 21.5.1945, and 'Interrogation of Dr. Theodor Netzer', 18.5.1945, in *The Story of Peenemünde*, 315 and 395–399; AVIA 54/1295, 'Scientists Employed Under DCOS Scheme by MoS. 30.8.1947'; AVIA 54/1827, TPA3 to Est.4(S), 22.4.1950; HO 213/2255, 'German Scientists whom the MoS would wish to retain provided they acquire British nationality', 13.1.1953.

41. TNA, AVIA 54/1295, 'Scientists Employed Under DCOS Scheme by MoS, 30.8.1947'; CAB 122/354, 'German scientists allocated to the United States whose services are no longer required' (1948); HO 213/2255, 'German Scientists whom the MoS would wish to retain provided they acquire British Nationality', 13.1.1953.

42. TNA, AIR 40/2832, 'Interrogation of Dipl. Ing. Hermann Zumpe at FIAT (Main)', 7.11.1946; AVIA 54/1827, 'German Scientists Recommended for Unestablished Appointments in the MoS in the Grade of PSO', 3.11.1950; Von Zborowski, 'BMW Developments', 297–324.

43. TNA, DEFE 43/18, record cards of German and other scientists and technicians, 1946–1964, WER–ZUS, 'ZUMPE, Hermann'; AIR 40/2832, 'Interrogation of Dipl. Ing. Hermann Zumpe at FIAT (Main)', 7.11.1946; AVIA 54/1295, 'Scientists Employed Under DCOS Scheme by MoS. 30.8.1947'; AVIA 54/1827, TPA3 to Est.4(S), 22.4.1950.

44. TNA, AVIA 54/1296, memorandum from TPA 3a to Est. 5.CI, 22.12.1948.

45. TNA, DEFE 21/14, memorandum from J. Gardiner, JIC, to various parties, 18.9.1952; Jonathan Hagood, 'Arming and industrializing Péron's "New Argentina": the transfer of German scientists and technology after World War II', *Icon* 11 (2005): 63–78.

46. TNA, AVIA 54/1826, 'Note on Long-term Employment of German Scientists under the DCOS Scheme', 7.1.1948; ibid, 'Note on the outcome of the JIC meeting held on 1 April 1948', 15.4.1948; ibid, DRPC, extract from minutes of meeting held on Tuesday, 20 April 1948; ibid, 'Long Term Employment of German Scientists', 22.10.1948.

47. TNA, AVIA 54/1295, memorandum from E. Haddon to D/RAE, 2.6.1948; AVIA 54/1826, TPA3 note, 28.9.1948; ibid, 'Long Term Employment of German Scientists', 22.10.1948; ibid, letter from Stafford Cripps to Winston Churchill, 13.6.1949; ibid, letter from (?) to Herbert Morrison, Lord President, 29.7.1949.

48. Morton, *Fire Across the Desert: Woomera and the Anglo-Australian Joint Project 1946–1980*, 189–193.

49. Baxter, 'Further Considerations on Selection of an Oxidant for Rocket Motors'.

50. TNA, AVIA 6/19787, P. Akass, 'A Short History of the Red Shoes Project', Technical Memo GW 390, RAE Farnborough (1955), 3–9; Twigge, *The Early Development of Guided Weapons in the United Kingdom, 1940–1960*, 28–30. The Thunderbird was apparently named after the numerous species of Australian birds that will call in response to sudden loud noises. Arthur Delbridge and J. R. L. Bernard (eds), *The Macquarie Concise Dictionary* (Sydney: The Macquarie Library, 1988), 1,053.

51. TNA, AVIA 54/1828, letter from A. Hall, D/RAE, to Dr. E. Jones, TPA1(d), MoS, 22.9.1952.

52. TNA, HO 334, naturalisation certificates: 370/27529, Karl Willi Kretschmer, 18.3.1953; AVIA 65/684, letter from A. Miers, Est. 4b, MoS, to J. Scholes, HM Treasury, 1.12.1955; 'Orenda-Nobel Staff List', Avroland, https://www.avroland.ca.

53. TNA, AVIA 54/1828, 'G. Fiedler (Revision of Salary)' (November 1952); HO 334, naturalisation certificate: 366/25927, Gustav Fiedler, 26.3.1953; AVIA 6/19787, P. Akass, 'A Short History of the Red Shoes Project', 15.

54. TNA, AVIA 54/1827, 'German Scientists Recommended for Unestablished Appointments in the MoS in the Grade of PSO', 3.11.1950; ibid, 'German Scientists', note for meeting on 13.11.1950; AVIA 54/1828, letter from J. Selby, RAE, to L. Haylor, AD/TPA3, MoS, 7.7.1952; AVIA 54/1828, 'Salaries of German Scientists Assimilated by White Paper Grades' (November 1952); HO 334, naturalisation certificates: 366/25925, Jurgen Diederichsen, 26.3.1953; 367/26215, Walter Müller, 16.4.1953; 374/29597, Ulrich Barske, 19.12.1953.

55. TNA, AVIA 54/1827, 'German Scientists Recommended for Unestablished Appointments in the MoS in the Grade of PSO', 3.11.1950; AVIA 54/1828, letter from AD/TPA3 to D/TPA, 17.4.1952; HO 334, naturalisation certificates: 366/25827, Karl Meier, 19.3.1953; 366/25858, Friedrich Jessen, 23.3.1953; 366/25879, Walter Koltermann, 23.3.1953; 367/26067, Hermann Treutler, 2.4.1953; 367/26204, Werner Schonheit, 18.4.1953; 370/27870, Hermann Zumpe, 25.8.1953; 373/29124, Walter Riedel, 18.11.1953; DEFE 21/14, letter from AD/TPA, MoS, to Scientific Advisor, United Kingdom Scientific Research Organisation, Germany, 24.1.1956.

56. TNA, AVIA 54/1826, 'Minutes of Meeting Held on Friday 4 March 1949 to Consider Proposals Made for the Long-term Employment of Selected German Scientists'.

57. 'Tridac the Unique', *Flight* (15 October 1954); D. Welbourne, *Analogue Computing Methods* (Glasgow: Pergamon Ltd, 1965), 84–85; James S. Small, 'Engineering, technology and design: The post-Second World War development of electronic analogue computers', *History and Technology* 11, no. 1 (1994): 33–48; Simon Lavington, *Moving Targets: Elliott Automation and the Dawn of the Computer Age in Britain, 1947–67* (London: Springer-Verlag, 2011), 133.

58. TNA, AVIA 54/1295, memorandum from André Kenny, GWRD 7, to Boyce, 30.6.1947; AVIA 54/1827, 'German Scientists Recommended for Unestablished Appointments in the MoS in the Grade of PSO', 3.11.1950; AVIA 54/1828, letter from AD/TPA3 to D/TPA, MoS, 17.4.1952; AVIA 54/1828, 'Position Status', 25.8.1952; AVIA 54/1828, letter from Elfers to A. Hall, D/RAE, 4.12.1952; AVIA 65/684, 'Employment of Aliens Return', 31.3.1954; AVIA 65/684, 'Employment of Aliens Return', 19.10.1953; DEFE 21/14, letter from AD/TPA, MoS, to Scientific Advisor, United Kingdom Scientific Research Organisation, Germany, 23.2.1954; AVIA 65/684, 'Employment of Aliens Return', 13.4.1954; DEFE 21/14, letter from AD/TPA, MoS, to Scientific Advisor, United Kingdom Scientific Research Organisation, Germany, 19.8.1954; HO 213/2255, letter from W. Jones, Nationality Division, HO, to A. Houghton, MoS, 14.10.1954.

59. TNA, DEFE 43/14, record cards of German and other scientists and technicians, 1946–1964, ROB–SCHLI, 'ROTHE, Paul'; AVIA 54/1827, 'German Scientists Recommended for Unestablished Appointments in the MoS in the Grade of PSO', 3.11.1950; DEFE 21/13, letter from TPA3(a), MoS, to FO (German Section), 25.6.1951; AVIA 54/1828, letter from AD/TPA 3 to D/TPA, MoS, 28.3.1952; AVIA 54/1828, 'List Showing Aliens Employed by MoS in R & D and Casualties during period from April 1952 – 30 September 1952'; AVIA 54/1828, letter from E. Jones, D/TPA, to D. Ballantine, MoD, 12.12.1952; HO 213/2255, 'German Scientists whom the MoS would wish to retain provided they acquire British Nationality', 13.1.1953; DEFE 21/14, letter from AD/TPA1, MoS, to DSI, MoD, 15.4.1953; HO 213/2255, 'Position as at 15.1.1954', regarding the applications for naturalisation by German specialists, attached to letter from W. Jones to A. Houghton, MoS, 18.1.1954; AVIA 65/684, 'Employment of Aliens Return', 31.3.1954.

60. TNA, AVIA 54/1827, 'German Scientists Recommended for Unestablished Appointments in the MoS in the Grade of PSO', 3.11.1950; AVIA 54/1828, letter from J. Selby, RAE, to L. Haylor, AD/TPA3, MoS, 7.7.1952; HO 334, naturalisation certificates: 372/28609, Karl-Heinz Schirrmacher, 21.10.1953; 372/28758, Hugo Ulrich, 28.10.1953; 377/31101, Siegfried Entres, 20.4.1954; AVIA 65/684, 'Employment of Aliens Return', 31.3.1954; Becklake, 'German Engineers: Their Contribution to British Rocket Technology after World War II', 157–172.

61. TNA, AVIA 54/1827, 'German Scientists Recommended for Unestablished Appointments in the MoS in the Grade of PSO', 3.11.1950; ibid, letter from PDSR (D) to D/TPA, MoS, 7.2.1951; DEFE 21/14, letter from AD/TPA3, MoS, to FO, 13.5.1952; AVIA 54/1828, copy of letter from Neunzig to Est. 3.d., MoS, 8.12.1952; DEFE 21/14, letter from AD/TPA1, MoS, to FO (German Section), 12.3.1953; HO 334/366/25839, naturalisation certificate, Hans Ziebland, 19.3.1953; AVIA 65/684, 'Employment of Aliens Return', 13.4.1953; Becklake, 'German Engineers: Their Contribution to British Rocket Technology after World War II', 157–172.

62. TNA, AVIA 54/1827, 'German Scientists Recommended for Unestablished Appointments in the MoS in the Grade of PSO', 3.11.1950; AVIA 54/1828, memorandum from TPA1, MoS to Est.4(S), MoS, 23.10.1952.

63. TNA, HO 213/2255, letter from D. Houghton, MoS to V. A. Jones, HO, 13.7.1954.

Chapter 6

1. CIOS report 1-I, 'Radar and Controlled Missiles, Paris Area' (1945); Philippe Jung, 'The True Beginnings of French Astronautics, 1938–1959, Part I', in *History of Rocketry and Astronautics, AAS History Series 28*, ed. Frank H. Winter (San Diego: American Astronautical Society, 2007), 75–107.

2. Jean de Lattre de Tassigny, *The History of the French First Army*, trans. Malcolm Barnes (London: George Allen and Unwin Ltd, 1952), 157–158; Carpentier, *Un demi-siècle d'aéronautique en France: Les missiles tactiques de 1945 à 1995*, 19; Huwart, *Du V2 à Véronique. La naissance des fusées françaises*, 90.

3. TNA, WO 202/777, signal from SHAEF Mission (France) to General Juin, Chief, CEC, National Defence, 10.10.1944; ibid, signal from SHAEF to CO, 6th and 12th Army Groups, HQ, 21st Army Group, CO, Communications Zone, US Army, and Supreme HQ, AEF Mission (France, for French Authorities), 25.10.1944.

4. TNA, WO 219/1028, SHAEF Staff Minute Sheet, draft letter from Lt. Col. Francis Miller, Deputy Chief, Special Sections, G-2 Division, to Commanding General, 12th Army Group, 5.3.1945; ibid, signal from Lt. Col. H. Allen, Asst. Adjutant General, SHAEF, to Commanding General, 12th Army Group, 6.3.1945; ibid, signal from Lt. Col. Allen, to Commanding General, 6th Army Group, 13.3.1945; De Lattre de Tassigny, *The History of the French First Army*, 413–420 and 460–466; Bower, *The Paperclip Conspiracy*, 141.

5. TNA, AIR 40/2874, ADI (K) Report No. 288/1945, 'Development of rotation-stabilized rockets in aircraft', 20.4.1945; ibid, ADI (K) Report No. 300/1945, 'Remotely controlled missiles (Ruhrstahl A.G. Bielefeld)', 4.5.1945; AIR 40/2875, ADI (K) Report No. 312/1945, 'Remotely controlled missiles (Henschel Flugzeugwerke A.G.)', 18.5.1945.

6. 'Bibliography on German Guided Missiles', 18.

7. TNA, FO 1031/5, signal from Agwar from the Combined Chiefs of Staff to SHAEF for General D. Eisenhower, United Kingdom Base HEC, and SHAEF Rear for Secretariat, CIOS, 8.5.1945; ibid, signal from Brigadier General T. J. Davis, Adjutant General, to Head, SHAEF Mission (France), 15.6.1945; WO 219/1580, signal from HQ, 6th Army Group to AC of S, G-2 (CI), 'French Sécurité Militaire Liaison Officers Attached to Sixth Army Group', 17.5.1945; ibid, signal from Lt. Col. H. Sheen, G-2 (CI), 22.5.1945.

8. TNA, FO 1031/5, 'British Investigators Visiting the French Zone', draft policy approved by France attached to signal from Brigadier B. J. Maunsell, Chief, FIAT, to 21st Army Group, 31.7.1945; ibid, signal from Maunsell to FIAT, CCG(BE), 25.8.1945; 'Note mensuelle d'information à l'usage des sections du CEPA', 1.2.1946 – SHD, Châtellerault AA 396 4H2 28; 'Le Centre National de la Recherche Scientifique (Notes documentaires et études)', 25.4.1947 – SHD Châtellerault AA 396 4H2 4; Frank Roy Willis, The French in Germany, 1945–1949 (Stanford: Stanford University Press, 1962), 71–73.

9. 'Renseignements divers d'ordre aérodynamique pour les projectiles type "TAIFUN"' – SHD, Châtellerault, AA 396 4H2 52; 'État des notes de renseignements remises par EMG/2 LR du 24 Octobre 1944 au 10 Janvier 1946', 10.1.1946 – SHD, Châtellerault, AA 396 4H2 206.

10. Marie-France Ludmann-Obier, 'Un aspect de la chasse aux cerveaux: les transferts de techniciens allemands en France: 1945–1949', Relations internationales 46 (1986): 195–208; Bower, The Paperclip Conspiracy, 218; Wegener, The Peenemünde Wind Tunnels: A Memoir, 119; John Gimbel, Science, Technology and Reparations: Exploitation and Plunder in Postwar Germany (Stanford: Stanford University Press, 1990), 40; 'Wind Tunnel Oral History', Part 1, 5–6, White Oak Laboratory Alumni Association, https://www.wolaa.org.

11. Wegener, The Peenemünde Wind Tunnels: A Memoir, 112.

12. Kraus and Jordan, 'Ultrarotstrahlung der Grenzschicht am bewegten Wasserfall-Aggregat', HAP 66/116, 15.7.1943; Wegener, 'Vorläufige Ergebnisse der Windkanalentwicklung von Rudern mit Hilfsrudern zur Steuerung des Geräts "Wasserfall"', HAP 66/153, 31.10.1944; Wegener and Eckert, 'Ruderentwicklung im Windkanal zur Raketensteuerung im Unter- und Ueberschall', HAP 66/165, 25.4.1945; Kraus and B. Herrmann, 'Berechnung und Diskussion der Hauttemperaturen TB1 für Projekt "Wasserfall" im Flach- und Steilschuß', HAP 66/168, 1.3.1945; Kurzweg, 'Die aerodynamische Entwicklung der Flakrakete "Wasserfall"', HAP 66/171, 15.3.1945 – SHD, Châtellerault, AA 910 6F1 11.

13. TNA, FO 1031/5, signal from Brigadier General T. Davis, Adjutant General, to Head, SHAEF Mission (France), 15.6.1945.

14. Huwart, Du V2 à Véronique. La naissance des fusées françaises, 100.

15. Ibid, 62.

16. F1 Kochel (Personal), Inventaires – List de personnels, 12.9.1945 – SHD, Châtellerault, AA 910 6F1 11; 'Rapport sur l'Emploi des Technicians Allemands de la Direction des Études et Fabrications d'Armement', 22.5.1947 – SHD, Châtellerault, AA 789 1H1 373; LRBA personnel dossiers, Eckart Finger, Elsbeth Hermann, Werner Kraus, Diethelm Schnapper – SHD, Châtellerault; TNA, DSIR 23/15963, BIOS Final Report 93, 'Brief Report on the Supersonic Wind Tunnel in Kochel', 10.1.1946, 8; Huwart, Du V2 à Véronique. La naissance des fusées françaises, 98–100.

17. Compte-rendu de missions: Au BEE, par le BEE, 11.1945 à 10.1950 – SHD, Châtellerault, AA 910 6F1 11.

18. TNA, AIR 20/8677, BIOS Group II report, 'Anti-Aircraft Rocket Rheintochter with Remote Control', Dr. Klein, 17.8.1945; FO 1031/5, 'Annex A: Limitations by Subjects on Russian Access to Information in the British Zone', attached to letter from Colonel Sanderson, Secretary, JIC (CCG), Joint Intelligence Co-ordinating Branch, CCG(BE), to JIC London, 19.11.1945; FO 1031/6, signal from British Military Mission to France to CCG(BE), 23.11.1945; ibid, signal from HQ Air Division (R & D Branch) to British Military Mission to France, CCG (Economic Division) and FIAT Main (British), 4.12.1945.

19. TNA, FO 1031/5, letter from Colonel Sanderson to the Secretary, JIC, 9.2.1946; ibid, amendment to 'Limitations on Allied Access to Information in the British Zone', attached to letter from Sanderson to seven British government departments, 21.2.1946; ibid, Amendment No. 2 to 'Limitations on Allied Access to Information on British Zone', 7.3.1946.

20. 'Rapport sur l'Emploi des Techniciens Allemands de la Direction des Études et Fabrications d'Armement', 22.5.1947 – *SHD, Châtellerault, AA 789 1H1 373.*

21. Proces-verbal, 18.4.1946 and 13.5.1946, compte-rendu de missions: Au BEE, par le BEE, 11.1945 à 10.1950 – *SHD, Châtellerault, AA 910 6F1 11.*

22. Huwart, *Du V2 à Véronique. La naissance des fusées françaises*, 103.

23. Stokes, Porter and Sharpe, 'Interrogation of Helmut Weiss', 21.5.1945 and 25.5.1945, in *The Story of Peenemünde*, 468; Proces-verbal, 18.4.1946 and 13.5.1946, compte-rendu de missions: Au BEE, par le BEE, 11.1945 à 10.1950 – *SHD, Châtellerault, AA 910 6F1 11*; LRBA personnel dossier, Helmut Weiss – *SHD, Châtellerault*; Dryden *et al.*, 'Technical Intelligence Supplement', 62; Huwart, *Du V2 à Véronique. La naissance des fusées françaises*, 103.

24. H. Liebhafsky, G. Gollen and J. Iball, 'Interrogation of Bringer', 22.5.1945, in *The Story of Peenemünde*, 270; 'Report on Operation Backfire', Volume 1, 12; LRBA personnel dossier, Heinz Bringer – *SHD, Châtellerault* (dossier non communicable).

25. LRBA personnel dossiers, Wolfgang Pilz and Robert Schabert – *SHD, Châtellerault*; 'Bibliography on German Guided Missiles', 43; Huwart, *Du V2 à Véronique. La naissance des fusées françaises*, 103–116.

26. Aktenvermerk, 16.9.1946, compte-rendu de missions: Au BEE, par le BEE, 11.1945 à 10.1950 – *SHD, Châtellerault, AA 910 6F1 11.*

27. 'Rapport sur l'Emploi des Techniciens Allemands de la Direction des Études et Fabrications d'Armement', 22.5.1947 – *SHD, Châtellerault, AA 789 1H1 373.*

28. Ibid.

29. L. M. Chassin, 'France's Missile Programme', *Survival* 2, no. 3 (1960): 117–121; Carpentier, *Un demi-siècle d'aéronautique en France: Les missiles tactiques de 1945 à 1995*, 32 and 67; Huwart, *Du V2 à Véronique. La naissance des fusées françaises*, 126–129.

30. Claude d'Abzac-Epezy, 'Avions ou missiles? L'armée de l'Air face au développement des "engins Spéciaux" après 1945', *Revue Historique des Armées* 178, no. 1 (1990): 94–101.

31. Carpentier, *Un demi-siècle d'aéronautique en France: Les missiles tactiques de 1945 à 1995*, 35.

32. 'Établissement de la liasse de définition de l'accélérateur de décollage, type Walter 109-502 modifié' – *SHD, Châtellerault, AA 711 1K1 743*; Philippe Jung, 'The SE-4100 Family: An Early French Experience in Rocketry', in *History of Rocketry and Astronautics, AAS History Series, Volume 17*, ed. John Becklake (San Diego: American Astronautical Society, 1995), 103–129; Carpentier, *Un demi-siècle d'aéronautique en France: Les missiles tactiques de 1945 à 1995*, 17 and 35; Huwart, *Du V2 à Véronique. La naissance des fusées françaises*, 86.

33. Carpentier, *Un demi-siècle d'aéronautique en France: Les missiles tactiques de 1945 à 1995*, 19–21.

34. Société MATRA, 'Étude de l'engin spécial R04 et de équipement' – *SHD, Châtellerault, AA 711 1K1 746.*

35. CIOS report XXVI-30, 7–8 and 12; 'Étude des combustibles liquides. Mélanges d'acide nitrique, triéthylamine et xylidine', 24.10.1946 – *SHD Châtellerault, AA 810-112 1*; SEPR, 'Étude du propulseur Allemand BMW 548 et la fabrication de 25 réacteurs de ce type' – *SHD, Châtellerault, AA 711 1K1 741*; SEPR, 'Étude et réalisation de moteurs à réaction, utilisant comme carburant, l'eau oxygénée concentrée' – *SHD, Châtellerault, AA 711 1K1 749*; J. Villain, 'The Evolution of Liquid Rocket Propulsion in France in the Last 50 Years', *AAS History Series, Volume 17*, ed. John Becklake (San Diego: American Astronautical Society, 1995), 87–101; Kyrill von Gersdorff, 'Transfer of German Aeronautical Knowledge After 1945', in E. H. Hirschel, H. Prem and G. Madelung, *Aeronautical Research in Germany: From Lilienthal Until Today*, 325–344.

36. CIOS report XXVI-30, 3–12 and 20; Villain, 'The Evolution of Liquid Rocket Propulsion in France in the Last 50 Years', 87–101; Philippe Jung, 'Postwar German Rocketry Influence in France: An Analysis (Part II)', Fifty-fifth IAC Congress, Vancouver, Canada (2004), AIAA, https://

www.aiaa.org; Carpentier, *Un demi-siècle d'aéronautique en France: Les missiles tactiques de 1945 à 1995*, 17–21; Walther Killy *et al.* (eds), *Dictionary of German Biography, Volume 10* (Munich: KG Saur Verlag, 2006), 657.

37. SEPR, 'Fourniture de 100 moteurs fusées prototype SEPR 4 de décollage d'engins spéciaux, 20 moteurs fusées SEPR 2, pour engins spéciaux NC-3500 et NC-3501' – *SHD, Châtellerault, AA 711 1K1 765*; SNCA du Centre, 'Étude de l'engin expérimental NC-3501' – *SHD, Châtellerault, AA 711 1K1 749*; Jung, 'The SE-4100 Family: An Early French Experience in Rocketry', 103–129; Carpentier, *Un demi-siècle d'aéronautique en France: Les missiles tactiques de 1945 à 1995*, 21; George Paul Sutton, *History of Liquid Propellant Rocket Engines* (Reston: AIAA, 2006), 786–787.

38. 'Projectiles de DCA Commandés à Distance Enzian' – *SHD, Châtellerault, AA 396 4H2 55*; SNCA du Centre, 'Étude de l'engin spécial NC-3510' – *SHD, Châtellerault, AA 711 1K1 765*; Carpentier, *Un demi-siècle d'aéronautique en France: Les missiles tactiques de 1945 à 1995*, 21.

39. Bower, *The Paperclip Conspiracy*, 141; Jean-Pierre Marec *et al.*, *Un demi-siècle d'aéronautique en France: Centres et moyens d'essais, Tome II*, 139–148.

40. Jung, 'The SE-4100 Family: An Early French Experience in Rocketry', 103–129; Philippe Jung, 'The SE-4200: First Ramjet Missile?', *AAS History Series, Volume 22*, ed. Philippe Jung (San Diego: American Astronautical Society, 1998), 115–156; Carpentier, *Un demi-siècle d'aéronautique en France: Les missiles tactiques de 1945 à 1995*, 20–44.

41. Carpentier, *Un demi-siècle d'aéronautique en France: Les missiles tactiques de 1945 à 1995*, 67–70.

42. Fiche de renseignement, *Dipl.-Ing.* Ewald Müller, 1949 – *SHD, Châtellerault, AA 396 4H2 35*.

43. TNA, AIR 20/8677, ADI (K) Report 331/1945, 18.6.1945; DSIR 23/15888, J. J. Henrici and D. Mandel, 'Aerodynamics of the Butterfly', 7.3.1946; Friedrich Georg, *Hitlers letzter Trumpf: Entwicklung und Verrat der 'Wunderwaffen'*, Volume 1 (Tübingen: Grabert, 2009), 562; Kevin Mahoney, *Fifteenth Air Force Against the Axis: Combat Missions over Europe during World War II* (Plymouth: Scarecrow, 2013), 347–373.

44. Philippe Masson, 'La Marine Française en 1946', *Revue d'histoire de la Deuxième Guerre mondiale* 110 (1978): 79–86.

45. 'Nombre et délimitation des Circonscriptions Régionales du Service de la Surveillance des Travaux et des Fabrications', 19.7.1946, modifiée 9.3.1948 et 26.2.1951 – *SHD, Châtellerault, AA 763 022 7*.

46. Ibid.

47. Mark Walker, *German National Socialism and the Quest for Nuclear Power, 1939–1949* (Cambridge: Cambridge University Press, 1993), 186; Winfried Engler, *Frankreich an der Freien Universität: Geschichte und Aktualität* (Stuttgart: Franz Steiner Verlag, 1997), 103 and 121–122; Olivier Huwart, *Sous-marins français, 1944–1954: La décennie du renouveau* (Rennes: Marines Éditions, 2003), 48; Carpentier, *Un demi-siècle d'aéronautique en France: Les missiles tactiques de 1945 à 1995*, 17, 20 and 34; Jean Robert, Jean-Jacques Serra and Philippe Jung, 'Maruca: An Early French Liquid Fuelled Rocket', presented at the 57th IAF Congress, 40th IAA History of Astronautics Symposium, Valencia, 5 October 2006; Norbert Desgouttes, *Les Commandements de l'Aéronautique Navale 1912–2013* (2013), Association pour la recherche de documentation sur l'histoire de l'aéronautique navale, https://www.aeronavale.org.

48. CIOS report XXXI-1, 'Establishments of the Forschungsanstalt der Deutschen Reichspost' (1945), 39–42; Marine Nationale État-Major Général, Deuxième Bureau Liaison Recherches, 'Armes nouvelles allemandes. Mise de feu infra-rouge PAPLITZ de la firme ELAC', Note de Renseignements No. 83, 22.2.1946 (translation of an Anglo-American report) – *SHD, Châtellerault, AA 396 4H2 225*; *Who's Who in Germany, A–L* (Who's Who Sutton's International Red Series, 1992), 923; Dominique Pestre, 'Guerre, renseignement scientifique et reconstruction, France, Allemagne et Grande-Bretagne dans les années 1940', in *De la Diffusion des sciences à l'espionnage industriel XVe–XXe siècle. Actes de la colloque de Lyon (30–31 mai 1996) de la SFHST,*

ed. André Guillerme (Fontenay Saint-Cloud: ENS Editions, 1999), 183–204; Huwart, *Sous-marins français, 1944–1954: La décennie du renouveau*, 48; Jung, 'Postwar German Rocketry Influence in France: An Analysis (Part II)'.

49. Lasby, *Project Paperclip: German Scientists and the Cold War*, 67.
50. At the Châtellerault archives in May 2015, I requested to read this dossier, but for reasons unknown I was denied access. The dossier is: "Engin radio-guidé allemand H.S. 117 ou Schmetterling (note de renseignements)" – *SHD, Châtellerault, AA 396 4H2 51*.
51. Carpentier, *Un demi-siècle d'aéronautique en France: Les missiles tactiques de 1945 à 1995*, 33 and 67; Robert, Serra and Jung, 'Maruca: An Early French Liquid Fuelled Rocket'; Jean-Pierre Marec *et al.*, *Un demi-siècle d'aéronautique en France: Centres et moyens d'essais, Tome I*, 31.

Chapter 7

1. Harlow, 'Alpha, Beta, and RTV 1: The Development of Early British Liquid Propellant Rocket Engines'; Neufeld, *Von Braun: Dreamer of Space, Engineer of War*, 282–295.
2. TNA, FO 371/109785, 'German proposals for collaboration with United Kingdom on guided weapons. Report by Controller of Guided Weapons and Electronics', MoS (1954).
3. David Miller, *The Cold War: A Military History* (London: John Murray, 1998), 287.
4. 'Missile, Surface-to-Air, Rheinmetall-Borsig Rheintochter R-1', Smithsonian NASM, https://www.si.edu/object/missile-surface-air-rheinmetall-borsig-rheintochter-r-i%3Anasm_A19660037000; 'Rheintochter R I Missile', Smithsonian NASM, https://www.airandspace.si.edu/collection-objects/rheintochter-r-i-missile/nasm_A19710756000; 'Hs 117 Schmetterling (Butterfly) Missile', Smithsonian NASM, https://www.airandspace.si.edu/collection-objects/missile-surface-air-henschel-hs-117-schmetterling/nasm_A19890595000.

Appendix 1

1. TLR/Flak-E 5, az. 101a and 140, Nr. 194/45, 1945, in *Story of Peenemünde*, 523–527; 'Burgund Control Equipment for the Rocket Schmetterling', Intelligence Report GDM-1, Signal Corps, USFET, 28.6.1945, and 'General Report on Guided Missiles', Intelligence Report GDM-2, Signal Corps, USFET, 11.7.1945, and 'German Cartesian Co-ordinate Control System for Supersonic Rockets', Intelligence Report GDM-5, Signal Corps, USFET, 23.7.1945, RG 38, National Archives College Park; US Naval Technical Mission in Europe Technical Report No. 236-45, 'General Survey of Rocket Motor Development in Germany', 5.9.1945; TNA, AIR 40/2167, H.F. King, 'German Jet-propulsion Units', AI2(g) Report No. 2373, 5.9.1945; DSIR 23/14848, 5-7; DSIR 23/15145, 48–50, 108–109 and 116; DEFE 15/216; DEFE 15/217; Dryden *et al.*, 'Technical Intelligence Supplement', 40–41; Gatland, *Development of the Guided Missile*, 114–123; Vüllers, 'Design and Development of the Solid Fuel Rocket', 253–262.

Appendix 2

1. See Dean Reuter, Colm Lowery and Keith Chester, *The Hidden Nazi* (Washington DC: Regnery History, 2019).

Appendix 3

1. Werrell, *Archie, Flak, AAA and SAM: A Short Operational History of Ground-Based Air Defense*, 74; Louis Brown, *A Radar History of World War II: Technical and Military Imperatives* (Bristol and Philadelphia: Institute of Physics Publishing, 1999), 452; Boris Chertok, *Rockets and People, Volume I* (Washington DC: NASA History Division, 2005), 176; Asif Siddiqi, *The Red Rockets' Glare: Spaceflight and the Soviet Imagination, 1857–1957* (New York: Cambridge University Press, 2010), 198–199.

2. TNA, AIR 20/8773, BIOS Final Report 1110, 5; AVIA 20/8640, H. Klein *et al.*, '"Rheintochter" Rocket Motor System of Drive with Liquid Propellants', 15.12.1946; Siddiqi, *The Red Rockets' Glare: Spaceflight and the Soviet Imagination, 1857–1957*, 198–199.

3. 'Notes on German Weapons Developments', Seventh Army Interrogation Centre, 3.6.1945; Norman Naimark, *The Russians in Germany: A History of the Soviet Zone of Occupation, 1945–1949* (Cambridge, Massachusetts: Harvard University Press, 1995), 177; Michael J. Neufeld, 'Rolf Engel vs the German Army: A Nazi Career in Rocketry and Repression', *History and Technology* 13 (1996): 53–72.

4. TNA, DSIR 23/15145, 68.

5. Siddiqi, *The Red Rockets' Glare: Spaceflight and the Soviet Imagination, 1857–1957*, 199–210.

6. Naimark, *The Russians in Germany: A History of the Soviet Zone of Occupation, 1945–1949*, 11.

7. Siddiqi, *The Red Rockets' Glare: Spaceflight and the Soviet Imagination, 1857–1957*, 217–231; Christian Lardier and Stefan Barensky, *The Soyuz Launch Vehicle: The Two Lives of an Engineering Triumph*, trans. Tim Bowler (New York: Springer, 2013), 13–14.

8. Boris Chertok, *Rockets and People, Volume II: Creating a Rocket Industry* (Washington DC: NASA History Division, 2006), 9–15. Special Committee No. 1 was the State Defence Committee Atomic Committee, which was set up in August 1945 after the atomic bombs were dropped on Hiroshima and Nagasaki; and Special Committee No. 3, which was created in June 1947, was formerly the State Defence Committee Radar Council that was set up in early June 1943. Ibid, 4–7.

9. TNA, AIR 40/2543, 'Technisches Büro No. 11, the former Institute Berlin', 3.1.1949. This is a British copy of a US Army report, dated 19.10.1948, which the Americans passed on to the STIB in Germany.

10. Ibid.

11. Ibid; ADM 260/140, 'Calculator "Schmetterling A"', British Naval Gunnery Mission, Special Technical Report Number 7 (1945). This report mentions the removal of the calculator from the *Kreiselgeräte* main works in Berlin by the Red Army.

12. TNA, DEFE 15/216; AIR 40/2832, 'Interrogation of Dipl. Ing. Hermann Zumpe at FIAT (Main)', 7.11.1946.

13. Chertok, *Rockets and People, Volume II: Creating a Rocket Industry*, 84–85 and 114.

14. TNA, AIR 40/2543; AIR 40/2832; Naimark, *The Russians in Germany: A History of the Soviet Zone of Occupation, 1945–1949*, 228–229.

15. TNA, AIR 40/2543.

16. 'Technical report on decimeter radio equipment for controlling bombs and rockets', EEIS, Signal Corps, US Ninth Army, 28.5.1945, RG 38, National Archives College Park.

17. TNA, AIR 40/2543.

18. Ibid.

19. Paul Maddrell, *Spying on Science: Western Intelligence in Divided Germany 1945–1961* (Oxford: Oxford University Press, 2006), 117.

20. Lardier and Barensky, *The Soyuz Launch Vehicle: The Two Lives of an Engineering Triumph*, 21.
21. Krag, 'Special Features of Anti-Aircraft Rockets with Swept or Low-Ratio-Aspect Wings', 545–614.
22. Chertok, *Rockets and People, Volume II: Creating a Rocket Industry*, 68–69 and 199–210.
23. Michal Fiszer, 'Moscow's Air Defense Network, Part 1: Foundations in Fear', *Journal of Electronic Defense* 27, no. 12 (2004): 41–48.
24. Lardier and Barensky, *The Soyuz Launch Vehicle: The Two Lives of an Engineering Triumph*, 21–24.

Glossary

1. Partly taken from Cagle, *Development, Production and Deployment of the Nike-Ajax Guided Missile System 1945–1959*, and *The Macquarie Concise Dictionary*.

Bibliography

Primary sources

The National Archives, Kew, Surrey

Admiralty

ADM 1/20374, 'Use by British industry of German scientific and industrial knowledge: recruitment of German scientists and technicians under the modified Darwin Panel Scheme', 1947–1952.

ADM 256/140, 'Calculator "*Schmetterling A*"', British Naval Gunnery Mission, Special Technical Report Number 7, 1945.

Air Ministry

AIR 20/8638, 'Anti-aircraft Rocket *Wasserfall:* Aerodynamic Development', 1945.

AIR 20/8640, '*Rheintochter* rocket motor system of drive with liquid propellants', Technical Information Bureau translation of report by H. Klein, J. Hennes, E. Prier and A. Weidmann dated 15 December 1946.

AIR 20/8677, ADI(K) report 331/1945, 'Remotely controlled AA projectiles "*Rheintochter*" and "*Taifun*"'; and ADI(K) report 359/1945, 'Radio control of German AA projectiles', 1945.

AIR 20/8678, 'Anti-aircraft rocket "*Rheintochter*" with remote control', Halstead Exploiting Centre translation of report by Dr. Klein dated 17 August 1945.

AIR 20/8773, British Intelligence Objectives Subcommittee Final Report No. 1110, 'Some aspects of German rocket developments', 1946.

AIR 34/702, 'Reports on German Anti-Aircraft Weapons', August 1943–May 1945.

AIR 40/1151, 'German Flak: Details of Disposition, Size, etc., MI15 Summaries of Flak Operations in Germany', May 1942–July 1945.

AIR 40/1310, 'Section IV – Part 4 – Controlled Missiles', MAP, 7 August 1945.

AIR 40/2162, AI2(g) reports 1751–2001, 1 March 1945–31 July 1946.

AIR 40/2167, AI2(g) reports 2284-2374, 1 November 1944-31 October 1945.

AIR 40/2458. 'Report on Interrogation of German Air Ministry Technical Personnel', 1945.

AIR 40/2532, 'German ground-to-air and air-to-air missiles: assorted photographs, drawings and negatives', 1945.

AIR 40/2542, 'Survey of Facilities in Germany for Development of Guided Missiles', 1945 (ALSOS report).

AIR 40/2543, '*Technisches Büro 11:* Department, Previously Named Institute Berlin', 1949 (US Army intelligence report).

AIR 40/2832, 'Interrogation of Dipl. Ing. Hermann Zumpe at FIAT (Main)', November 1946.

AIR 40/2874, 'ADI(K) Reports: Nos 151–300 (1945): information obtained from German POWs', 1945.

AIR 40/2875, 'ADI(K) Reports: Nos 300–399 (1945): information obtained from German POWs', 1945.

Ministry of Supply

AVIA 6/13225, R. C. Murray, 'Developments in liquid bi-fuel rocket technique at the MOS Establishment at Trauen, Germany. Jan–June 1946', August 1946.

AVIA 6/15499, 'History of Development of RTV 1', 1950.

AVIA 6/15882, 'A Review of Rocket Developments at BMW Until November 1944 (*Über die R-Entwicklung bei BMW November 1944*)', translated by R. C. Murray, January 1947.

AVIA 6/19787, 'Short History of the Red Shoes Project', 1955.

AVIA 12/82, 'Operation Surgeon: Memorandum', 1946.

AVIA 40/4466–4516, *Schmetterling* engineering drawings.

AVIA 40/5206, *'Modellflügel'*, 1944 (*Rheintochter*).

AVIA 40/4812, 'Missile 1' (*Enzian*).

AVIA 41/388, 'Chemical Analysis of German Rocket Propellants: Tentative Results', 1946.

AVIA 48/1, 'Coordinating Committee on Guided Missiles and Projectiles: Organisation', 1945–1947.

AVIA 48/3, 'Brakemine: Minutes of First to Tenth Meetings', 1946–1947.

AVIA 49/10, 'BIOS Working Party: classification of Halstead Exploitation Centre Documents', 1946–1948.

AVIA 54/1295, 'Employment of Germans in United Kingdom Under DCOS Scheme: General', 1946–1953.

AVIA 54/1296, 'Employment of Germans Under DCOS Scheme: Families', 1946–1949.

AVIA 54/1403, 'Employment of German Scientists and Technicians: Denial Policy', 1946–1950.

AVIA 54/1404, 'Halstead Exploiting Centre: review of activities and eventual closure; security gradings and lists of documents', 1946–1950.

AVIA 54/1407, 'Plans to Exploit German Work on Guided Projectiles and Rockets (Unterlüss)', 1946–1948.

AVIA 54/1826, 'Long Term Employment of German Scientists', 1947–1950.

AVIA 54/1827, 'Long Term Employment of German Scientists', 1950–1952.

AVIA 54/1828, 'Long Term Employment of German Scientists', 1952–1953.

AVIA 65/684, 'German Scientists: Conditions of Service and Promotion', 1948–1956.

Cabinet Office

CAB 122/350, 'German Scientists (Defence)', 1947.

CAB 122/354, 'German Scientists (Defence)', 1948.

CAB 176/3, 'Joint Intelligence Sub-Committee, Secretariat: minutes (1944) 5–1002, 1 January 1944–17 July 1944'.

CAB 176/4, 'Joint Intelligence Sub-Committee, Secretariat: minutes (1944) 1003–1675, 17 July 1944–30 December 1944'.

Ministry of Defence

DEFE 15/194, 'Examination of German *Enzian* Rocket Motor', July 1945.

DEFE 15/216, 'Examination of German *Wasserfall* (C2) Guided Anti-Aircraft Rocket', 1946.

DEFE 15/217, 'German Non-Guided Flak Rocket *Taifun*', 1946.

DEFE 21/13, 'German Scientists', 1950–1952.

DEFE 21/14, 'German Scientists', 1952–1957.

DEFE 41/62, 'Joint Intelligence Committee (JIC) Germany: Minutes', 1946–1947.

DEFE 43/3, 'Foreign Office (FO) and Ministry of Defence (MoD): Scientific and Technical Intelligence Branch (STIB) and related bodies: Record Cards of German and other Scientists and Technicians, 1946–1964, BEY-BY'.

DEFE 43/10, 'FO and MoD: STIB and related bodies: Record Cards of German and other Scientists and Technicians, 1946–1964, LA-MAN'.

DEFE 43/13, 'FO and MoD: STIB and related bodies: Record Cards of German and other Scientists and Technicians, 1946–1964, PLA-RIX'.

DEFE 43/14, 'FO and MoD: STIB and related bodies: Record Cards of German and other Scientists and Technicians, 1946–1964, ROB-SCHLI'.

DEFE 43/15, 'FO and MoD: STIB and related bodies: Record Cards of German and other Scientists and Technicians, 1946–1964, SCHLO-SEI'.

DEFE 43/17, 'FO and MoD: STIB and related bodies: Record Cards of German and other Scientists and Technicians, 1946–1964, THI-WEN'.

DEFE 43/18, 'FO and MoD: STIB and related bodies: Record Cards of German and other Scientists and Technicians, 1946–1964, WER-ZUS'.

Department of Scientific and Industrial Research

DSIR 23/14848, CIOS report XXVII-67, 'Aerodynamics of rockets and ramjets research and development work at *Luftfahrtforschungsanstalt Hermann Göring*', 1945.

DSIR 23/15140, 'Visit to *Messerschmitt* Design and Development Department, Oberammergau, Bavaria, 18–25 June', 1945.

DSIR 23/15145, CIOS report XXXII-125, 'German Guided Missile Research', 1945.

DSIR 23/15888, 'Aerodynamics of the Butterfly (*Schmetterling*)', 1946.

DSIR 23/15963, 'Supersonic Wind Tunnel at Kochel (BIOS interrogation)', 1946.

Foreign and Colonial Office

FO 371/109785, 'Proposal for collaboration with West Germany on development of guided weapons', 1954.

FO 1031/5, 'Liaison with Russians, French and Other Allies: Policy', 1945–1946.

FO 1031/6, 'Liaison with Russians, French and Other Allies: Individual Cases', 1945–1947.

FO 1031/12, 'Operation Abstract', 1947.

FO 1031/51, 'British Intelligence Objectives Subcommittee – General', 1945.

FO 1031/236, 'Lists of Documents found at Wesermünde', 1947.

FO 1050/1417, 'Combined Intelligence Priorities Committee (CIPC) Black List', 1944.

Home Office

HO 213/2255, 'German Scientists Working for Ministry of Supply: Exemption from Naturalisation Restrictions', 1952–1955.

HO 334, 'Home Office: Immigration and Nationality Department, Certificates of Naturalisation', 1870–1987.

War Office

WO 202/777, 'French Liaison', 1 October 1944–31 July 1945.

WO 219/1028, 'T-Force: Miscellaneous Papers', 1945.

WO 219/1580, 'Organisation of the French Military Security Service: Requirements of French Liaison Officers', 1944–1945.

Imperial War Museums, London

CIOS report XXI-1, 'Organization of *Telefunken*', 1945.

CIOS report XXVI-83, '*Bayerische Motor Werke*', 1945.

CIOS report XXVIII-41, '*Institut für Physikalische Forschung Neu Drossenfeld*', 1945.

CIOS report XXVIII-53, '*Walterwerke* Kiel', 1945.

CIOS report XXX-80, 'Bavarian Motor Works (BMW) – A production survey', 1945.

CIOS report XXXI-2, 'Research work undertaken by the German universities and technical high schools for the *Bevollmachtigter für Hochfrequenztechnik*; independent research on associated subjects', 1945.

CIOS report XXXII-66, '*Deutsche Forschungsanstalt für Segelflug* Ainring', 1945.

CIOS report XXXII-88, '*Stassfurter Rundfunk*, Stassfurt', 1945.

National Archives of Australia

D874, NG592, '*Enzian* – German World War II surface-to-air missile at Store 2 Salisbury', 1954.

D879, DB16, '*Schmetterling* plans: *Flak Sicht Gerät* (A) and *Flak Folge Gerät* (B)'.

D879, DB17, '*Schmetterling* plans: A *Bodo* – B aerial array with *Kehl* – C transmitter *Kehl*'.

D897, K61/95, 'Woomera – Thunderbird, 1961'.

US National Archives and Records Administration, College Park, Maryland, United States

Record Group 38 US Naval Technical Mission in Europe. A collection of 12 technical intelligence reports by US military intelligence organisations based on interrogations of Dr. Theodor Sturm.

Memorandum from Commander in Chief, United States Fleet and Chief of Naval Operations, to various parties, subject: US Naval Technical Mission in Europe, 26 December 1944.

'Report on Operation Backfire'. Volume 1, United Kingdom War Office, January 1946.

Smithsonian National Air and Space Museum

'Missile, Surface-to-Surface, Hermes A-1, Experimental'.

Deutsches Museum, Munich

'*Projekt Wasserfall*', 6 September 1943.

Centre des Archives de l'Armement et du Personnel Civil (CAAPC), Châtellerault, Service Historique Défense

Serie 396/4H2, '*Collection de Documents Concernant les Armements et les Techniques*', 1808–1972.

Serie 711/1K1, '*Marches Passes par la Direction des Constructions Aéronautiques*', 1945–1963.

Serie 763/022, '*Cabinet Delegation Générale pour l'Armement*', 1934–1971.

Serie 789/1H1, '*Archives du Cabinet de la Direction des Armements Terrestres*', 1911–1993.

Serie 810/112, '*Notes Technique du Laboratoire Balistique*', 1944–1968.

Serie 910/6F1, '*Laboratoire de Recherches Balistiques et Aérodynamiques de Vernon, Archives Allemandes sur les V2*', 1915–1951.

Laboratoire de Recherches Balistiques et Aérodynamiques (LRBA) personnel dossiers.

The Foundation for German Communications and Related Technologies

CIOS report I-1, 'Radar and Controlled Missiles Paris Area', 1944.

CIOS report XXII-19, '*I.G. Farbenindustrie A.G.*, Leuna', 1945.

CIOS report XXIII-15, '*I.G. Farbenindustrie A.G.* Frankfurt/Main', 1945.

CIOS report XXIV-12, '*I.G. Farbenindustrie*-Oppau Works Ludwigshafen', 1945.

CIOS report XXVI-3, '*Seefliegerhorst Wesermünde* (Evacuation from *Erprobungsstelle der Luftwaffe*, Karlshagen)', 1945.

CIOS report XXVI-30, 'Gas turbine development at BMW', 1945.

CIOS report XXXI-1, 'Establishments of the *Forschungsanstalt der Deutschen Reichspost*', 1945.

CIOS report XXXI-71, 'Interrogation of Helmut Gröttrup, Dipl. Ing, *Elektromechanisch Werke*', 1945.

CIOS report XXXII-38, 'Explosives summary of capacity and production in Germany', 1945.

BIOS Final Report 160, '*Luftfahrtforschungsanstalt Hermann Göring* Völkenrode, Brunswick', 1945.

BIOS Final Report 163, 'Some German aircraft armament projects with particular reference to fire control developments', November 1945.

BIOS Final Report 867, 'Television development and application in Germany', 1946.

Lawrence, Lovell. 'General survey of rocket motor development in Germany'. US Naval Technical Mission in Europe Report 236-45, September 1945.

Other documents

American Power Jet Company. 'Analysis and Evaluation of German Attainments and Research in the Liquid Rocket Engine Field'. Central Air Documents Office (Army-Navy-Air Force), Dayton, 1952. DTIC.

Baxter, A. 'Further Considerations on Selection of an Oxidant for Rocket Motors'. RPD Technical Note 11, February 1949. DTIC.

'Bibliography on German Guided Missiles'. Headquarters Army Air Forces Air Materiel Command, Dayton, July 1946. Hathitrust.

Bragg, James W. *Development of the Corporal: The Embryo of the Army Missile Program, Volume I.* Huntsville: Army Ballistic Missile Agency, 1961. DTIC.

Broughton, L. and Frauenberger, H. 'A Comparison of Liquid and Solid Propellant Boost Rocket Motors'. RPD Technical Note 25, January 1950. DTIC.

Bullard, John W. *History of the Redstone Missile System.* Huntsville: Army Missile Command, 1965. DTIC.

Cagle, Mary T. *Development, Production and Deployment of the Nike-Ajax Guided Missile System 1945–1959.* Huntsville: US Army Rocket and Guided Missile Agency, 1959.

Cagle, Mary T. *Loki Anti-aircraft Free Flight Rocket December 1947–November 1955.* Huntsville: US Army Ordnance Corp, 1957. DTIC.

De Lattre de Tassigny, Jean. *The History of the French First Army.* Translated by Malcolm Barnes. London: George Allen and Unwin Ltd, 1952.

Dornberger, Walter. *V-2.* London: Hurst and Blackett, 1954.

Dryden, H. L., Tsien, H. S., Wattendorf, F. L., Williams, F. W., Zwicky, F. and Pickering, W. H. 'Technical Intelligence Supplement: A Report of the AAF Scientific Advisory Group'. Headquarters Army Air Forces Air Materiel Command, Dayton, 1946. DTIC.

Dryden, H. L., Morton, G. A. and Getting, I. A. 'Guidance and Homing of Missiles and Pilotless Aircraft: A Report of the AAF Scientific Advisory Group'. Headquarters Army Air Forces Air Materiel Command, Dayton, 1946. DTIC.

Fricke, Werner. 'Test Firing of R-I Guided Missile Rheintochter'. August 1947. (English translation of German report, *Schuss von Gerät R-I*, July 1944.) DTIC.

'German Mechanical Time Fuzes'. US Naval Technical Mission in Europe Technical Report No. 491-45, September 1945. DTIC.

'Ground to Air Pilotless Aircraft'. Air Technical Service Command, 1 October 1945. DTIC.

'Historical data on US Naval Technical Mission in Europe: first narrative'. November 1945. Fischer-Tropsch Archive.

Holder, D.W. 'The High-Speed Laboratory of the Aerodynamics Division, NPL'. Aeronautical Research Council Reports and Memoranda No. 2560, December 1946. London: Her Majesty's Stationery Office, 1954. AERADE.

House, William C. 'Project Squid: Liquid Propellant Rockets Field Survey Report'. Volume II Part 2, Princeton University, 30.6.1947. DTIC.

'Interrogation of Albert Speer, former Reich Minister of Armaments and War Production'. SHAEF G-2, 29 May 1945. Cornell University Law Library.

Long, W. 'The Design of Emplacements for Rocket Motor Testing'. RPD Technical Note 16, June 1949. DTIC.

Mair, W. A. (ed.). 'Research on High-Speed Aerodynamics at the Royal Aircraft Establishment from 1942 to 1945'. Aeronautical Research Council Reports and Memoranda No. 2222, September 1946. London: His Majesty's Stationery Office, 1950. AERADE.

'Nike: the US Army's Guided Missile System'. Western Electric book rack service for Employees. Date of publication unknown.

'Nitroglycerin, Diethylene Glycol Dinitrate and Similar Explosive Oils – Manufacture and Development in Germany'. US Naval Technical Mission in Europe Technical Report No. 257-45, September 1945. DTIC.

'Notes on conference on 14.10.44 at TLR/Flak-E about the organisation and the presumed supply points for *Flak-Rakete "Schmetterling"*'. TLR/Flak-E. 5/II, Az. 144 Nr. 703/44, 16.10.1944; and TLR/Flak-E. 5/II B, Az. 103 Nr. 745/44, 22.10.1944. *Deutsche Luftwaffe.*

'Notes on German Weapons Developments'. US Seventh Army Interrogation Centre, 3 June 1945. Cornell University Law Library.

'Project Hermes Interim Report Number 1'. General Electric Company. Film, date unknown. Online Archive.

'Semi-Annual Report of Bumblebee Project July–December 1949'. Johns Hopkins University Applied Physics Laboratory, March 1950. DTIC.

Simon, Leslie E. *German Scientific Establishments*. New York: Mapleton House, 1947. Hathitrust.

Singelmann, D. and Müller, H. 'BMW Rocket-engine Development'. Headquarters Air Materiel Command, Wright-Patterson Air Force Base, May 1948. DTIC.

The Story of Peenemünde, or What Might Have Been (also known as Peenemünde-East, through the Eyes of 500 Detained at Garmisch). US Army Ordnance Department, 1945. Online Archive.

'Trainers for Operators of Guided Missiles'. US Naval Technical Mission in Europe Technical Report No. 229-45, September 1945. DTIC.

United Kingdom Government, 'Patents, Designs, Copyright and Trade Marks (Emergency) Act, 1939', https://www.legislation.gov.uk/ukpga/Geo6/2-3/107/enacted.

US Department of the Interior and Building Technology Incorporated. 'Historic Properties Report. White Sands Missile Range, New Mexico, and Subinstallation Utah Launch Complex, Green River, Utah'. Final Report, July 1984. DTIC.

Von Kármán, Theodore. 'Where We Stand: A Report of the AAF Scientific Advisory Group'. Headquarters Army Air Forces Air Material Command, Dayton, 1946. DTIC.

White, L. D. 'Final Report, Project Hermes V-2 Missile Program'. General Electric Guided Missiles Department, September 1952. Smithsonian Libraries.

Wiseman, L. 'The Suitability of the Oxidants Liquid Oxygen, Hydrogen Peroxide and Nitric Acid in Liquid Propellant Systems for Operational Use'. ERDE Technical Memorandum No. 13/M/48, October 1948. DTIC.

Secondary sources

American Men and Women of Science: The Physical and Biological Sciences, Volume 2. Jacques Cattell Press, 1982.

Avroland Canada, https://www.avroland.ca.

Babcock, Elizabeth. History of the Navy at China Lake, California, Volume 3, Magnificent Mavericks: Transition of the Naval Ordnance Test Station From Rocket Station to Research, Development, Test, and Evaluation Center, 1948–1958. Washington DC: Naval Historical Center and the Naval Air Systems Command, 2008.

Baucom, Daniel. 'Eisenhower and Ballistic Missile Defence: The Formative Years, 1944–1961'. Air Power History 51, no. 4 (2004): 4–17.

Becklake, John. 'German Engineers: Their Contribution to British Rocket Technology after World War II'. In History of Rocketry and Astronautics, AAS History Series, Volume 22, edited by Philippe Jung, 157–172. San Diego: American Astronautical Society, 1998.

Becklake, John. 'German Rocket Engineers in Britain – Their Influence Revisited'. Acta Astronautica 59 (2006): 510–515.

'BOMARC: Boeing's Long-range A.A. Missile'. Flight, 24 May 1957.

Boog, Horst, Rahn, Wernher, Stumpf, Reinhard and Wegner, Bernd. Germany and the Second World War, Volume VI: The Global War. Oxford: Oxford University Press, 2001.

Boog, Horst, Krebs, Gerhard and Vogel, Detlef. Germany and the Second World War, Volume VII: The Strategic Air War in Europe and the War in West and East Asia 1943–1944/5. Oxford: Oxford University Press, 2006.

Bower, Tom. The Paperclip Conspiracy: The battle for the spoils and secrets of Nazi Germany. London: Michael Joseph, 1987.

Brown, Louis. A Radar History of World War II: Technical and Military Imperatives. Bristol and Philadelphia: Institute of Physics Publishing, 1999.

Burgess, Eric. 'German Guided Missiles and Rockets'. The Engineer, no. 184 (October 1947).

Campbell, Estelle. 'The V2 rocket – how it worked and how we acquired it'. Australian War Memorial, https://www.awm.gov.au/articles/blog/v2rocket.

Carpentier, René. Un demi-siècle d'aéronautique en France: Les missiles tactiques de 1945 à 1995. Comité pour l'histoire de l'aéronautique. Academie de l'air et de l'espace, https://www.academie-air-espace.com.

Chassin, L. M. 'France's Missile Programme'. Survival 2, no. 3 (1960): 117–121.

Chertok, Boris. Rockets and People, Volume I. Washington DC: NASA History Division, 2005.

Chertok, Boris. Rockets and People, Volume II: Creating a Rocket Industry. Washington DC: NASA History Division, 2006.

Clark, John D. Ignition! An Informal History of Liquid Rocket Propellants. New Jersey: Rutgers University Press, 1972.

Cochrane, Rexmond C. The National Academy of Sciences: The First Hundred Years 1863–1963. Washington DC: National Academy of Sciences, 1978.

Collier, Basil. The Defence of the United Kingdom. London: HM Stationery Office, 1957.

Converse III, Elliott V. Rearming for the Cold War 1945–1960. Washington DC: Historical Office, Office of the Secretary of Defense, 2012.

Cronich, L. L. "Aerodynamic Development of Fleet Guided Missiles in the Navy's Bumblebee Program'. Presented at the Seventeenth Aerospace Sciences Meeting, 15–17 January 1979.

d'Abzac-Epezy, Claude. 'Avions ou missiles? L'armée de l'Air face au développement des engins spéciaux apres 1945'. Revue historique des Armées, no. 178 (1990): 94–101.

Delbridge, Arthur, and Bernard, J. R. L. (eds). *The Macquarie Concise Dictionary*. The Macquarie Library, 1988.

Desgouttes, Norbert. *Les Commandements de l'Aéronautique Navale 1912–2013*. Association pour la recherche de documentation sur l'histoire de l'aéronautique navale, 2013. https://www.aeronavale. org.

Dorril, Stephen. *MI6: Inside the Covert World of Her Majesty's Secret Intelligence Service*. New York: Simon & Schuster, 2002.

Dougherty, Kerrie. 'A German rocket team at Woomera? A lost opportunity for Australia'. *Acta Astronautica* 55 (2004): 741–751.

Edmaps, https://www.edmaps.com.

Engler, Winfried. *Frankreich an der Freien Universität: Geschichte und Aktualität*. Stuttgart: Franz Steiner Verlag, 1997.

Faber, Harold (ed.). *Luftwaffe: An Analysis by Former Luftwaffe Generals*. London: Sidgwick and Jackson, 1979.

Fiszer, Michal. 'Moscow's Air Defense Network, Part 1: Foundations in Fear'. *Journal of Electronic Defense* 27, no. 12 (2004): 41–48.

Gatland, Kenneth. *Development of the Guided Missile*. London: Iliffe, 1952.

Georg, Friedrich. *Hitlers letzter Trumpf: Entwicklung und Verrat der 'Wunderwaffen', Volume 1*. Tübingen: Grabert, 2009.

Gerrard-Gough, J. D. and Christman, Albert B. *History of the Naval Weapons Center, China Lake, California, Volume 2: The Grand Experiment at Inyokern*. Washington DC: Naval History Division, 1978.

Gimbel, John. 'Project Paperclip: German Scientists, American Policy, and the Cold War'. *Diplomatic History* 14, no. 3 (1990): 343–365.

Gimbel, John. *Science, Technology and Reparations: Exploitation and Plunder in Postwar Germany*. Stanford: Stanford University Press, 1990.

GlobalSecurity.org, https://www.globalsecurity.org.

Goodman, Michael S. *The Official History of the Joint Intelligence Committee, Vol. 1, From the Approach of the Second World War to the Suez Crisis*. New York and London: Routledge, 2014.

Gudgin, Peter. *Military Intelligence: A History*. Stroud: Sutton, 1999.

Guillerme, André (ed.). *De la diffusion des sciences à l'espionnage industriel, XVe–XXe siècle*. Saint-Cloud: ENS Editions, 1999.

Gunston, Bill. *The Illustrated Encyclopedia of the World's Rockets and Missiles*. London: Salamander Books, 1979.

Hagood, Jonathan. 'Arming and industrializing Péron's "New Argentina": the transfer of German scientists and technology after World War II'. *Icon* 11 (2005): 63–78.

Hall, Charlie. *British Exploitation of German Science and Technology, 1943–1949*. Abingdon and New York: Routledge, 2019.

Harlow, J. 'Alpha, Beta, and RTV 1: The Development of Early British Liquid Propellant Rocket Engines'. In *History of Rocketry and Astronautics, AAS History Series, Volume 22*, edited by Philippe Jung, 173–201. San Diego: American Astronautical Society, 1998.

Hirschel, E. H., Prem, H. and Madelung, G. *Aeronautical Research in Germany: From Lilienthal until Today*. Berlin: Springer-Verlag, 2004.

Höhne, Heinz. *The Order of the Death's Head: The Story of Hitler's SS*. London: Penguin, 2000.

Huwart, Oliver. *Sous-marins français, 1944–1954: La décennie du renouveau*. Rennes: Marines Éditions, 2003.

Huwart, Olivier. *Du V2 à Véronique. La naissance des fusées françaises*. Rennes: Marines Éditions, 2004.

Huzel, Dieter. *Peenemünde to Canaveral*. Englewood Cliffs, N. J.: Prentice Hall, 1962.

Jones, Reginald Victor. *Most Secret War*. London: Hamish Hamilton, 1976.

Jung, Philippe. 'The SE-4100 Family: An Early French Experience in Rocketry'. In *History of Rocketry and Astronautics, AAS History Series 17*, edited by John Becklake, 103–129. San Diego: American Astronautical Society, 1995.

Jung, Philippe. 'The SE-4200: First Ramjet Missile?'. In *History of Rocketry and Astronautics, AAS History Series 22*, edited by Philippe Jung, 115–156. San Diego: American Astronautical Society, 1998.

Jung, Philippe. 'Postwar German Rocketry Influence in France: An Analysis (Part II)'. Presented at the Fifty-Fifth IAC Congress, Vancouver, Canada, 2004.

Jung, Philippe. 'The True Beginnings of French Astronautics, 1938–1959, Part I'. In *History of Rocketry and Astronautics, AAS History Series 28*, edited by Frank H. Winter, 75–107. San Diego: American Astronautical Society, 2007.

Keisel, Kenneth M. *Wright Field*. Charleston: Arcadia Publishing, 2016.

Killy, Walther, *et al.* (ed.). *Dictionary of German Biography, Volume 10*. Munich: K G Saur Verlag, 2006.

Kit, Boris and Evered, Douglas S. *Rocket Propellant Handbook*. New York: Macmillan, 1960.

Krag, Bernd. 'Special Features of Anti-Aircraft Rockets with Swept or Low-Ratio-Aspect Wings'. In *German Development of the Swept Wing: 1939–1945*, edited by Hans-Ulrich Meier, 545–614. Reston: American Institute of Aeronautics and Astronautics, 2010.

Kroener, Bernhard R., Müller, Rolf-Dieter, and Umbreit, Hans. *Germany and the Second World War, Volume V, Part II: Organisation and Mobilization in the German Sphere of Power: War Administration, Economy, and Manpower Resources 1942–1944/5*. Oxford: Clarendon Press, 2003.

Ladas, Stephen Pericles. *Patents, Trademarks, and Related Rights: National and International Protection, Volume 1*. Cambridge, Massachusetts: Harvard University Press, 1975.

Lardier, Christian and Barensky, Stefan. *The Soyuz Launch Vehicle: The Two Lives of an Engineering Triumph*. Translated by Tim Bowler. New York: Springer, 2013.

Lasby, Clarence G. *Project Paperclip: German Scientists and the Cold War*. New York: Atheneum, 1971.

Lavington, Simon. *Moving Targets: Elliott Automation and the Dawn of the Computer Age in Britain, 1947–67*. London: Springer-Verlag, 2011.

Lee, R. G. *et al. Guided Weapons*. London and Washington: Brassey's, 1998.

Ludmann-Obier, Marie-France. 'Un aspect de la chasse aux cerveaux: les transfers de techniciens allemands en France: 1945–1949'. *Relations internationales* 46 (1986): 195–208.

Maddrell, Paul. *Spying on Science: Western Intelligence in Divided Germany 1945–1961*. Oxford: Oxford University Press, 2006.

Mahoney, Kevin. *Fifteenth Air Force Against the Axis: Combat Missions over Europe during World War II*. Plymouth: Scarecrow, 2013.

Marec, Jean-Pierre, *et al. Un demi-siècle d'aéronautique en France: Centres et moyens d'essais, Tome I*. Comite pour l'histoire de l'aéronautique. Academie de l'air et de l'espace, https://www.academie-air-espace.com.

Marec, Jean-Pierre, *et al. Un demi-siècle d'aéronautique en France: Centres et moyens d'essais, Tome II*. Comite pour l'histoire de l'aéronautique. Academie de l'air et de l'espace, https://www.academie-air-espace.com.

Masson, Philippe. 'La Marine Française en 1946'. *Revue d'histoire de la Deuxième Guerre mondiale* 110 (1978): 79–86.

McGovern, James. *Crossbow and Overcast*. London: Hutchinson and Co., 1965.

Miller, David. *The Cold War: A Military History*. London: John Murray, 1998.

Mills, James and Johansen, Graeme. 'Project Abstract: an Anglo-American intelligence operation in 1947 to recover guided weapon technical documentation buried in Germany'. *Intelligence and National Security* 34, no. 1 (2019): 129–148.

Mills, James. 'Pandora's box closed: the Royal Air Force Institute of Aviation Medicine and Nazi medical experiments on human beings during World War II'. *Studies in History and Philosophy of Science Part C: Studies in History and Philosophy of Biological and Biomedical Sciences*, 79 (2020).

Mills, James. 'The transfer and exploitation of German air-to-air rocket and guided missile technology by the Western Allies after World War II'. *The International Journal for the History of Engineering & Technology* 90, no. 1 (2020): 75–108.

'Missiles 1959'. *Flight*, 6 November 1959.

Morton, Peter. *Fire Across the Desert: Woomera and the Anglo-Australian Joint Project 1946–1980*. Canberra: Australian Government Publishing Service, 1989.

Naimark, Norman N. *The Russians in Germany: A History of the Soviet Zone of Occupation, 1945–1949*. Cambridge, Massachusetts: Harvard University Press, 1995.

Neufeld, Michael J. 'Hitler, the V-2, and the Battle for Priority, 1939–1943'. *Journal of Military History* 57, no. 3 (1993): 511–538.

Neufeld, Michael J. *The Rocket and the Reich*. New York: The Free Press, 1995.

Neufeld, Michael J. 'Rolf Engel vs the German Army: A Nazi Career in Rocketry and Repression'. *History and Technology* 13 (1996): 53–72.

Neufeld, Michael J. 'Overcast, Paperclip and Osoaviakhim: Looting and the Transfer of German Military Technology'. In *The United States and Germany in the Era of the Cold War, 1945–1990, A Handbook, Volume I: 1945–1968*, edited by Detlef Junker, 197–203. New York: German Historical Institute, Washington DC and Cambridge University Press, 2004.

Neufeld, Michael J. *Von Braun: Dreamer of Space, Engineer of War*. New York: Vintage Books, 2008.

Ordway, Frederick I. and Sharpe, Mitchell R. *The Rocket Team*. London: Heinemann, 1979.

Ordway, Frederick I. and Wakeford, Ronald C. *International Missile and Spacecraft Guide*. New York: McGraw-Hill Book Company, 1960.

'Our Latest on Show. Impressive Display at Farnborough: Newest British Aircraft and Engines Alongside Germany's Best: An Outline Description of an Outstanding Exhibition'. *Flight*, 1 November 1945.

Pestre, Dominique. 'Guerre, renseignement scientifique et reconstruction, France, Allemagne et Grande-Bretagne dans les années 1940'. In *De la Diffusion des sciences à l'espionnage industriel XVe–XXe siècle. Actes de la colloque de Lyon (30–31 mai 1996) de la SFHST*, edited by André Guillerme, 183–204. Fontenay Saint-Cloud: ENS Editions, 1999.

Pocock, Rowland F. *German Guided Missiles of the Second World War*. London: Ian Allan, 1967.

Postan, M. M., Hay, D. and Scott, J. D. *Design and Development of Weapons: Studies in Government and Industrial Organisation*. London: HM Stationery Office and Longmans Group Limited, 1964.

Quick, A. W., and Benecke, Theodore (eds). *History of German Guided Missiles Development*. Brunswick: Verlag E. Appelhans and Co., 1957.

Rankin, Nicholas. *Ian Fleming's Commandos: The Story of 30 Assault Unit in WWII*. London: Faber and Faber, 2011.

Reuter, Dean, Lowery, Colm and Chester, Keith. *The Hidden Nazi: The Untold Story of America's Deal with the Devil*. Washington DC: Regnery History, 2019.

Robert, Jean, Serra, Jean-Jacques and Jung, Philippe. 'Maruca: An Early French Liquid Fuelled Rocket'. Presented at the Fifty-Seventh IAF Congress, Fortieth IAA History of Astronautics Symposium, Valencia, 5 October 2006.

Rosenberg, Max. *The Air Force and the National Guided Missile Program 1944–1950*. USAF Historical Division Liaison Office, 1964.

'Rocketry at Westcott. A Visit to the Ministry of Supply Rocket Propulsion Establishment'. *Flight*, 21 August 1947.

Siddiqi, Asif. *The Red Rocket's Glare: Spaceflight and the Soviet Imagination, 1857–1957*. New York: Cambridge University Press, 2010.

Small, James S. 'Engineering, technology and design: The post-Second World War development of electronic analogue computers'. *History and Technology* 11, no. 1 (1994): 33–48.

Smithsonian Institution, National Air and Space Museum, https://www.airandspace.si.edu.

Stever, Guy. *In War and Peace: My Life in Science and Technology*. Washington DC: Joseph Henry Press, 2002.

Sutton, George Paul. *History of Liquid Propellant Rocket Engines*. Reston: American Institute of Aeronautics and Astronautics, 2006.

The Conference at Yalta and Malta, 1945, Volume 1. Washington DC: United States Department of State (Historical Division), 1955.

The Nike Historical Society, https://www.nikemissile.org.

'The Royal Naval Scientific Service'. *Nature* 154, no. 3,908 (1944): 398–399.

'Tridac the Unique'. *Flight*, 15 October 1954.

Twigge, Stephen. *The Early Development of Guided Weapons in the United Kingdom, 1940–1960*. Chur, Switzerland: Harwood Academic Publishers, 1993.

Uttley, Matthew. 'Operation "Surgeon" and Britain's Post-War Exploitation of Nazi German Aeronautics'. *Intelligence and National Security* 17, no. 2 (2002): 1–26.

Uziel, Daniel. *Arming the Luftwaffe: The German Aviation Industry in World War II*. Jefferson: MacFarland and Company, 2012.

Villain, J. 'The Evolution of Liquid Rocket Propulsion in France in the Last Fifty Years'. In *History of Rocketry and Astronautics, AAS History Series 17*, edited by John Becklake, 87–102. San Diego: American Astronautical Society, 1995.

Von Braun, Wernher. 'The Redstone, Jupiter, and Juno'. *Technology and Culture* 4, no. 4 (1963): 452–465.

Walker, Mark. *German National Socialism and the Quest for Nuclear Power, 1939–1949*. Cambridge: Cambridge University Press, 1993.

Webster, Charles and Frankland, Noble. *The Strategic Air Offensive Against Germany 1939–1945*. London: HM Stationery Office, 1961.

Wegener, Peter. *The Peenemünde Wind Tunnels: A Memoir*. New Haven and London: Yale University Press, 1996.

Welbourne, D. *Analogue Computing Methods*. Glasgow: Pergamon Press Ltd., 1965.

Werrell, Kenneth P. Archie, *Flak, AAA and SAM: A Short Operational History of Ground-Based Air Defense*. Alabama: Air University Press, 1988.

Westermann, Edward B. *Flak: German Anti-aircraft Defenses, 1914–1945*. Lawrence, Kansas: University Press of Kansas, 2001.

White Oak Laboratory Alumni Association, https://www.wolaa.org.

Who's Who in Germany, A–L. Who's Who Sutton's International Red Series, 1992.

Willis, Frank Roy. *The French in Germany, 1945–1949*. Stanford: Stanford University Press, 1962.

Winter, Frank H. and James, George S. 'Highlights of 50 Years of Aerojet, A Pioneering American Rocket Company, 1942–1992'. In *History of Rocketry and Astronautics, AAS History Series 22*, edited by Philippe Jung, 53–104. San Diego: American Astronautical Society, 1998.

Zaloga, Steven J. 'Defending the capitals: The first generation of Soviet strategic air defense systems 1950–1960'. *The Journal of Slavic Military Studies* 10, no. 4 (1997): 30–43.

Zimmerman, David. *Top Secret Exchange: The Tizard Mission and the Scientific War*. Montreal: McGill-Queen's University Press, 1996.

Index